FOR LIBRARY USE ONLY

LIBRARY
COLLEGE of the REDWOODS
EUREKA
7351 Tompkins Hill Road
Eureka, California 95501

REF S 561 .J36 2000
James, Sydney C.
Economic & business
principles in farm planning

Economic & Business Principles In Farm Planning & Production

EuRef

Economic & Business Principles In Farm Planning & Production

by Sydney C. James & Phillip R. Eberle

Iowa State University Press / Ames

Sydney C. James was raised in Utah where he earned an MS degree in agricultural education before receiving a PhD in agricultural economics from Oregon State University. He was on the faculties of New Mexico State University and Iowa State University before retiring from Brigham Young University. He has international experience as a consultant in Peru and Portugal. During his 30-year career as a college professor, he taught undergraduates the principles of farm and ranch management and did research to support his teaching. He is author of two other books widely used by farm and ranch managers.

Phillip R. Eberle was raised on a farm in Washington State. He received an MA degree in agricultural economics from Washington State University and received his PhD in economics from Iowa State University. Dr. Eberle's teaching and research focus is in farm management and farm real estate values. He is presently an associate professor in the Department of Agribusiness Economics at Southern Illinois University, Carbondale.

© 2000 Iowa State University Press
All rights reserved

Iowa State University Press
2121 South State Avenue, Ames, Iowa 50014

Orders: 1-800-862-6657
Office: 1-515-292-0140
Fax: 1-515-292-3348
Web site: www.isupress.edu

Authorization to photocopy items for internal or personal use, or the internal or personal use of specific clients, is granted by Iowa State University Press, provided that the base fee of $.10 per copy is paid directly to the Copyright Clearance Center, 222 Rosewood Drive, Danvers, MA 01923. For those organizations that have been granted a photocopy license by CCC, a separate system of payments has been arranged. The fee code for users of the Transactional Reporting Service is 0-8138-2880-5/2000 $.10.

∞ Printed on acid-free paper in the United States of America

First edition, 2000

Library of Congress Cataloging-in-Publication Data

James, Sydney C.
Economic and business principles in farm planning and production / by Sydney C. James and Phillip R. Eberle.—1st ed.
p. cm.
Includes bibliographical references (p.) and index
ISBN 0-8138-2880-5
1. Farm management. I. Title: Economic and business principles in farm planning and production. II. Eberle, Phillip R. III. Title.

S561 .J36 2000 00-025254
630'.68'1—dc21

The last digit is the print number: 9 8 7 6 5 4 3 2 1

CONTENTS

FOR LIBRARY USE ONLY

CHAPTER NINE
Organization and Ownership of the Farm Business, 239

CHAPTER TEN
Financial Planning for Ownership and Operation, 269

CHAPTER ELEVEN
Acquisition of Farm Machinery Services, 315

CHAPTER TWELVE
Acquiring and Managing Labor, 349

CHAPTER THIRTEEN
Income Tax Management, 383

This book is written for students in colleges of agriculture and for practicing farmers and ranchers and those who advise them. Although it could be used as a beginning text, juniors and seniors or readers with some practical experience probably will better understand it. The concepts are not particularly difficult, but courses or experience in accounting, microeconomic principles, and business management would be helpful. These topics are covered in this book but are generally better understood the second time around. Because applications are always easier to understand by those with the technical background, introductory courses in agronomy, animal science, range management, and other related disciplines give special meaning to the topics covered.

There are few new concepts or ideas not contained in other good farm management texts. After teaching farm management for many years and experiencing practical solutions to farm business problems, it is difficult to remember just when and where certain concepts and principles were learned. We are indebted to many excellent teachers, both in and out of the classroom, and to the authors of textbooks selected for our classes. Although there are other books that do a good job in teaching farm management, we suggest that this one excels in every chapter. Each major topic is illustrated with practical examples.

We have selected 10 topics. There are books written on each of them, thus, our discussions can only be introductory in nature. Nevertheless, we have tried to include subject matter most useful and vital for the financial success of the farm business. Thorough coverage was a major goal.

Following the introduction are three chapters on farm business accounting. Financial information and analyses are not only necessary for controlling the farm business but also for business planning and implementing decisions. Budgeting must be done within the framework of existing parameters, and economic principles are generally applied using coefficients gleaned from physical and financial records of the business.

Chapters on budgeting and economic principles follow accounting. The next chapter is on marketing. We included marketing management because marketing decisions are inseparable from production and risk management when making major business decisions. Our purpose was to integrate marketing management into choices about production and risk management. We concentrated on price outlook and marketing plans. Risk management follows marketing. This topic can best be treated in the context of the total farm business, including production, marketing, and finance.

The next four chapters have to do with acquiring the resources to farm with, beginning with business organization and ownership. Following this chapter are chapters on financial planning, acquiring machinery services, and acquiring and managing labor. The last chapter is on income tax management.

Most agricultural students are required to take one or two accounting courses, taught either in the business college or by agricultural economics faculty. Even though fewer agricultural accounting courses are being taught, most instructors of farm management and finance include accounting material in their courses. The three financial accounting chapters in this book fit both situations. Students not having any experience in accounting should be able to understand the material in this book. Single-entry accounting procedures are explained, but double-entry procedures with complete profit center accountability are emphasized. Those having had instruction in commercial accounting will soon realize that the principles are the same even though the applications are different. Farmers and farm managers need to integrate double-entry accounting into their businesses to take advantage of its flexibility and ability to measure enterprise profitability. Those acquainted with farming and familiar with accounting principles can glean the important material in these chapters much faster than those who have not had this kind of background.

Even though we recommend that instructors follow the chapter sequence of this book, there are other potentially successful approaches. It would be perfectly logical to start with business organization and ownership followed by financial planning. Some instructors like to begin with economic principles because this topic provides a major contribution of agricultural economists. Course outlines that do not follow the sequence of chapters in this book should pose no problems. Even though we tried to tie the book together in the introductory chapter and follow a logical sequence throughout the text, any chapter should be understandable by itself. During the winter semester, when interest in taxes is high, you may wish to move income tax management forward to follow accounting principles.

Rather than cite all of the major farm management texts at the end of every chapter, we have elected to list them here. Those we know of that were published in the last 15 years are shown. Any omissions are regrettable. We do not attempt to list the many texts reviewed in business management and associated disciplines. Other references are listed within the text and in footnotes.

Kay, Ronald D., and William M. Edwards, *Farm Management.* McGraw-Hill, 1994.
Herbst, John, and D. E. Erickson, *Farm Management: Principles, Budgets, Plans.* Stipes Publishing Co., 1993.
Castle, Emery N., Manning H. Becker, and A. Gene Nelson, *Farm Business Management.* MacMillan, 1987.
Kadlec, John E., *Farm Management—Decisions, Operation, Control.* Prentice-Hall, 1985.
Boehlje, Michael D., and Vernon R. Eidman, *Farm Management.* Wiley, 1984.

Economic & Business Principles In Farm Planning & Production

CHAPTER 1

The Role of Farm Management

The Concept of Farm Management

A farm manager's job is to decide how to use available resources to obtain what is wanted. Resources include labor, land, buildings, machinery, and other capital. Management not only controls the acquisition of factors of production but also determines the proportions in which they are combined. Different combinations of factors create different products, and the proportions and amounts used affect the quantity produced. The development of new products and yield expectations from differing levels of factor inputs is discovered by agricultural scientists, engineers, and technicians. But equally important is the manager who applies scientific information to each peculiar set of resources to reach a selected set of goals and objectives. Some product lines are more profitable or desirable than others, and we all can benefit from improved methods of production. Even though the prices received for products sold and factor input purchased are determined in the marketplace, management can modify these prices and inputs to increase farm net income. A manager's job is to apply his or her knowledge of production and markets to the factors or resources that he or she controls in a manner that fulfills the wants, goals, and objectives of self, family, and business associates most efficiently. Figure 1.1 depicts the interaction of science and management in helping farmers reach their desired goals from available resources.

Farm planning begins by defining the resources currently owned and rented. Resources include real property (e.g., land, buildings, and improvements) and personal property (e.g., machinery and equipment, livestock, crops, supplies, and cash). Resources need to be specifically defined and are identified in greater detail in subsequent chapters. Resources having market value are listed in the balance sheet or net worth statement. Liability claims upon asset resources are likewise specified. The detail and accuracy of specifying resources may determine the success of the farming activity. Successful crop production depends upon soil characteristics, erosion hazards, fertilizer needs, moisture, varieties, and other modifying agents. Successful livestock production depends upon breed, weaning percentage, birth rate, rate of weight gain, feed efficiency, and animal health. Building and machinery investments may be limiting factors in livestock and crop production. It is important to note that all production activity begins where you currently are situated and any changes must recognize this beginning.

FIGURE 1.1

The Relationship of Management to Resources and Operator Goals

Goals, too, are important and must be clearly defined. Without goal specification, management is without direction. It is impossible to reach an unknown target. Goals are not of equal importance and may not be achievable in the same time period. Some goals relate to the "good life" (e.g., health, leisure, education, retirement, and security). Goals important to the business may be farm ownership, farm expansion, reduction of debt, an increase in profit or income, and stability of income or avoidance of large losses. To achieve these longer-run goals, shorter ones need to be identified. Goals important to next year's operations may include buying a new piece of machinery, improving the swine farrowing facility, changing to narrower crop rows, adding more grain storage, and even taking a family vacation in August. Goal specification relates to the subjects of religion, philosophy, sociology, psychology, history, and economics. Not to be forgotten are family traditions.

Goals change with time and must be reevaluated periodically. Also, some goals conflict with other goals and these conflicts must be resolved. By using maximization principles, managers seek to convert available resources into goal achievements most efficiently and expeditiously. Goal achievement is the center of all economic activity.

The technology of converting resources into products limits the reaching of goals. We now enjoy a much higher standard of living because of newly discovered technologies. Much more will be done in the future. A successful manager must know and use those technologies relating to the products that he or she produces.

As mentioned, farm management is primarily concerned with deciding how to allocate farm resources most efficiently. The following paragraphs outline principal questions a farm manager may need to address.

How can the factors of production be obtained? The acquisition of factors of production continually plagues farmers. Obtaining farmland by inheritance, purchase, lease, parent-and-child agreement, or other formal arrangement precedes any agricultural production. Credit to finance real- and personal-property investments and purchase fertilizer, feeds, livestock, and machinery may be necessary to produce profitable levels of production. Contracting for production rights to raise some livestock and specialty crops may be a required means for producing some products.

What form of business organization should be selected? Most farm businesses are organized as single proprietorships, but partnerships and corporations are increasing in number and importance. Determining which organization best suits the needs, desires, and abilities of the owners, managers, and operators is of utmost importance. The organizational structure affects the acquisition and use of the factors of production and the size of the business. It also affects family relationships and business continuation. Each form has its advantages and disadvantages, and the lists are not the same for all people. Selecting the form of business organization is truly one of the very important considerations of management.

Which products should be produced? Most farms are capable of producing more kinds of products than are found there. Enterprise choice depends upon factor require-

TABLE 1.1

Net Income and Return on Investment of Specialized Farms in Iowa (1987–1996)

	Hog Raising		Beef Feeding		Dairy		Cash Grain		Beef Raising	
Year	**$**	**%**	**$**	**%**	**$**	**%**	**$**	**%**	**$**	**%**
1987	77,369	18.0	52,996	11.0	42,772	13.0	43,355	8.0	57,615	15.0
1988	53,019	11.0	54,423	9.0	47,373	14.0	57,418	13.0	46,008	14.0
1989	47,616	10.0	39,188	6.0	—[a]	—[a]	40,716	9.0	28,099	6.0
1990	80,669	17.0	103,428	13.0	57,057	15.0	36,378	6.0	30,060	5.0
1991	46,371	8.0	5,517	(2.0)	25,581	3.0	35,118	2.0	35,248	6.0
1992	52,493	7.0	89,330	10.0	47,205	11.0	44,295	4.0	38,578	7.0
1993	43,743	7.0	27,977	3.7	38,625	7.3	28,255	5.0	23,766	3.9
1994	20,347	3.4	36,634	4.7	39,690	6.7	44,934	8.2	28,339	4.5
1995	38,845	11.4	56,825	7.3	38,353	6.0	62,326	12.1	33,822	3.9
1996	91,360	11.3	77,759	8.6	40,143	6.4	61,689	10.2	38,874	4.3
1987–96 Average[b]	58,183	10.4	54,405	7.1	41,867	9.2	45,648	7.8	36,059	7.0

Source: *1996 Iowa Farm Costs and Returns,* FM-1789. Iowa State University Extension, Ames.

[a] Not estimated, insufficient observations.

[b] Averages for dairy exclude 1989.

ments for the different products, factor and product markets, and the desires of the operator or landlord. Some enterprises are selected only at the exclusion or reduction of others. A few enterprises complement each other, thus adding to the output of both. Still other enterprises can be added or increased without causing a change in existing enterprises. The choice of enterprise may well determine the success of a farm operation. The importance of enterprise selection is illustrated in Table 1.1 where the return on investments is shown for several enterprise specializations.

How should the products be produced? There are many possible ways of producing products. For example, corn can be produced with little labor and much machinery or vice versa, with commercial fertilizer or without, or by cultivating several times or with minimum tillage and the use of chemicals. Corn can be harvested for storage in gastight silos, or dried naturally or with drying equipment. Livestock rations combine forages, concentrates, protein, and vitamins and minerals in various proportions. The degree of quality or finish may affect the market price received. Most producers are interested in determining the method of production that will give the kinds and amounts of products they want at a desirable time and with the least cost. Importantly, there is a least-cost way to carry out any production activity; however, it may not result in the highest profit. The amount of profit received from any enterprise is a function of production costs and sales receipts.

The choice of enterprises is not unrelated to the method of production. The introduction of new technology that changes the production cost structure may cause new enterprises to replace old ones.

What should the level of production be from land and other fixed factors? Level of production refers to yields and not to the selection of products to produce or to the size of the farm business. Yield levels for most products can be controlled by the farm manager. For example, corn yields can be made to vary from 100 to 175 bushels per acre by changing the method of production, variety, fertilizer, chemicals, or the number of plants per acre. In some areas, irrigation becomes extremely important in regulating yields. Butcher hogs can be sold at almost any weight between 180 and 240 pounds. Dairy cattle can be fed to produce different levels of milk, as can hens to produce more or fewer eggs. If profit is a farm goal, then farmers would want to compare the profitability of different levels of input and production. There is an optimal level of inputs to use for each product produced.

When and where is it optimal to buy and sell? Prices paid for the factors of production and prices received for products sold vary throughout the season and from market to market. Many products, because of climatic conditions, can be produced naturally only in one season of the year; thus, the marketing period is limited. But there are many other products with extended or flexible seasons of production, such as for some livestock. However, the costs of production may change with the season. This change may be due to input factor costs or to production requirements. Seasonal production patterns cause seasonal price patterns.

Our competitive market system assumes price bargaining among buyers and sellers. If activity on the buyer's side is stronger than on the seller's side, prices rise and vice versa. The level of activity on the seller's side is determined by the amount of supplies on hand, and the activity on the buyer's side is determined by the demand for the product or factor being traded. In addition, government actions affect prices for inputs and products. Farm producers need to understand market forces and how prices are established and what causes them to change, both nationally and internationally.

Thus, when to buy and sell becomes very important to the financial success of any business. Price forecasting, adjusting production to expected price changes, and buying and selling activities may be the most important work of a farm manager.

How large should the business be? The total volume of business largely determines the amount of profit that can be made. It is necessary to do a large volume of business to have a high net income or profit, albeit an insufficient condition. The size of the production plant is measured by the number of acres farmed, the number of head of livestock, the number of fruit trees, the value of assets managed, and the volume of sales. The size of the business activity is important for gaining the benefits from economies of scale. Size also allows for specialization of labor and management.

How should the business office be organized? Not to be overlooked are farm accounts and records, including income statements, balance sheets, production records, farm business analysis, credit accounts, and tax reporting. Historical records form the basis for major decisions and business improvements because they mark the beginning point and show the direction for change. The analysis points out not only weaknesses but also strengths. Balance sheets provide security information for credit transactions that may be necessary for production efficiency and growth. Business growth and equity control can be obtained only if income is greater than needed for personal and family consumption. Budgeted plans can do little more than project future receipts, expenses, and net income. Tax reporting must be accurate and within legal limits; tax management can save considerable money for use by the family and business.

Several management areas affect farm profits (Figure 1.2). It is appropriate that the bail, or handle, be labeled "General Economy." The health of the general economy both domestic and international determines the general price level. The general price level measures the total economic cadence of our economy and may be the most important factor in determining financial success in farming. Seldom has farming been profitable during a general depression.

FIGURE 1.2
Factors Affecting Profits in Farming

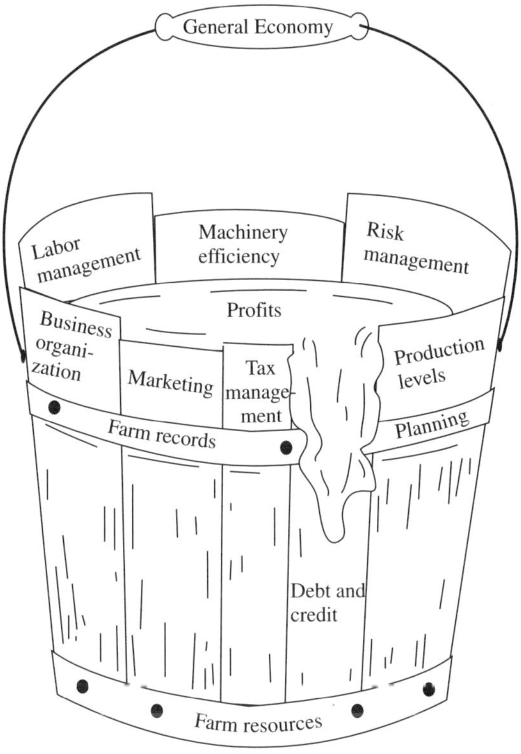

Importance of Management in Farming

The importance of management to farm earnings is illustrated in Table 1.2 that compares net farm income and other measures of financial success with various measures of farm production efficiency. The low-profit farms are the bottom third of all farms measured, with the high-profit farms being the high third. The large differences in net income ($118,499) and returns to management ($107,850) can be attributed to the size of the business and to the other factors previously discussed. Consider crop production efficiency. Corn yields were 6 bushels per acre higher and the price received $0.08 more per bushel for the high third compared with the low third of farms. The price differential alone amounts to about $12 per acre. Yield and price ($25 per acre combined) could amount to $7,500 on a farm with 300 acres of corn. Considering all crops, the value per acre was $22 higher for the high third of farms compared with the low third and machinery costs were $16 lower, for a combined difference of $38. For livestock the returns per $100 feeds fed were $25 higher for the high third compared with the low third. Assuming feed costs about $190 to finish a steer calf, this differential in feed efficiency amounts to $48 per head. The combined effect of farming at a high level of efficiency returned an 18.2 percent to assets for the high third of farmers compared with 2.0 percent for the low third. Following these farmers through the years reveals that many of the same persons remain in the same classifications year after year. The logical conclusion is that some farmers are better managers than others and it shows in their financial performances.

TABLE 1.2

Comparison of High- and Low-Profit Farms in Iowa, 1996

Comparison	High Third	Middle Third	Low Third
Farm Income:			
Livestock production less feed	92,191	20,639	14,948
Value of crop production ($)	243,691	147,630	135,455
Miscellaneous ($)	30,716	13,561	11,169
Crop inventory gain or loss ($)	27,813	11,414	4,279
Value of farm production ($)	394,412	193,245	165,851
Operating expenses ($)	147,291	78,714	79,701
Fixed expenses ($)	104,529	60,391	61,553
Capital gain or loss ($)	2,095	840	1,591
Net farm income, accrual ($)	144,687	54,980	26,188
Operator and family labor ($)	21,858	19,808	19,019
Charge for equity capital ($)	37,004	21,396	29,196
Return to management ($)	85,824	13,777	–22,026
Crop Efficiency:			
Acres in crops (acre)	847	581	507
Crop acres per person (acre)	473	386	392
Crop value per acre ($)	337	340	315
Corn yield (bu/acre)	151	151	145
Average price received for corn ($/bu)	2.32	2.32	2.24
Total economic cost for corn ($/bu)	2.41	2.61	3.05
Machinery cost per acre ($)	45	53	61
Machinery investment per crop acre ($)	163	151	172
Livestock Efficiency:			
Value of livestock production ($)	215,017	58,266	59,457
Feeds fed ($)	122,826	37,627	44,457
Livestock returns per $100 feed fed ($)	168	151	143
Financial Efficiency:			
Value of farm production per person ($)	244,393	166,246	153,263
Value of farm production per $1 expense ($)	1.68	1.51	1.21
Asset turnover ratio	0.53	0.56	0.35
Return to owned assets (%)	18.2	12.5	2.0
Return to equity (%)	26.0	15.6	–1.9
Debt to asset ratio	0.30	0.23	0.25

Source: 1996 *Iowa Farm Costs and Returns,* FM 1789. Iowa State University Extension, Ames.

Decision-Making Functions

Farm management is both an art and a science. Farmers are continually confronted with timing problems, such as when to plant, harvest, irrigate, cultivate, fix the barn, arrange for a new loan, buy inputs, and sell products. Some farming activities can only be learned from experience. These decisions fall largely under the art of planning. Decision making involves a priority system, part of which can be determined scientifically, but on a day-to-day basis relies heavily on intuition, tradition, judgment, and experience. Some things must be learned by doing. For example, plowing too early after a rain may waste time and affect the soil adversely, but plowing too late may result in poor seed germination. Bred animals do not usually give birth on the day planned.

Farm managers must adjust for weather, accidents, breakage, and other unplanned events if resources are to be used to maximize returns.

Farm management, as a science, is concerned with decision making. The significance of science is in the methods used rather than the results obtained. The scientific method substitutes objectivity for the subjectivity found in the following three methods:

Tenacity—holding fast to habits, traditions, and opinions.

Authority—unquestionably following the advice of another because of his or her position or status.

Intuition—following the "self-evident."

Scientific objectivity, however, implies interpersonal validity; thus, the methods used and the results obtained are independent of the researcher. The scientific approach is involved in the decision-making process outlined below.

1. Recognition of a problem
2. Observation of relevant facts
3. Specification of alternatives and hypotheses
4. Analysis and test of hypotheses
5. Choice of alternatives
6. Taking action or implementation
7. Bearing responsibility for action taken and control

The first four items are particularly relevant to the scientific method. The activities that distinguish the researcher from the manager are contained in items 5, 6, and 7. The ability to make choices and take action may well be the most important difference between good and poor management.

An understanding of the principles of management is basic to an appreciation of the importance of change and uncertainty. Without change, the need for management would gradually but largely disappear. It is conceivable that without change and uncertainty, master planners could develop a blueprint for each business and for the whole economy. Then, a hired supervisor could carry out the plan according to exact specifications.

Change comes primarily from two sources: the action of individuals and the natural environment. Management itself is largely responsible for new technology that may free productive factors or create new products. Advertising companies can even change the tastes and preferences of individuals. To enumerate all the changes brought about by management would require a lengthy discussion on how our whole economic system operates. Ours is not a planned economy but rather one of checks and balances through the market system and governmental regulation. Natural forces continually upset the plans of management, particularly in agriculture where production to a great extent is controlled by the weather.

Uncertainty, then, is a function of two major forces: time and knowledge. Time allows change to take place. Knowledge relates to the amount of information about a problem situation. It is very difficult to have complete and accurate information about all the variables involved with any problem.

Planning under risk and uncertainty is discussed in Chapter 8. It is sufficient here to recognize the importance of uncertainty to the management process. Each step in the decision-making process is discussed below.

1. *Recognition of a problem.* Problem perception is probably the most difficult and important step in the decision-making process. "Everybody has problems," is a

truism. But all problems are not of equal importance and not even what they may seem. A problem is said to exist if pain or difficulty is felt. The problems of what, how, how much, when to produce, and how to obtain the factors of production were discussed previously. Other problems have been implied and are covered throughout the text.

Many problems grow out of conflicts between purported facts and goals—facts versus facts, goals versus goals, and facts versus goals. Often the information received about a product or situation is contradictory. The problem is to determine which of the purported facts are true. Advertised claims for machinery, fertilizers, feeds, seed, and new practices need to be verified before being broadly adopted. There also may be conflicting goals, such as those between the farm and family. A new family auto may conflict with the purchase of a new tractor, a community project with a family outing, leisure time with family security, and religious worship with crop harvest. The ordering of goals thus becomes very important. Lastly, goals and facts are not always compatible. Having the latest tractor, a beautifully kept farmstead, a purebred beef herd, or following certain conservation practices may not generate maximum net farm income. Dairying as a means of producing income may not allow for a family vacation. Without careful evaluation of the facts and a clear understanding of the family and farm goals, a problem is likely to be poorly conceived.

A consideration of facts and values as they relate to a problem causes one to arrive at certain beliefs or hypotheses about the problem. The hypothesis may be that adopting a certain practice or buying a new machine will (or will not) increase income (or some other goal). The formulation of ideas is largely a deductive process; that is, drawing tentative conclusions about the unknown from certain information.

It has been said that a problem is half solved once it is properly defined. For a real problem to exist it should be important, timely, feasible, and unique. Importance keeps management off the trivia—items that time will usually solve without an expenditure of research effort. Timeliness says get to it while the answer can do some good. Market-price advantage requires the adoption of output-increasing technologies prior to the masses. Feasibility says there is no need to tackle a problem for which the solution is impossible or uneconomical. If the cost of a solution is greater than the benefit from it, dismiss it. Uniqueness means, "don't do work on something for which someone else already has the solution."

Farm records are of great benefit in identifying farm problems. Record keeping and analysis are discussed in Chapters 2-4. The importance and ordering of problems can be ascertained more easily through the consideration of economic principles discussed in Chapter 6.

A decision tree may be helpful in visualizing problem identification (Figure 1.3). Illustrating the importance of identifying the problem is the college student who has grade difficulty. Troubles with concentration, motivation, restlessness, and class skipping are symptoms that may bring about a visit to the school psychologist. The symptoms soon dictate the seriousness of the problem and the need for a remedy. The hard part is in discovering the root of the problem. The root may be a carryover from youth, a traumatic experience, girlfriend or boyfriend problems, body malfunctions, or drug abuse. Once discovered, the prescription for health may be relatively easy to give. The medicine, however, could be nasty to take.

2. *Observation of relevant facts.* This phase involves data gathering as it relates to the problem under study—the learning process. The decision maker must become

FIGURE 1.3

A Decision Tree Illustrating the Various Choices a Farmer May Need to Make

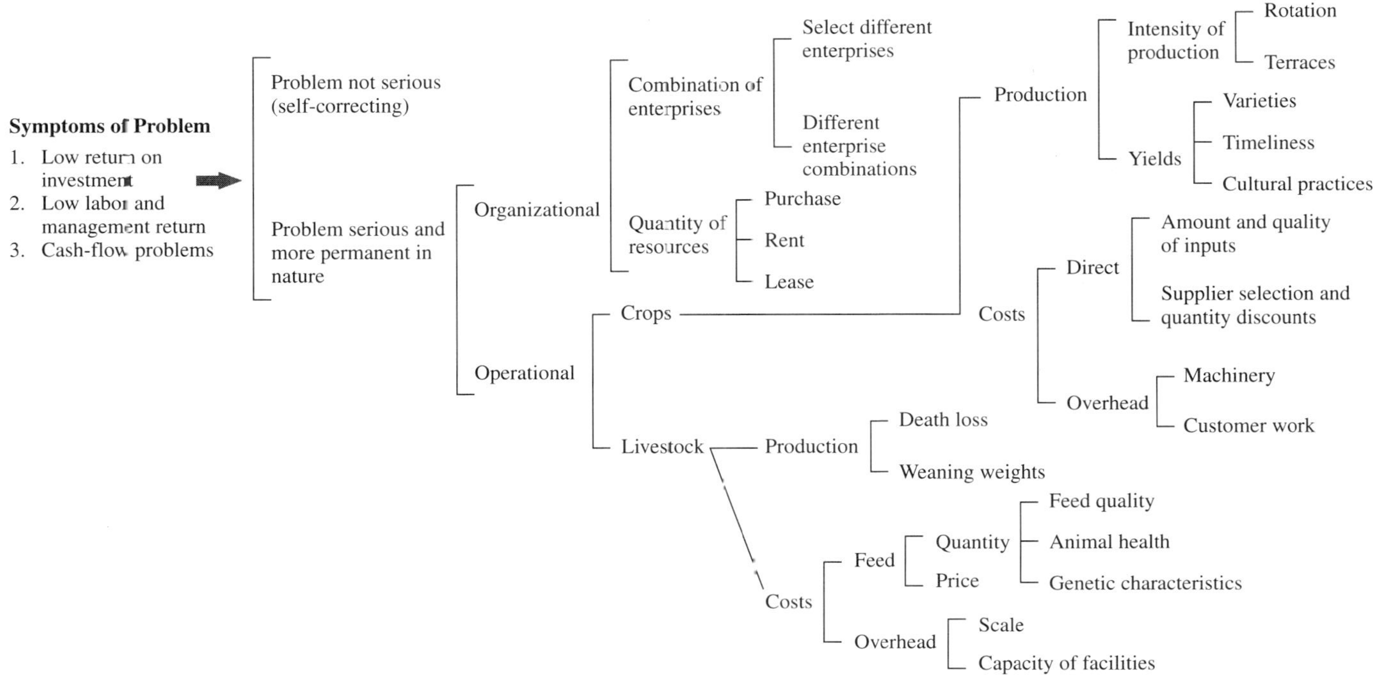

Symptoms of Problem

1. Low return on investment
2. Low labor and management return
3. Cash-flow problems

Problem not serious (self-correcting)
Problem serious and more permanent in nature
Organizational
Combination of enterprises
Select different enterprises
Different enterprise combinations
Quantity of resources
Purchase
Rent
Lease
Operational
Crops
Production
Intensity of production
Rotation
Terraces
Yields
Varieties
Timeliness
Cultural practices
Costs
Direct
Amount and quality of inputs
Supplier selection and quantity discounts
Overhead
Machinery
Customer work
Livestock
Production
Death loss
Weaning weights
Costs
Feed
Quantity
Price
Feed quality
Animal health
Genetic characteristics
Overhead
Scale
Capacity of facilities

engaged in fact-finding to fit his or her beliefs and hypotheses to the true state of nature. Many farm managers make provision for a continual flow of information into their business through magazines, newspapers, agricultural experiment station and cooperative extension service publications, as well as radio, television, and the Internet. For these data to be useful, a manager must develop some system of classification or catalog by subject. The categories might include prices, production requirements for crops and livestock, building plans, machinery and equipment, leasing arrangements, credit sources, insurance rates, new products, and the like.

The general types of data found in farm magazines is not sufficient for many problems that will arise. For some problems the manager will need to seek the advice of specialists in such disciplines as soils, crops, botany, entomology, animal nutrition, veterinary medicine, engineering, economics, and law. Many people from private companies, as well as those in public institutions, also can help.

Certainly one's knowledge need not, and possibly cannot, be perfect before one is willing to, or forced to, act. Suffice it to say here that managers engage in fact-finding only until they are sufficiently satisfied with the answer to be willing to make a decision to act. Observations may be made all through the analysis phase.

It is clear that all decisions are not of the same urgency. Consider, for example, two boys dipping in the farm pond when one of them who cannot swim finds himself in water over his head. The other boy does not have time to gather information about the depth of the water, measure his own swimming ability, or compare strengths. He must act now to save his buddy. Situations like this happen often in farming. The decision maker must act before he or she is ready and without good information about the consequences. Forced decisions are generally more uncertain than if there were time for gathering substantiating data.

Again, the importance of complete farm records and accounts should be stressed. These records furnish the background against which the manager makes many, if not most, decisions. Records provide data about the resource base, production performance rates, and marketing practices that tell a lot about the managerial ability of the operator. Economic theory is important in telling which of the multitude of facts are most relevant and useful for the problem under study.

3. *Specification of alternatives and hypotheses.* Specifying the alternatives takes place somewhere in the problem-identification and data-gathering phases. Decision making is selection among choices. However, choices must be distinguishable in measurable ways. Specifying alternatives is similar to developing hypotheses about alternative outcomes. A hypothesis is a contention based upon preliminary observations of what appear to be facts, which may or may not be true. To test the hypothesis is to compare the contention with the facts.
4. *Analysis and testing of hypotheses.* The analysis tests the hypotheses listed as alternatives. If the facts support a hypothesis or contention it is accepted as being correct or true. If not, it is rejected as false. Rejection of a hypothesis may be as important as its acceptance. Farm managers use computers, data banks, and decision models (i.e., forward-planning techniques such as budgeting, linear programming, marginal analysis, and business techniques) to substantiate their contentions. Decision making is seldom without risk because there are nearly always uncertainties. Once again, classification makes the job easier. The decision-making tools available to farm managers are the subject matter of this book and are presented in the chapters that follow.

5. *Choice of alternatives.* Alternatives are selected based upon the analysis, values of the operator, and means of implementation. Most decisions involve welfare criteria. The consequences and benefits must both be weighed. Two types of errors are possible: something that is true can be rejected or something that is false can be accepted.

   Which type of error is the more serious is not argued here. The consequences of being wrong are important to consider in addition to the benefits of being right. Consider, for example, that a decision to buy cattle was wrong; that is, the price received is below the expected level and there is a loss. For the established feeder, this may be just another loss in a series of profit–loss years. There may be other products to support the loss. However, such a loss could wipe out the beginning farmer. But suppose the established farmer rejects the purchase of liability insurance only to discover too late that a large legal claim has been filed against the business. Even an established farmer may not be able to withstand such a large loss. The following generalizations about errors in decision-making may be useful:

   - The accuracy required depends upon the chance (probability) of making an error and upon the size of the error.
   - An increase in accuracy increases the amount of facts and data required.
   - The importance of accuracy and the seriousness of mistakes vary with the decision and the decision maker.
   - As equity and assets increase individuals and organizations can sustain greater absolute loss and spend more on attaining accuracy and security.
   - Because in any decision the consequences of the two types of errors are different, it is important to consider them separately.

6. *Taking action or implementation.* This phase involves carrying out or putting into practice the chosen plan. Implementation requires the acquisition and coordination of the factors of production. Loans may need to be negotiated, lands rented, machinery leased or purchased, buildings remodeled, livestock purchased, and workers hired.

   Taking action may well be the most difficult step because there is always the chance of being wrong or not wanting to be committed. However, if action is postponed, it may be too late and opportunities lost. The problem can be well defined, the data complete, and the analysis clever, but all to no avail because of inaction. Many brilliant persons fail because they can't carry out a plan of action.

7. *Bearing responsibility for action taken and control.* Part of being a manager is accepting responsibility for the action taken. Success may be met with disbelief and failure with ridicule. Society and the individual are both important here. Each person has preconceived notions of how things are and should be. Time is the proving ground.

   Records and accounts are again important. These can be used to measure progress or change. Comparisons are useful not only with prior time periods but also with budgeted plans. Successes and failures are both important to identify and evaluate. Even the best of plans go wrong through no fault of management. Re-implementation could be the very best next action. Those plans that seemed to go right possibly could be improved. The control aspects involve correcting deviations from plans, and modifying, redirecting, and making necessary improvements.

Classification of Decisions

Classifying decisions is helpful to the manager for budgeting time wisely and keeping away from trivia. The following system outlined by Castle, Backer, and Nelson[1] may be helpful:

Importance—a measure of the size of potential gain or loss. It is obvious that all decisions are not of the same magnitude. Consider the importance of buying land compared with determining the type of fence to enclose it, or the decision to invest in a dairy operation compared with whether or not to use artificial insemination for the cows. Some decisions, such as buying liability insurance, are very important, not so much from the cost or gain potential, but by potential loss.

Frequency—how often the same or a similar decision is made. A farm may be purchased only once in a lifetime, a tractor every 5–10 years, fertilizer once each year, and feed several times a month. Livestock are fed several times each day. The cumulative effect of often-repeated decisions may be great, but it may be possible to develop guidelines that greatly minimize the process. A new decision may be required only when certain elements change a stipulated magnitude.

Imminence—the penalty for waiting. Some decisions can be delayed without penalty and there is plenty of time for tabulation and deliberation, whereas other decisions require immediate attention. Consider the decision of building or repairing a fence compared with the decision to combine grain when a severe storm front has been forecast. The planting season is not the time to check the planter out and make necessary repairs or to make arrangements to purchase seed and fertilizer.

Revocability—capability to reverse a decision or change one's mind. Once made, some decisions can be revoked only at considerable cost. Consider the decision to plant an orchard compared with purchasing a 6-bottom plow. If the plow does not plow well, or if the tractor cannot pull it, it may be difficult to trade it or even sell it without great loss. A farmer who cannot afford drastic, costly changes in operation may be wise to select a type of farming that will permit him or her flexibility in planning.

Available alternatives—possible choices. Some situations permit a variety of choices, whereas other situations allow only one or two. A soil in the Willamette Valley in Oregon may support more than 20 different crops for which there are ready markets. For most soils in the Midwest, only three or four crops are viable alternatives. Some desert areas may only be useful for cattle or sheep, and public sentiment may not even allow these animals. Many farmers fret and waste time over things they cannot change.

Fields of Study in Farm Management

Undergraduate curricula in agricultural economics are usually divided into the following specializations: farm management or production, marketing and agribusiness management, agricultural finance, and farm and food policy. Although other courses may be taught, these four are considered to be core courses. This textbook concentrates upon production principles but recognizes the importance of marketing and finance in the management functions just presented. Planning requires integrating these four areas into the planning and implementation phases.

Production or farm management. Production and planning are used somewhat synonymously in this context. One could say that production is the implementation of plans, but normally that is too specific. Production also includes planning, organizing, and control. Operations management might be a term that separates production from marketing and finance. This book is mainly about the economic and business principles and practices of production.

Marketing. There is little one can do in planning production that does not require pricing before alternatives can be economically selected. Indeed, the prices paid for production inputs and prices received for products sold may be more important than are production practices. This book includes a discussion about marketing to help complete the planning process. The treatment is that of application and not discovery.

Finance. Finance is part of the implementation process. The size of the business investment, and associated operating costs, necessary to provide full employment to a farm family at an acceptable level of income are so large that only a very few do not need outside financing. In fact, even if one could self-finance the total business, it probably would not be in the economic best interest to do so. Financing is a means for using borrowed capital more productively than it costs, thus multiplying the returns to equity. Therefore, financial considerations need to be integrated into production plans. The discussion in this book does not include sources of borrowed money, how interest rates are determined, or how the national money markets work.

Policy. Policy discussions are not part of this text. But you should be in a much better position to deal with policy questions and with various state and federal policies that affect your business.

Relationship of Farm Management to Other Fields of Study

Farm management is broader than most fields of study in agriculture. It includes not only agricultural physical sciences such as agronomy, horticulture, entomology, meat animal production, dairy production, poultry production, and agricultural engineering but also social sciences and humanities such as economics, sociology, philosophy, ethics, and psychology. Like any other business, a successful farm operation must be built upon sound business practices. The principles involved are mainly those from economics.

Economics, in a narrow sense, is the study of the allocation of scarce resources among competing ends to maximize goals. The goals might be anything (e.g., health, education, security, family living comforts, or leisure). The goal emphasized most throughout this text is maximum net farm income or profits. It is realized that this goal is not a final goal for most people, but it does allow for the achievement of some higher goals, assuming that the farm is primarily a business unit.

The elements involved under this definition of economics are as follows:

- Insatiable human wants
- Limited resources
- Techniques of production
- Principles of maximization

This definition of economics does not differ greatly from the characterization of the farmer's job as used in the beginning of this chapter.

The insatiability of human wants is well demonstrated in American society. There seems to be no end to the capacity to consume goods and services. New technology continues to be developed. Obsolescence is built into many of the items purchased, such as automobiles, household appliances, and clothing. New goods and services are being placed on the market each day. Also, two cars, several television sets, air-conditioning, and many other past luxuries are among the "necessities" for many of today's families. Tomorrow it will be something else. This insatiability gives vitality and stimulus to the whole system. The farmer is part of this great American society.

Most people are keenly aware of resource limitations. Critical resources to farmers are now known to be in short supply. Production costs and limitations cause the forces of supply and demand to operate and a marketplace to be established. For most factors, this price distributes the resources to where they can add the most to productive value. Price is the center of determining what, how, how much, and when to produce as previously discussed. Farmers, too, are concerned with limited resources and prices. Limited resources on the farm might include land acreage of certain qualities, buildings and their arrangements, labor, machinery, numbers of livestock on hand, money in the bank, and available credit. Thus, the farm manager must take these beginning resources as a base or starting point and organize the farm business around them in such a way as to maximize the returns from them.

Techniques of production almost always limit how resources are converted into usable products, and the associated level of profitability or other desirable goals. Farmers must understand the technical aspects of the crops and livestock being produced, and the tools used in their production. They cannot hope to be successful without this knowledge. Often, managers must select between competing technologies.

The function of agricultural economists is not to generate new technology. This aspect is left to physical, chemical, and biological scientists in public and private research, and to innovative practitioners. It may be necessary to search the literature and to contact specialists to determine just what the factor-to-product conversion ratios are to evaluate them. Much of the research done by agricultural economists in farm management is of this nature. A farmer manager also may need to predict the future regarding new technology and consumer behavior when forward planning. Anticipating technological change should not be confused, however, with the creation of new technology. Thus, the approach generally used in this book is to assume given technologies in their particular fields. Certain technologies referred to are used only to illustrate the management principles discussed. Management principles are more enduring than the technologies they evaluate.

The relationship of farm management to psychology and sociology can be illustrated by the involvement of the farm operator with the human agent. The nature and requirements of hired labor are changing in the direction of nonagricultural industries. Government labor laws, the influence and control of labor unions, and the demands of labor for shorter and more regular working hours, along with security and benefits are illustrative of these changes.

Not to be considered lightly is the constant flow of service agents contacting farmers to sell seed, fertilizer, feed, chemicals, insurance, and the like. These agents are skilled, educated salespeople from whom valuable information can be obtained. They not only have products for sale but also often are technicians in their specialized fields. Farmers generally obtain their supplies from several different businesses, have membership in one or more cooperatives, obtain credit from the local banker, and market their produce through a large number of firms. Each of these human agents is different with respect to education, personality, character, and company policies.

Any discussion of fields of study related to farm management must include mathematics and statistics. These fields are disseminated through most other areas of learning, including technical agriculture, business, and economics. Mathematics and statistics are used to interpret and evaluate farm business records and accounts, evaluate the claims of salespeople, and interpret findings from research. Business computers are now available to farmers at a reasonable price. Some farmers have computers already and others will soon. Computer programs are available and the decade to follow will see many more.

Thus, over the period of a year, the farm manager may have made decisions on how much fertilizer to apply (agronomy), whether to feed plant protein or urea to the livestock (animal science), whether to dry the corn artificially (agricultural engineering), whether to keep his or her own records or submit them to electrical data processing (business administration), where and when to market livestock and crops (marketing), whether to try the new product that a salesperson described (sales psychology), how to conserve farm resources (ethics), and how much to pay in taxes (law). All of these decisions involve economics. This book is designed to integrate these various subject areas as they relate to the management of a farm business.

Economics must rest its existence on principles of optimization and maximization of the human goods and services necessary in fulfilling goals under limited resources through techniques of production. Agricultural economics, including farm management, is an applied field; hence, the principles discussed are applied in solving farm problems.

The Functions of Management

The activities in which a farm business manager is involved generally fit into one of the following categories: planning, organization, implementation, and control. Many management textbooks are divided into these categories. Even though this book is not so divided, the framework is a convenient one to describe the function of management. The following paragraphs briefly describe each function, and they are referred to periodically in the book.

Planning. This chapter began by identifying a dependency between the goals and objectives of the manager and his or her resources. Planning begins with identifying the long-run, intermediate, and short-term goals of management. Goals and objectives were briefly mentioned, and others are listed at the beginning of Chapter 9. Resource identification was likewise discussed earlier in this chapter and is the subject of Chapter 2. The types of problems management deals with were classified in this chapter and procedures for dealing with them are discussed in Chapters 5 and 6 and in other chapters. Decision-making functions were likewise categorized in this chapter. Because planning seldom materializes as envisioned, it is important to have contingency plans. These plans are discussed in Chapter 8. The major contents of this book, as its title suggests, are about planning.

Organization. Organization has to do with the structure of the business and its workforce. The ownership and operational forms that the business may take are discussed in Chapter 9. Obtaining and managing labor is the subject of Chapter 12. Writing job descriptions, negotiating contracts and pay rates, employee benefits, policies and procedures, and performance standards are very important. Labor management is the subject matter of organizational behavior.

Implementation. Taking action was one of the decision-making functions. Once the manager has adopted a plan of action and an organizational structure, it is time to take action. Making it work is implementation. First, the resources need to be acquired. Some may be on hand, such as land, improvements, machinery and equipment, livestock, and supplies. Other resources will need to be rented or purchased using savings or credit. "Financial planning for ownership and operation" is discussed in Chapter 10. Acquiring the services of machinery and labor are the subjects of Chapters 11 and 12, respectively. Many implementation tasks require a study of other disciplines such as soils and agronomy, horticulture, livestock science, and range science. Some aspects of putting the plan into action must be learned through experience. This experience is part of the art of farming previously discussed.

Control. Control is the disciplinary function of management. Control requires the comparison of results with goals and objectives and formulated plans. Procedures for monitoring finances are discussed in Chapters 2–4. These procedures provide more than control features in that the analysis of a business also identifies its strong and weak parts that may affect new plans. Thus, financial and physical records are necessary for identifying problems needing corrective "control" action, as well as being part of the planning phase. Seldom are plans fully realized, so it is necessary to set deviations, or ranges, around expectations within which corrective action is not considered necessary. The setting of these ranges is part of the discussion in Chapter 8. Farm accounting and business analysis are placed at the beginning of this text, even though controls are considered as coming later in the management process because of the structure and organization of the financial accounts form, the format for budgeting and economic principles. Also, the process of analysis often suggests problems and opportunities for testing in the planning phase.

Summary Rules for Success

- Know and be able to do well all the practical jobs connected with the enterprises on the farm you are managing. Learn the art of farming.
- Know the scientific principles of crop and livestock production and machinery technology required. Learn the technology of farming.
- Understand the legal procedures and governmental regulations for organizing and operating the farm business, including the ownership and organization of the business and legal requirements for paying taxes and maintaining the environment. Learn governmental rules and regulations pertaining to your business.
- Know and use the basic business and economic principles in accordance with which the common farm practices and scientific knowledge should be applied. Learn management and supervisory skills.

This book concentrates on the last two rules.

Notes

1. *Farm Business Management,* Macmillan Publishing Co., 3rd ed., 1987.

CHAPTER 2

Financial Accounts—Introduction and Balance Sheet

INTRODUCTION TO ACCOUNTING PRINCIPLES

Importance and Uses of Records and Accounts

Accounting is defined as an activity that communicates business financial information for reporting and decision making. The means of communication is typically the balance sheet, income statement, cash-flow statement, and statement of owner equity. Although each of these reports has a well-defined general format they can be modified to meet specific purposes. Reports may relate to the total business or to only a part of it, such as a livestock or crop profit center. Through these statements accountants communicate with individuals outside the firm (financial accounting) or to individuals within the firm (managerial accounting). The accounts and procedures discussed in this section pertain to both types of accounting.

Records and accounts are major tools for sound business management. They are not only useful when making decisions but also for controlling the implementation of decisions and evaluating successes and failures. Many problems are recognized because of some felt difficulty (e.g., lack of cash flow, insufficient time to get things done, slow growth rates in animals, and low crop yields). But other problems go unrecognized because they are hidden or masked by more noticeable items. The detection of difficult management problems poses a challenge to all farm accountants. Data from farm records are often necessary for evaluating changes in business organization and operation. Sometimes errors are made in planning because the data used were not relevant or realistic. A problem is never really solved until the facts are gathered and an evaluation is made.

This chapter is a revision of material found in James, Sydney C. and Everett Stoneberg, *Farm Accounting and Business Analysis,* 3rd ed., Iowa State University Press, 1986.

Farm accounting is a subset of the generally accepted accounting principles (GAAP) followed by certified public accountants. Farmers, however, have deviated from GAAP due to the nature of their businesses, lack of training, and requirements of the Internal Revenue Service (IRS). Farms are generally single proprietorships whose owners are both managers and laborers. Some items produced often are inputs to other production activities. Historically, accounting typically was done with cash journals. The IRS emphasized single-entry cash accounting. Because farms have become larger and more complicated, with many alternative ownership and operating organizations, the need for financial and managerial accounting has increased. Accounting by computers has generally modeled double-entry systems. Thus, there arose a need for standardized methods of accounting for agricultural producers. To this end the Farm Financial Standards Task Force (FFSTF) was organized in 1989. Their corporate name now has been changed to Farm Financial Standards Council (FFSC). They have released "Financial Guidelines for Agricultural Producers II," December 1997.[1] Agricultural accountants, particularly those in professional work, should study GAAP and recommendations of the FFSC. If there are differences in this book they are for clarity and to broaden applications. The construction of the statements proposed by the FFSC should pose no difficulty. Hopefully, GAAP are not compromised.

Major uses of farm records and accounts follow:

A management tool. Farm records and accounts allow farmers to measure their success in using the factors of production to produce agricultural products for a profit. When combined with production data, they point out the strong and weak points in the farm business. Problems are identified for improvement. Records and accounts help control the implementation of decisions and measure the success of change.

Preparing income tax reports. All taxpayers must keep records that will enable them to accurately prepare an income tax return, and that will permit the IRS to determine whether the law has been correctly applied. Although most farmers see the need of records for tax reporting, only a small number recognize the opportunities for tax management. The objective of tax management is to maximize after-tax income, not necessarily to minimize the amount of taxes paid. Experience suggests that farmers who keep poor records usually pay relatively more taxes.

Basis for credit. According to the American Bankers Association, "The banker who has records of the borrower's business is able to compare the borrower's past performance against standards for the area. These records also become a basis for projecting and evaluating the future profitability and loan repayment capacity of the business. Records, properly and accurately kept, provide the banker with the financial information needed for prompt handling of credit requests."[2] Records and accounts furnish information for planning the credit needs and repayment schedules of the farmer. Credit worthiness not only is assessed but also evidence of loan security is provided.

Additional uses. Records and accounts also provide the basis for farm lease arrangements and other contracts, farm insurance programs, participation in government programs, and conformation to legal requirements and limits.

Components of a Farm Records and Accounting System

Even though an unlimited number of financial records and reports are used in farming, the following are typical:

Asset and liability accounts. These accounts are physical and financial accounts of all farm resources (assets) and the claims against them (liabilities). The proper separation and ordering of the assets and liabilities produces the balance sheet or net worth statement of the business.

Income and expense accounts. These accounts document the financial flows into (income) and out of (expenses) the business over a period of time. Included are cash and noncash transactions. Subtracting expenses from income gives net income, which measures the profitability of operating the business.

Capital accounts. These accounts are for recording the purchases and sales of breeding livestock, machinery and equipment, buildings and improvements, and land. Capital assets last longer than 1 year so their purchase costs cannot be debited fully as expenses in the year purchased. But, many of these assets are depreciable and do affect future expense accounts. Also, gains and losses from sales of capital assets need to be shown in income statements. Purchases and sales of capital assets directly affect the listing of assets in the balance sheet. Over time, the value of these assets changes to reflect general economic conditions.

Credit accounts. These accounts are for recording new loans and keeping track of principal and interest payments on old loans. Unpaid operating and business accounts are included in this definition as accounts payable and receivable.

Production and statistical records. These records relate to the production of crop and livestock enterprises or profit centers and the resources used. Labor, machinery, feed, seed, fertilizer, pesticides, and yield recordings are examples of statistical records.

Employee records. Federal and state withholding taxes, insurance and retirement benefits provided by the employer, and employee withholdings need to be accounted for and reported. Records of labor allocations to the various profit and cost centers are also important and necessary in business management. Accounting for labor use is elusive and requires daily monitoring. However, with consistency and dedicated effort reliable results can be obtained.

Enterprise records and accounts. An accountant can record for individual enterprises or profit centers the same information as for the total farm. Detailed enterprise analyses often can be made with only a few additional records.

Farm business analyses. The balance sheet and receipt and expense accounts are combined with production records to probe for strong and weak areas within the business. These analyses are commonly called efficiency measures. The information they provide is useful in identifying problems and directing future farm management decisions. Growth and progress can be measured over time by a comparison among years.

Figure 2.1 shows many of the possible records and accounts that can be kept and their relationships to each other. Note that business accounts have been separated from personal and family financial accounts and those of other businesses.

Accounting Period

The accounting period normally corresponds to the calendar year. However, a fiscal year accounting period is more useful for some farmers. A fiscal year does not begin on January 1. The beginning month may correspond to the beginning of a lease arrangement, such as March 1. Some government agencies operate on a fiscal year beginning July 1.

The production period should be considered in selecting an accounting period. It may be useful to adopt an accounting period that corresponds to the flow of receipts and expenses from major enterprises. For example, a broiler producer may wish to balance the accounts after each batch of broilers sold. Some cropping seasons end in midsummer, and this time might be the best to balance accounts. The time period selected between summaries should be meaningful for analyzing the activities of the business.

FIGURE 2.1

Major Components of Farm Records and Accounts

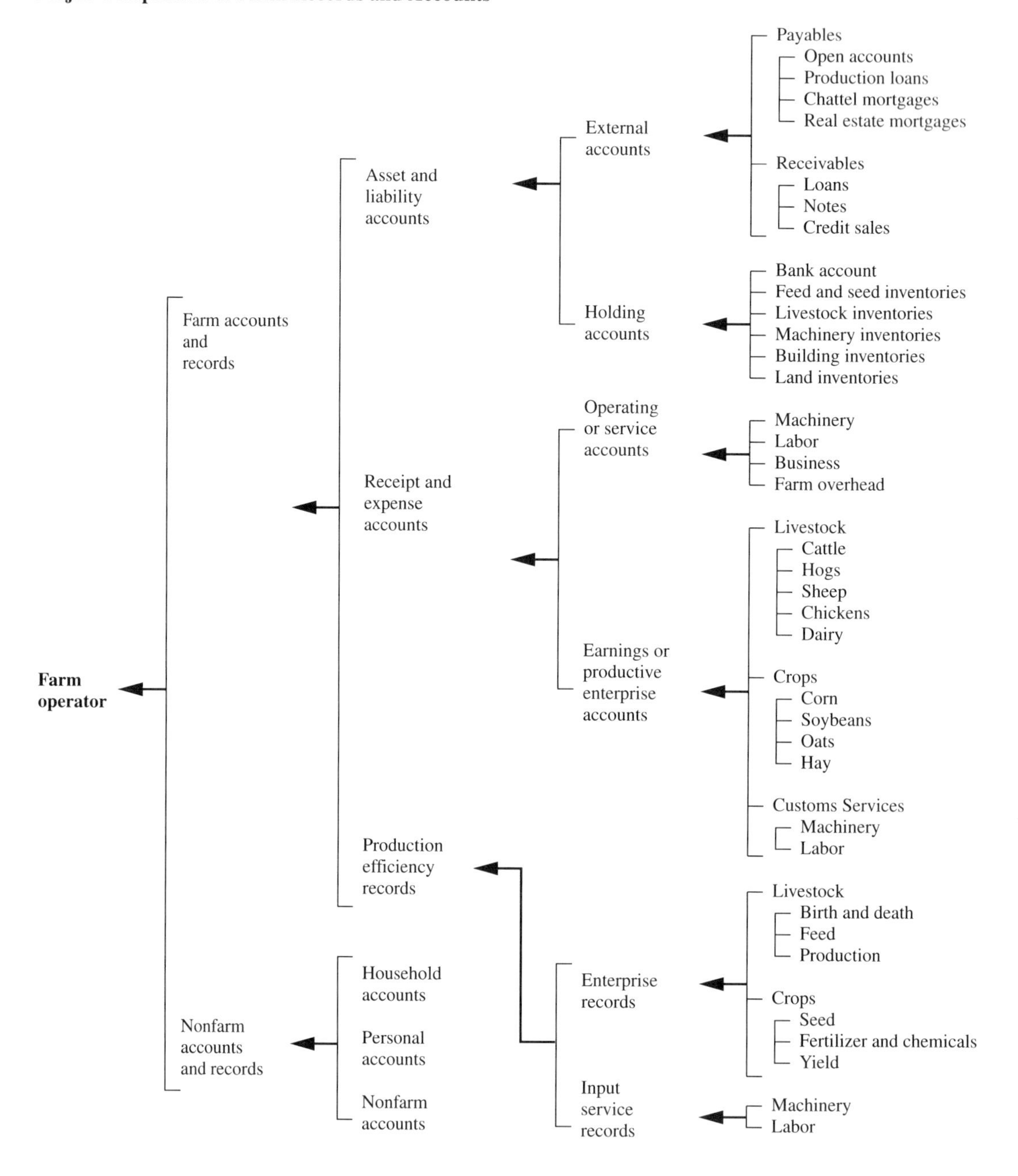

FIGURE 2.2
The Accounting Period

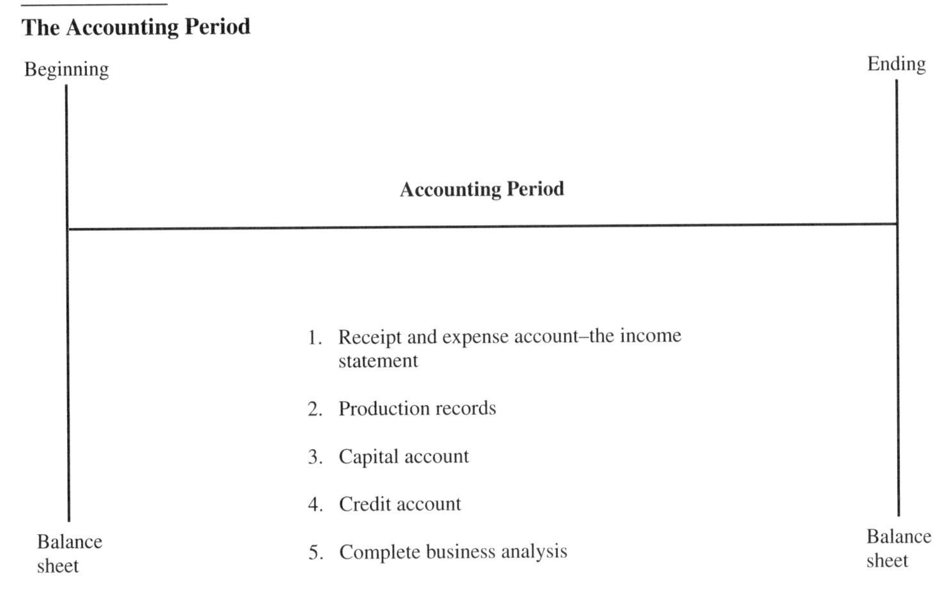

In this book a calendar year generally is used for discussion purposes. Figure 2.2 illustrates how the records and accounts previously listed relate to each other and fit into the accounting period.

The balance sheet, or net worth statement, gives the financial position of the business at a specified time, usually at the beginning of the accounting period and again at the end. Between the beginning and ending times are many changes (e.g., production activities, purchases and sales, loan payments, and new loans). These changes are recorded in the income statement and other accounts and records defined previously. Whereas the balance sheet statement is static in concept, the income and expense accounts, production records, and capital and credit accounts are dynamic and explain the changes reflected by a comparison of balance sheets. It should be recognized that all production activities, purchases, and sales affect in some way the balance sheet, but they may only be recorded periodically.

Annual summaries may not be sufficient to guide the activities of a business. Quarterly and sometimes monthly summaries of major or key segments of the business may provide necessary and useful information for efficient management. Through the use of electronic data processing, these summaries are available as frequently as necessary.

A tabulation of net income just prior to the end of the calendar or fiscal year can be very useful for income tax management. For example, farmers who know approximately what their taxable incomes will be may be able to plan sales and purchases to shift tax payments between years.

The Accounting Entity

It is important to determine which financial transactions belong to the farm business and which belong to some other business or family activity. The area to which each account applies is called an accounting entity. The home and family form a separate accounting entity from the farm as do other business ventures. A farmer should keep separate records and accounts of each of these three levels of activities. Because there may be frequent transfers of funds and resources among these entities, it may not be easy to identify where each item belongs. Expenses that may need to be divided include the telephone, automobile, banking, magazines, and business office. Identification of the financial entities makes the allocation process easier and more accurate.

A major purpose of most farm business ventures is to produce income for family consumption, currently and in the future. Generally, there is not a wage agreement so income must be transferred from the business to the family as needed. Similar transfers may exist between the farm and other business interests (e.g., farm profits may be used to make nonfarm investments). Likewise, nonfarm income may be transferred to the farm. A high percentage of farms receive income from nonfarm employment.

Even though each of these entities should be separated, it may be useful to bring them together into a composite account. Balance sheets can be joined to determine total net worth and to measure growth. Income statements likewise can be joined to determine total net income and to balance farm, nonfarm, and family activities. Cash-flow statements often include farm and nonfarm income and expenses, thus showing the combined financial needs of the business and family.

BALANCE SHEET

The balance sheet is a summary of the financial assets of the business and the claims upon them. Assets not claimed by creditors are the owner's residual claim or equity. The form of the balance sheet is illustrated as follows:

Balance Sheet

Assets		Liabilities
$		$
$		$
$		$
$		$$$
$$$$	Owner's Equity	$
		$$$$

Owner's equity often is referred to as net worth or capital. Likewise, the balance sheet is referred to as a net worth statement. Net worth may better be used to refer to the composite of all business and personal assets of the farmer and his family, whereas equity is reserved for the business entity. Also, net worth is more meaningful when tabulated using market prices. Capital is applied to firms that are financed with sales of stock. In this book equity is used to refer to the difference between assets and liabilities, except when a market-based balance sheet is illustrated. The general accounting formula is assets = liabilities + owner's equity (net worth). This balance or equality is always maintained because owner's equity is a residual claim (Fig. 2.3).

The assets of a business are its productive resources that have economic value. To have economic value, resources must be scarce and capable of rendering future services to the business. As such, these items can be bought or sold on the market. Examples of farm assets are land, fences, wells, buildings, machinery, equipment, livestock, crops, fuel, fertilizer, receivables, and cash. To list the assets is to describe the farm in

FIGURE 2.3
Basic Accounting Formula

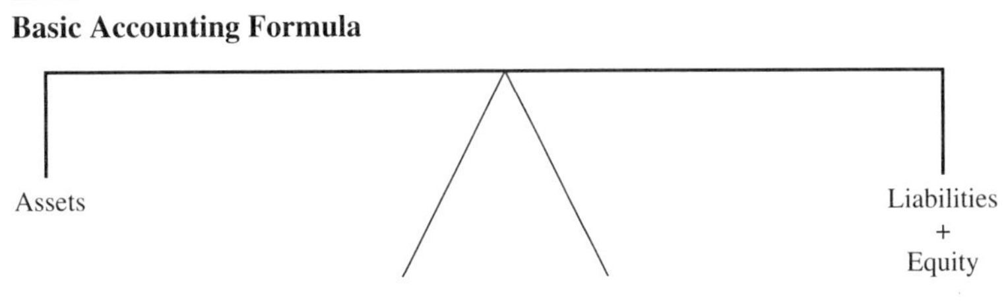

terms of its financial worth. These assets are typically listed on the left-hand side or on the top of the balance sheet. The categories into which they usually are grouped are current assets and noncurrent assets. The latter is often divided into intermediate assets and fixed assets.

Current assets are cash, savings, accounts receivable, investments in growing crops, and inventory. Inventory may include supplies, raw materials, and merchandise being held for sale such as feeder livestock and crops. Looking at it another way, current assets are cash or near cash and generally are either consumed in the production process or sold. These assets provide liquidity.

Intermediate-term assets (sometimes called working assets) provide services to production activities but are not depleted in one accounting period, and are not directly and legally attached to land. As they wear out and grow old they lose their value and thus become an expense to the business. Examples are machinery, equipment, tools, and breeding livestock. Most of these assets can be found on the business depreciation schedule (discussed later in this section).

Fixed assets (sometimes called long-term assets) are the real property and include land and attachments to land, such as roads, wells, terraces, fences, and buildings. Land normally does not depreciate if properly cared for. However, land improvements may depreciate if there is a wearing out, using up, or obsolescence.

A typical farm balance sheet is shown in Table 2.1. (The farm accounting and business analysis [FABA] farm is used in illustrations throughout the financial accounting chapters.) Compare the asset and liability classes and consider the types of items included in each.

Many textbooks follow GAAP and define only two classes of assets. Intermediate assets and fixed assets are combined under noncurrent assets. Classification of liabilities follows this pattern. (FFSTF followed GAAP.) This book continues the traditional three-category classification to separate real from personal property and to facilitate the business analysis. Real property is separated from personal property for financial and taxation reasons. The financing of land is considerably different from financing machinery or livestock and often is handled by separate agencies.

The decision of where to classify an asset is not always clear and may even seem arbitrary. But there is something to be said for consistency and comparability. Unless the same definitions are used from one accounting period to the next, the financial changes cannot be accurately measured. If records of investment, income, and production are to be compared among farms, they must have the same basis. Two useful tests can be applied when classifying assets, the convertibility test and the use or service test.

The convertibility test relates to the ease with which the asset can be converted into cash at its market value. Many items have well-established market outlets and the price is broadly known. Grain and feeder livestock are examples. Other items, such as farm machinery, are regularly traded, but the markets are not so well established and

TABLE 2.1

Balance Sheet for the FABA Farm, Beginning YRXX

Assets	Value	Liabilities	Value
Current:		*Current:*	
Cash	$ 12,200	Feed bills	$ 1,635
Life insurance (cash value)	4,500	Cattle note	30,000
Savings	10,050	Production loans	18,000
Feed	58,738	Accrued unpaid interest	13,811
Nonfeed crops and supplies	8,086	Principal payments due:	
Feeder hogs	41,093	Intermediate term	2,130
Feeder cattle	36,518	Long-term loans	22,524
Total current assets	$ 171,185	*Total current liabilities*	$ 88,100
Noncurrent:		*Noncurrent:*	
Intermediate:		Intermediate:	
Machinery and equipment	$ 72,962	Auto loan (farm share)	$ 2,279
Breeding hogs	19,660	Total	$ 2,279
Portable buildings	8,192	Long term:	
Total	$ 100,814	Silo loan	$ 17,000
Fixed:		Farm mortgage	318,674
Farmland	$ 832,800	Total	$ 335,674
Buildings and improvements	80,556	*Total noncurrent liabilities*	$ 337,953
Total	$ 913,356		
Total noncurrent assets	$ 181,370	**Total Liabilities**	$ 426,053
		Equity	$ 759,302
Total assets owned	**$1,185,355**	Liabilities + Equity	$1,185,355
Total assets rented	$ 180,000		
Total assets managed	$1,365,355		

prices are less well known or understood. Thus, it may take considerably more time and effort to find the highest price for a used machine. The land market, which includes improvements, is the least established, and there may not be a separate market for some improvements. Many farms are on the market for a year or more waiting for a buyer who is seeking those peculiar resources and is able to come to terms with the seller.

The use test has to do with the nature of the service the assets render to the business. Most current assets are completely used up in the production process within the accounting period, or they are sold. Most working assets provide services over several accounting periods, thus becoming less valuable. Fixed assets such as soil provide services indefinitely and may even become more valuable in the process.

Most farm lenders require an accurate accounting of current assets because the returns from the use and sale of these assets indicate the potential cash flow into the business to service debt payments. There are demands on cash income for family living expenses, debt retirement and interest payments, and operating expenses of the business that must be met. Lenders do not like to carry over short-term debt.

The intermediate-term assets are also of concern to lenders. First, these assets are not as readily available for sale in the accounting period. Their purchase cost is normally recovered over a period of years through a depreciation expense. Even though there are sales of intermediate assets, such as breeding livestock and used machinery, they normally are replaced. For instance, if breeding animals are sold, young breeding stock normally is transferred from current assets to replace them, or replacements are purchased. Likewise, replacement of depreciated machines usually involves large expenditures of money.

Liabilities represent the financial claims upon the owned assets. Normally, these liabilities are the unpaid debts of the business, such as accounts payable, notes payable, mortgages, accrued unpaid interest, and wages payable. Liabilities are classified according to their respective assets: short-term, intermediate-term, and long-term liabilities.

Current (also called short-term) liabilities correspond to the current assets and include notes, unpaid accounts, interest accrued at the time the balance sheet was made, and principal payments falling due within the following accounting period.

When a comparison of current assets with current liabilities is made, its meaning is enhanced if the principal payments on noncurrent liabilities falling due within the following accounting period are included with short-term liabilities. This approach gives a more accurate accounting of the liabilities that are to be paid out of current assets.

Intermediate-term liabilities correspond to the intermediate assets and include notes and security interest agreements with principal payments running for more than one accounting period—usually more than 1 year and less than 10 years. If intermediate-term assets were not separated from fixed assets then intermediate-term liabilities would be classified with long-term liabilities.

Long-term liabilities are mortgages and contracts on real property. They usually run 10 to 30 years from their initiation.

The residual claim belongs to the owner(s) and is referred to as the owner's equity or net worth. It is shown after liabilities. If the assets are looked at from a corporate sense the owner's equity is similar to the stockholder's claim on the assets.

For the time spent, the balance sheet provides a greater amount of financial information than any other account a farmer may keep. It is a static account, much like a photo that records everything at a moment in time, and forms a benchmark against which all financial progress is measured. It is the foundation of most credit transactions. Timed comparisons are meaningful whenever measured, during the year and between years.

Asset Valuation

The term inventory can be used as a noun or a verb. As a noun it refers to property. In nonagricultural business, the property is restricted to goods held for sale, goods in the process of production, and raw materials. Durable goods such as depreciable property and land normally are not included. In farm accounting the items in inventory are typically livestock, harvested crops, and supplies. However, other personal and real property are sometimes referred to as inventory.

As a verb, inventory refers to the process of counting and valuing resources. A working definition of inventory is an accounting of all property owned or controlled by the farm operator, including a listing of all liabilities. The process of listing and valuing is concerned mostly with crops, livestock, and productive supplies.

Taking the inventory includes two processes: a physical count and a valuation. Making the physical count is a practical matter that does not need lengthy discussion. The only equipment required is a notebook, pencil, and measuring device such as a rule and scale. Each piece of property is listed and adequately described; however, the following points should be kept in mind:

- It is important to measure quantities of items in units that are commonly used and meaningful and that aid in establishing values (e.g., bushels for small grains, hundredweight for market livestock, gallons for fuel).
- Like items should be grouped according to their contribution to the business or their place in it. For example, all livestock should be grouped under one classification and further divided by type, age, or purpose. Crops should be grouped and divided by type and differentiating properties. Farm machinery should be grouped and divided by purpose or use. It may be desirable to separate powered from non-powered machinery, or the livestock equipment from the crop equipment, etc. Divisions to be meaningful should be dictated by the business analyses the operator wishes to make. The groupings should fall into the categories established for the balance sheet.

Converting measured quantities into meaningful units is sometimes troublesome. For example, it may be necessary in the inventory process to measure the quantity of grain on hand in cubic feet, to count the number of bales of hay or measure its cubic feet, or to measure the depth of grain in the bin or silo. Tables and conversion formulas that facilitate this process are available. Even though measurements under these methods are not exact, they are usually accurate enough for practical purposes.

Common methods used in valuing farm properties are as follows:

- Market cost
- Net market price
- Farm production cost
- Unit livestock price
- Cost minus depreciation
- Cost minus depletion
- Capitalization

The cost referred to in the listed methods may include more than just the price at the point of purchase. The cost at the farm may include transportation, site preparation, and installation, which must be added to the purchase price of the item, particularly when the service life of the asset is more than one accounting period.

Market cost refers to the purchase price of an item. This method works particularly well for items that have been recently purchased and will be used up within a relatively short period of time, such as feed, fuel, small tools, fertilizer, seed, and feeder livestock. For feeder livestock it may be necessary to adjust the cost for changes in value after purchase to reflect value added after purchase.

Net market price is the anticipated value of an asset at the marketplace minus transportation and selling charges. Thus, it is the money the farmer could bring home from the sale of an asset. This method works particularly well for items that are held primarily for sale, or for which a market price is well established, such as market livestock and farm-produced crops. The net market price method is useful for all assets where the purpose is to establish a net worth statement based upon market values. Farms are often appraised on the basis of comparative market value.

Cost or market, whichever is lower, is a method closely related to the first two methods. The cost of producing an item, or its original purchase price, is compared with the current market price, and the lower of the two amounts is chosen. The major advantage of this method is its conservatism. However, the user should recognize that the lower value might be too conservative when prices are increasing and not conservative enough when they are falling.

Farm production cost relates to the cost of producing an asset on the farm. The production cost should not include the opportunity cost of the operator's labor and management because of the unrecognized income involved. The same may be said of unpaid interest costs. Generally, this method is most useful in valuing assets that have been produced on the farm to be used in other farm-production activities and to not be sold. A good example is feed grain and breeding livestock.

The unit livestock price or base value method, as prescribed by the IRS, has its basis in the costs of production but is not tied to it over time. This method classifies livestock according to age (i.e., calves, yearlings, 2-year-olds, mature animals) and each is valued by its production cost. The IRS requires that if this method is adopted all raised livestock must be valued by it, and once established the values cannot be changed without IRS consent. Example values are the following:

- Calves (from 3 to 9 months) $300
- Yearling heifers (from 10 to 15 months) 400
- Two-year-old heifers (16 months to calving) 500
- Mature cows 600

It should not be used for purchased breeding livestock. These animals are better valued by using a cost minus depreciation approach.

Cost minus depreciation is used for investment properties that provide services to the business for a period longer than one year and become less valuable in doing so. Depreciation is a regular adjustment to value. The balance sheet value of a depreciable asset is its beginning value minus the sum of the depreciation amounts claimed in prior periods. The initial purchase price, or value, of the asset is in the form of a prepaid expense, in that not all of the cost is used up the first year but is allocated over the life of the asset. Depreciation is a deduction on the income statement. Assets valued by this method include breeding livestock (particularly those purchased), machinery and equipment, and buildings and improvements. A discussion of specific depreciation methods follows later in this chapter.

The *cost minus depletion* method is used for valuing a stock asset such as a gravel pit. The value of the asset changes in accordance with the amount of resource displaced or removed during the accounting period. Timber stands, mineral deposits, oil wells, and organic deposits are examples of such resources.

Capitalization as a valuation method is appropriate for assets whose value is based upon a flow of costs or income. The present value has a time dimension to it. Land, fruit orchards, and breeding livestock are examples. Land provides income over an infinite period of time. Orchards produce only after several years of sequential investments. The value of a beef or dairy cow is based on the future income stream from her production. The cost of time enters into such valuation problems. The right to collect $100 one year from now is not worth $100 today, and an expenditure made several years ago is now more costly than when originally made—waiting is an expense. Moreover, investing money one way prohibits it from being invested in another. The cost of investing in the first way is the return that could have been received by investing in the second. Capitalization methods are useful for handling these time-dimension kinds of valuation problems. Capitalization methods and procedures are discussed in Chapter 10.

The listing of rented properties should be considered but not as owned resources. Conceptually, rented and leased resources are not greatly different from assets purchased with borrowed money. Having a value of rented assets makes it possible to tabulate the value of total assets managed and the return to assets managed as discussed in Chapter 4. Note in the example balance sheet (Table 2.1) that the rented assets are separated from owned assets and shown only as an addendum on the asset side. Values for rented assets probably need to be estimated.

Valuation Method Selection

Table 2.2 lists assets commonly found in the balance sheet and indicates which methods may be most useful when placing a value on them. More than one method of valuation can be applied for many assets. The valuation methods selected are not the same for all farm businesses, and some businesses may have more than one balance sheet, each using a different method of valuation.

The following factors should be considered when selecting a method of valuation:

- *The values established are entered in the balance sheet.* If the objective is for this statement to reflect current net worth, market value methods should be selected.
- *The values established are used in evaluating the efficiency of the business.* For example, asset and liability ratios measure financial stability and changes in assets and equity show business growth. Also, inventory values for crops may be used in tabulating the value of crops produced per acre and value of feed fed, which are used in figuring livestock returns per $100 feed fed.
- *The values used in the balance sheet may be used when tabulating taxable income.* Most farmers do not keep separate inventories for business analysis and tax reporting. Neither do they keep separate depreciation schedules, one for tax reporting and one for internal management.
- *The principles of conservatism and consistency should be considered.* Balance sheet analyses can only be meaningful if the same procedures are used from one period to the next.

TABLE 2.2

Useful Methods to Consider When Valuing Various Kinds of Farm Assets

Assets	Purchase Cost	Net Market Price	Lower of Cost or Market	Farm Production Cost	Cost Minus Depreciation	Unit Livestock Price	Capitalize Costs or Returns
Cash		×					
Marketable securities	×	×					
Life insurance		×					
Accounts receivable	×	×[a]					
Prepaid expense	×						
Growing crops				×			
Harvested crops		×	×	×			
Purchased feed	×						
Feeder livestock		×	×	×			
Breeding livestock		×	×		×	×	
Machinery		×			×		
Buildings		×			×		
Land	×	×					×
Orchard		×			×		×

[a]Less an allowance for doubtful accounts.

- *The principles of the going concern and stable monetary unit should be considered, especially for land.* The valuation procedures for a farm going out of business may be different from one that will continue in operation. And, the fluctuations in the general economy if incorporated into asset values could render efficiency measures meaningless.
- *Changes in the balance sheet from one period to the next are registered in the income statement,* unless adjusted internally through equity.

Cost- or Market-Basis Valuation of the Balance Sheet

Some balance sheets have two columns, one for a cost basis of valuation and the other for a market basis. The cost basis provides for a more stable monetary unit, and thus real growth is reflected in changes that occur. The market basis is more usable for finance and credit purposes. Also, market-based values are more accurate when calculating income returns on the capital invested. However, it should be recognized that market-based values, particularly for noncurrent assets, can be no more than an estimate of tangible asset values. The true value of any asset can only be determined through sale to a third party. A balance sheet showing both the cost basis of value and the market basis of value is shown in Table 2.3.

Market-based liabilities must include deferred or contingent taxes. If the assets were marketed there would be income taxes due. These liabilities are called deferred liabilities. Net worth or equity tabulated using market values is a more accurate representation of the financial position of the owner after all debts have been paid, including taxes. Some credit agents require the market approach to value when evaluating a

TABLE 2.3

Balance Sheet for the FABA Farm; Cost and Market Values; End YRXX

Assets	Cost Basis	Market Basis	Liabilities	Cost Basis	Market Basis
Current:			Current:		
Cash	$ 10,814	$ 10,814	Accounts payable	$ 885	$ 885
Savings and investments	15,540	15,540	Feeder cattle note	40,000	40,000
Feed crops	67,234	67,234	Crop production loan	16,000	16,000
Nonfeed crops	8,978	8,978	Accrued unpaid interest	15,238	15,238
Feeder livestock	80,616	80,616	Principal payments due	35,202	35,202
Total current assets	$ 183,187	$ 183,187	Total current liability	$107,325	$107,325
Intermediate:			Intermediate:		
Machinery and equipment	$ 115,864	$ 135,000	Vehicle loan	$ 0	$ 0
Breeding livestock	22,625	25,000	Tractor loan	24,731	24,731
Portable buildings	8,880	8,000	Harvest machine loan	16,487	16,487
Total intermediate assets	$ 147,369	$ 168,000	Total intermediate liabilities	$ 41,218	$ 41,218
Fixed:			Long term:		
Buildings and improvements	$ 73,968	70,000	Silo loan	$ 12,680	$ 12,680
			Farm mortgage	298,853	298,853
Farm land	832,800	900,000	Total long-term liabilities	$311,533	$311,533
Total fixed assets	$ 906,768	$ 970,000	Deferred taxes		$ 63,000
Total assets	**$1,237,324**	**$1,321,187**	**Total liabilities**	**$460,076**	**$523,076**
			Equity	**$777,248**	**$798,111**

new loan and their security position. The amount of deferred tax liability can be estimated as follows:

1. Estimate net farm profit by subtracting expenses from income at the time of valuation.
2. Estimate capital gains income from selling nondepreciable real property.
3. Estimate depreciation recapture including first-year allowances.
4. Add items 1, 2, and 3, and multiply by the marginal income tax rate.

Net farm profit for the specific date of the balance sheet is estimated by adding the following, if income tax is tabulated by the cash method:

Market value of raised crops on hand and/or
Market value of raised market livestock on hand
Plus: market value less cost of purchased market livestock on hand
Plus: value of unharvested crops
Plus: accounts receivable
Plus: year-to-date net farm income from operations (this may be a negative figure)
Less: accounts payable
Less: accrued unpaid interest

If income tax is tabulated using the accrual method, then those assets that were on hand when the last tax payment was tabulated already have been taxed. Thus, any market value established for crops and livestock would be reduced by the last inventory taken to tabulate taxable income. Likewise, it would be the difference between receivables and payables to be recorded.

Capital gains income is estimated by adding the following, if taxable income is estimated by the cash method:

Market value of raised breeding livestock
Plus: market value of land less its purchase cost as adjusted for nontaxed improvements
Plus: market value of securities less purchase cost

For the accrual taxpayer, the market value for raised breeding livestock is reduced by the last inventory value used to estimate taxable income. Furthermore, if purchased livestock are not on depreciation, then they are considered in the same manner as raised breeding livestock.

Depreciation recapture for the cash and accrual taxpayer is estimated for all personal and real properties on depreciation by subtracting from estimated market values their undepreciated or residual values as listed on the depreciation schedule or ledger. For this purpose first-year-depreciation (section 179 deduction) is treated like depreciation.

The marginal tax rate is difficult to estimate. The tax rate to use is that required by federal and state income tax laws on the added income. The appropriate rate to use cannot be determined until the amount of taxable income has been estimated. If the sale of the farm is for cash then capital gains income may be large and the tax rate relatively high. But if the proceeds from sale are spread out over future years, as with a seller-financed sale, the marginal tax rate may not be much different from the current tax rate. Perhaps a tax rate near the historical average for the taxpayer is as good as any other rate to use. The alternative minimum tax could become a factor and should be considered when estimating the contingent tax liability.

Even when using the market basis, changes in land values due to price fluctuations probably should not be made between the opening and closing dates of successive balance sheets. Capital gains on landholding are realized only when the land is sold. This gain may be realized only once in the lifetime of a farmer. Up until this time any gains or losses are only on paper because they have not been converted to cash. It is good practice to value land and similar assets conservatively and to hold these values nearly constant over several years. Changes should be made between the closing inventory of one period and the opening inventory of the next and should be reflected as capital gains or losses, not as farm income. This change can be shown as a separate line item in the asset list and under equity.

Depreciation Methods

Depreciation, as defined previously, is a systematic method for arriving at the value of a working asset based on its cost, as well as a method of prorating its cost over its productive life. Thus, there are three major purposes for depreciating an asset:

- To estimate current value
- To calculate business expense
- To adjust taxable income

These three purposes should be kept in mind in the discussion that follows. The three major causes of depreciation are as follows:

- Wear and tear—the wearing out of an item through use.
- Obsolescence—a time consideration as new technical developments render the old machines less useful or desirable compared with newer ones.
- Deterioration—a lessening in value caused by elements of nature. For machinery this might be rust; for buildings, decadence; for livestock, aging.

Obsolescence and deterioration are changes that take place even if the item is not productively used. Thus, they are fixed-cost elements that affect the farmer after purchase whether the item is used or not. Wear and tear is a variable cost, one the farmer has only when the item is used. Because these fixed-cost elements are usually greater and more significant than the variable-cost elements, depreciation is generally considered a fixed cost.

Many depreciable assets have residual value, commonly called salvage value, after they are no longer useful in the business. Salvage value may be an item's scrap value, its market value when the farmer plans to dispose of it, or its value when it ceases to depreciate. For most machinery and buildings, salvage value is generally scrap value. For breeding livestock, it is normally the anticipated cull market value. The amount to be depreciated, then, is the difference between an asset's cost (or basis, as defined by the IRS) and its salvage value. The IRS sets the salvage value at zero in its cost recovery system.

Before calculating annual depreciation, it is necessary to estimate an asset's useful or depreciable life. This lifetime is mainly an individual matter, determined by the farmer's own experience or sometimes by the experience of other farmers. Guidelines and rules established by the IRS often dictate the asset lives farmers apply, but these values may distort the real values of the assets. It should be remembered that the lives used by the IRS are tax recovery periods and not an estimate of how long a machine or other depreciable asset will last.

Four methods of depreciation commonly have been used:

- Straight-line method
- Declining-balance method
- Sum-of-the-years-digit method
- IRS cost recovery system

Any method that gives a reasonable depreciation allowance consistent with the purposes previously discussed and within legal limits is acceptable. Businesses typically use the method approved by the IRS, but the IRS has not been consistent over the years in prescribing uniform methods. Furthermore, the methods prescribed were more concerned with tax adjustment than with market value. IRS procedures have been modifications of historical methods. Thus, basic and traditional methods of depreciation are presented followed by a brief discussion of IRS allowances. To illustrate these four methods, assume the following situation involving a crop harvest machine:

- Purchased new in YRXX for $24,000
- Lifetime of 10 years
- Salvage value of $3,000
- Full year's depreciation the first year

The straight-line (SL) method is the old standby used by many farmers. Under this method the cost or other basis of the property less its estimated salvage value is deducted in equal amounts over its estimated useful life.

$$\text{Annual depreciation} = \frac{\text{Cost (or basis) minus salvage value}}{\text{estimated useful life}}$$

The annual depreciation deducted each year is constant.

EXAMPLE 2.1

Tabulations using the SL method follow:

Annual depreciation = ($24,000 – $3,000)/10 = $2,100
Value at end of first year = $24,000 – $2,100 = $21,900
Value at end of second year = $21,900 – $2,100 = $19,800
Value at end of third year = $19,800 – $2,100 = $17,700
Value at end of tenth year = $3,000

If the machine were purchased midyear, say April 1, the first year depreciation would be as follows: depreciation first year = $2,100 × 9/12 = $1,575 and depreciation last year = $2,100 × 3/12 = $525.

The declining-balance (DB) method is a fast write-off method. The largest depreciation deduction is taken the first year and becomes smaller in following years. Under this method a constant percentage of the remaining undepreciated balance is taken as depreciation each year as follows: annual depreciation = constant percent × undepreciated balance at beginning of year.

The DB method uses no salvage value and never reaches a zero balance. Thus, it becomes necessary to make adjustments in the last years of the assets life. It is common to switch to the SL method part way through the asset's life. Another method is to stop depreciation once the asset reaches salvage value.

Two methods are commonly used to specify the constant percentage to apply. The first method is called the double-declining balance method (DDB) or 200% DB method. The percentage to be multiplied is tabulated by dividing 200 by the asset's specified life. (This rate is twice that of the straight-line method.) The second method is called the 150% DB method. The annual percentage rate of depreciation is calculated by dividing 150 by the asset's life.

EXAMPLE 2.2

Applying the DDB method to the example problem, the following depreciations and values are obtained:

Constant rate of depreciation = 200/10 = 20% or 0.20
First-year depreciation = 0.20 × $24,000 = $4,800
End-of-first-year value = $24,000 – $4,800 = $19,200
Second-year depreciation = 0.20 × $19,200 = $3,840
End-of-second-year value = $19,200 – $3,840 = $15,360

If, instead of a full year, a part of a year were involved, say an April 1 purchase, these computations would be as follows:

First-year depreciation = 0.20 × $24,000 × (9/12) = $3,600
End-of-year value = $24,000 – $3,600 = $20,400
Second-year depreciation = 0.20 × $20,400 = $4,080
End-of-year value = $20,400 – $4,080 = $16,320

The sum-of-the-years-digits (SYD) method falls between the two methods previously discussed. It is calculated as follows:

$$\text{Annual depreciation} = (\text{cost} - \text{salvage value}) \times \frac{\text{remaining life in years at beginning of year}}{\text{sum of digits (total years of life)}}$$

The sum-of-years-digits may be tabulated by the formula SYD = L [(L + 1)/2], where L = life.

EXAMPLE 2.3

Applying the SYD method to the example problem, the following depreciations and values are obtained:

$$\text{First-year depreciation} = (\$24{,}000 - \$3{,}000) \times \left(\frac{10}{(1+2+3+\ldots+10) = 55} \right) = \$3{,}818$$

End-of-year value = $24,000 – $3,818 = $20,182

$$\text{Second-year depreciation} = (\$24{,}000 - \$3{,}000) \times \left(\frac{9}{55} \right) = \$3{,}436$$

End-of-second-year value = $20,182 – $3,436 = $16,746

If, instead of a full year, part of a year was involved, say an April 1 purchase, the first-year depreciation would be the full-year amount times the proportion of the year used as follows:

First-year depreciation = $3,818 × 9/12 = $2,864
Second-year depreciation = ($3,818 × 3/12) + ($3,436 × 9/12) = $3,532
Third-year depreciation = ($3,436 × 3/12) + [($24,000 – $3,000) × 8/55 × 9/12] = $3,150

FIGURE 2.4
Methods of Depreciation

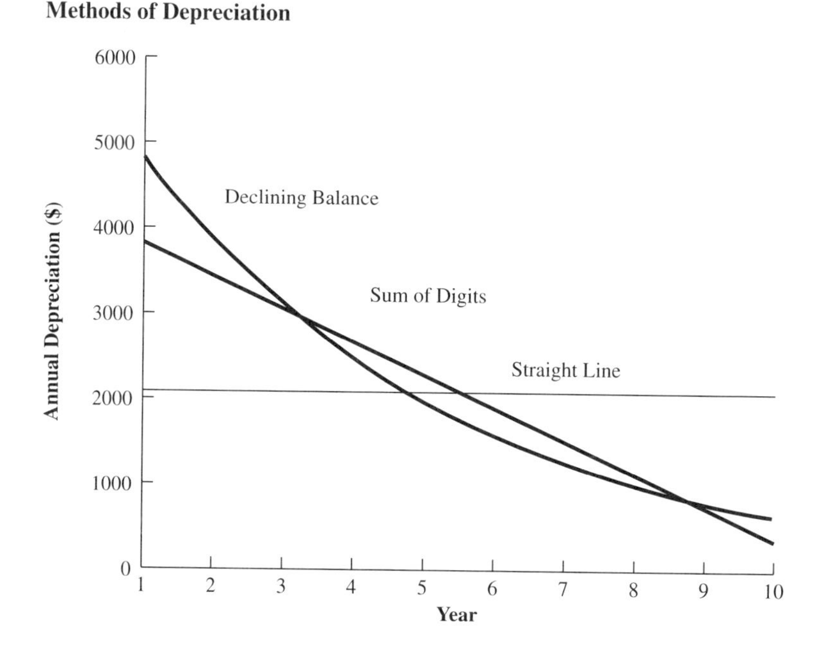

The SL, DB, and SYD methods are illustrated graphically in Figure 2.4.

The 1981 Tax Reform Act introduced the accelerated cost recovery system (ACRS) that is still in existence in a modified form. Its name was changed to the modified ACRS in 1986. It is now referred to as the MACRS or general depreciation system (GDS). The ACRS prescribes a schedule for writing off qualified depreciable investments as business expenses when tabulating taxable income. Even though this schedule is not intended to be used to estimate asset values, the unrecovered balances are being used in balance sheets by accountants, loan agents, and others. Thus, it is important to understand the GDS to be able to interpret the values given for depreciable properties. The IRS also allows using a straight-line depreciation method called ADS. The recovery period is longer for the ADS than the GDS. There must be consistency in the use of either method.

The GDS separates property into one of several recovery (depreciation) periods (e.g., 3-year, 5-year, 7-year, 10-year, 15-year, and 20-year periods). For example, farm trucks purchased in 1998 are 5-year property, machinery and equipment are 7-year property, fruit and nut trees are 10-year property, drainage facilities are 15-year property, and farm buildings are 20-year property. Currently, the 150% DB method is applied with the first year of ownership assumed to be 6 months. Thus, the depreciation recovery period is really 1 year longer than those shown.

The IRS also has a first-year depreciation allowance for direct-expense qualifying purchases.[3] This is section 179 deduction. Originally, this allowance was set at a relatively small amount but had risen to $18,000 in 1997, and is scheduled to reach $25,000 in year 2002. The allowance is treated like depreciation and shown on the same IRS form. Although farmers can benefit from this deduction for tax purposes, its use in tabulating values for the balance sheet are questionable. There is an accumulation effect if used for several years. The basis of a new asset is adjusted for the section 179 deduction before regular depreciation tabulations. Thus, it is possible to have the 179 deduction and a regular depreciation deduction the same year.

Deciding which method to use is troublesome if the purposes of the valuation are not emphasized. Keeping in mind the purposes of establishing a realistic market value as well as a business expense, the following considerations are useful in deciding which method to apply:

- Powered machinery (trucks, tractors, combines, silo unloaders, etc.) normally change in value faster in the early years than in the later years of service life, so faster write-off methods usually fit these items best.
- Nonpowered machinery (plows, planters, manure loaders, wagons, etc.) normally change in value at the same rate throughout the service life, so the SL method usually fits these items best.
- Farm buildings do not have an established market value, so the depreciation method selected should be related to the services provided. Usually the SL method approximates these items.
- New untried machines do not have an established length of service life, so it may be wise to select a faster write-off method.
- Small tools and equipment (hammers, wrenches, shovels, feed pans, buckets, etc.) whose purchase price is relatively small and whose service life is undetermined should be direct expenses and not depreciated. Normally, annual replacement costs are about equal to the depreciation. It seems more realistic to treat the replacement cost as a current expense and hold the inventory value nearly constant.
- The method selected should meet the legal requirements of the IRS if no other depreciation is being tabulated.

The method selected may not greatly affect the total farm depreciation. The accumulated depreciation of an asset cannot be any greater than 100 percent of its initial depreciable value. If the amount of depreciation is greater in the early years of life, it is less in later years. If equal amounts of new depreciable properties are added to the inventory each year, the selection of a method makes little difference. However, farmers tend to group their purchases of new properties according to fat-and-lean business years, and the selection of a method is important as an income leveler and tax adjuster.

There is an investment income advantage for faster write-off methods of depreciation as allowed by the IRS in addition to the income-averaging benefits of purchasing depreciable properties in years of higher-than-average income. This advantage, of course, is only present when depreciation from new investment purchases is not offset by smaller depreciation amounts claimed for purchases in earlier years. That is, if regular and equal replacement investments are made, the method of depreciation makes little difference because only 100% of the depreciable amount can be claimed as a business expense. But given the lumpy nature of new investments and the need for new purchases to reduce taxable income in high-income years, there is an investment advantage in selecting a faster write-off method. The money saved in taxes can be invested and reinvested, thus earning interest income credits during the catch-up years of the slower-rate method. This result is illustrated in Example 2.4, using a farm machine under GDS (150% declining-balance) with a 7-year life versus the same machine with ADS (straight-line) and a 10-year life.

EXAMPLE 2.4

Illustration of the income advantage of using the faster write-off method under GDS and ADS. The following information is given:

Item purchased for $12,000, estimated life 10 years, no salvage value, opportunity interest return 10 percent, the individual purchasing the item pays taxes at 35 percent at the margin.

	Year										
	1	2	3	4	5	6	7	8	9	10	11
Depreciation schedule:											
ADS, 10-year life	600	1,200	1,200	1,200	1,200	1,200	1,200	1,200	1,200	1,200	600
GDS, 7-year life	1,285	2,296	1,804	1,470	1,470	1,470	1,470	736	0	0	0
Income tax reductions on amount of depreciation claimed											
ADS[a]	210	420	420	420	420	420	420	420	420	420	210
GDS[a]	450	804	631	515	515	515	515	258	0	0	0
Difference($)	240	384	211	95	95	95	95	–162	–420	–420	–210
Investment advantage:											
Available money ($)[b]	240	624	835	930	1,024	1,118	1,212	1,050	630	210	0
Interest at 10% ($)	24	62	84	93	102	112	121	105	63	21	0

Accumulated advantage = $787
NPV[c] of annual earnings = $472

[a]Annual depreciation times tax rate.
[b]The first year $240 were saved in tax payments that could be invested. At 10% return these $240 would earn $24.
[c]Net present value (NPV) of the investment advantage at a discount rate of 10%.

In Example 2.4 it is assumed there are not any offsetting smaller depreciations in prior years that will nullify the effects of faster write-offs. That is, if the same investments were made each year and a faster method of depreciation was used, the same total depreciation would result as if the slower straight-line method were used. In the example the entire tax savings are available for investment, which is not usually the case. Money earned through the investment of taxes saved earns income (compounding principal) in the following period. In the second year, $240 is available for investment, but in the third year $648 is available ($240 plus $24 investment earnings plus $384 savings) for investment. The discounted value of $264, assuming 10% is also the discount rate, is $240. Discounting the annual earnings for the total period gives a present value for this product of $472. Now, consider the present-value advantage of writing off the total investment of $12,000 the first year by using IRS section 179 compared with either of these other methods.

It is good practice to separate items in the depreciation schedule according to their use or to some other meaningful allocation pattern rather than just listing them in date-of-purchase sequence. Following is an example:

- Powered machinery that has general use
- Specialized crop machinery
- Specialized livestock machinery
- General-purpose machinery
- Livestock by kind
- Buildings and improvements

Form 2.1 illustrates how a depreciation schedule is organized and transactions are posted.

FORM 2.1

Machinery, Equipment, and Moveable Building Depreciation Schedule

Item	Date Acquired	Cost or Basis $	Business Use %	Depreciation Method	Section 179 Deduction $	Life or Rate	Salvage Value $	Depreciation Amount $	YRXX-1 Depreciation $	YRXX-1 Value $	YRXX Depreciation $	YRXX Value $
1 Tractor, IH	3/16/XX-7	9,200	100	SL	0	10 yr	1,200	8,000	800	4,000	200	Traded
2 Tractor, JD	3/27/XX	24,200	100	GDS	10,000	7 yr	0	14,200			1,521	12,679
3 Plow, 4b	9/25/XX-7	1,800	100	SL	0	8 yr	200	1,600		680	150	Traded
4 Plow, 5b	9/20/XX	6,600	100	GDS	0	7 yr	0	6,600			707	5,893
5 Manure spreader	4/30/XX-5	2,500	100	200DB	0	7 yr	200	2,500	276	763	83	Sold
6 Auto, 4d	6/15/XX-3	12,000	25	GDS	800	5 yr	0	2,200	331	1,212	270	942
7 Crop harvester	1/10/XX	24,000	100	GDS	0	7 yr	0	24,000			2,570	21,430
8												
9												
10 Moveable buildings	Various	17,000								8,192	2,143	8,880
11												
12												
13												
Above totals										14,847	9,922	67,146
32												
33 All other items of machinery and equipment not shown										58,115	54,343	48,718
Column totals										36,962	10,356	62,006

Note: Items are illustrative and do not follow any particular system. Even when depreciation methods do not specify a salvage value it is good practice to show one.

Difficult Valuation Problems

Valuation of a depreciable asset when a trade is involved. The depreciable value of a new asset is the basis the owner has in it. The IRS defines basis to mean the cost or other original basis assigned to the property when first acquired. The basis of property purchased for money is ordinarily its cost. If property is acquired in a taxable trade, the basis is a combination of the cash difference paid and the adjusted basis of the asset traded. The adjusted basis is defined to mean the basis increased by any improvements or alterations that would raise its value or decrease its value by depreciation, depletion, or amortization deductions, or by deductions for losses such as casualty loss that would lower its value.[4] For most property the adjusted basis is the undepreciated balance or end-of-year value of a depreciable asset.

EXAMPLE 2.5 Calculation of Basis Where a Gain Is Received

Given:

Cash difference (boot) paid	$14,000
Trade-in allowance on item traded	+6,000
Market price of item purchased	$20,000
Adjusted basis (undepreciated balance) of item traded	4,000
Gain on item traded ($6,000 – $4,000)	2,000
Basis of item purchased ($20,000 – $2,000)[a] or ($14,000 + $4,000)[b]	$18,000

[a]Market price minus gain on item traded.
[b]Undepreciated balance of item traded plus cash difference.

Thus, it can be seen that the gain received on the item traded was used to reduce the cost of the item purchased. It may be useful to consider the above in terms of recovered and unrecovered costs. Purchase costs of depreciable assets are normally recovered through depreciation or sale.

Unrecovered cost of item purchased	$20,000
Unrecovered cost of item traded	+4,000
Total unrecovered cost	$24,000
Recovered cost on item traded	–6,000
Net unrecovered cost	$18,000

The newly purchased item is entered on the depreciation schedule at $18,000. The closing inventory reflects a $4,000 loss over the beginning inventory with regard to the item traded. No taxable gain is realized.

EXAMPLE 2.6 Calculation of Basis Where a Loss Is Sustained

Given:

Cash difference (boot) paid	$14,000
Trade-in allowance on item traded	+6,000
Market price of item purchased	$20,000
Adjusted basis (undepreciated balance) of item traded	$8,000
Loss on item traded ($6,000 – $8,000)	–2,000
Basis of item purchased ($14,000 + $8,000)	$22,000

From an unrecovered and recovered cost standpoint:

Unrecovered cost of item purchased	$20,000
Unrecovered cost of item traded	+8,000
Total unrecovered costs	$28,000
Recovered costs on item traded	–6,000
Net unrecovered costs	$22,000

Whether trades are treated as unlike or like exchanges makes little difference to the long-run balance sheets or income statements, disregarding tax considerations. Gains or losses on items traded are realized over a period of years (the life of the asset purchased) rather than in the year the transaction was made. Realization is through depreciation adjustments. The intent of the IRS is to encourage more realistic depreciation schedules. This type of schedule involves both the asset's life and the depreciation method and ties them more closely to market values.

Valuation of silage and standing crops. Because there generally is no market established for these items, one must be estimated. The production cost method is a possibility but often the costs of production are hard to determine. It may be possible to value silage in terms of equivalents of other crops for which a price is established; for example, 1 ton of silage is approximately equivalent to 1/4 ton of hay and 3.5 bushels of corn. Another method is to value the corn silage on the basis of the estimated corn grain equivalent adjusted for the difference in harvesting costs.

Standing crops may be valued in one of two ways. One method considers the costs of production, whereas the other works backward from the expected market value. There are accounting benefits from accumulating production costs for seed, fertilizer, irrigation, machinery operations, and labor in the current asset account. These accounts might be called production-in-progress accounts. At the time of harvest they could be moved into the expense account. The accumulated production expense is a reasonable approach to valuing standing crops. The market approach would estimate a yield and a price to tabulate expected gross income, then subtract from this amount the costs of harvesting and other unmet costs of production. Particularly in the market approach, an adjustment needs to be made for the uncertainty of harvest.

Valuing buildings whose original function has changed. Many buildings are being used for entirely different purposes than for which they were originally constructed. Examples include horse barns being used to house cattle, cattle sheds for machinery, and chicken houses for hogs. For the original owner, changes in use poses no problems because the same depreciation schedule can be continued. Conversion costs may need to be added to the depreciation schedule. The difficulty in valuing buildings comes mainly when the building changes ownership. The new owner considers the old building in terms of its new intended use. The value of a former chicken house that will be used to farrow sows is in terms of a farrowing house. Thus, a building's new value is in what it would cost to replace it with a building that would provide the same intended use, and depreciated to its present condition. Replacement cost minus depreciation may be an appropriate method to use when giving value to a newly purchased old building.

Valuing major repairs. Repairs are either current expenses or capital investments. There are no strict guidelines to apply when deciding how to handle major repair costs but the following considerations may be useful:

- Does the repair increase the life of the asset being repaired?
- Does the repair cost have a residual value that can be recovered through sale at the end of the current accounting period?
- How large is the repair expenditure relative to the size of the business?

If the repair increases the life of a depreciable asset, it probably should be capitalized and depreciated. Likewise, if the repair cost increases the market value of the asset, even though it does not increase the asset's life, it probably should be capitalized and depreciated. The third question is not so easy to answer; it relates to the materiality principle and suggests a practical approach. A small expenditure may not justify the accounting cost of adding it to the investment account and depreciation schedule. For a small farmer, retiring the tractor may be a major investment, whereas for a large farmer this action may be an annual event and thus be considered a routine expense.

Change in the general price level. General price-level changes affect asset values and equity. Agricultural prices are in continual fluctuation, not only because of changes in the general economy but also do to natural price cycles within agriculture. These price-level changes, if entered into the balance sheet, cause valuation changes that are not generated by management decisions. Management decisions are reflected in net income. One method for handling price-level changes so they do not affect net income is to keep particular asset prices constant within the accounting period. Changes in asset values are then made between the ending of one accounting period and the beginning of the next. Changes in value are then separately recorded as capital gains or losses, both for assets and equity. Shifts in prices should be reflected in the balance sheet only when the new prices are considered permanent. Land, breeding livestock, assets not being held for sale, and depreciable assets are the major assets affected. It is important to keep in the balance sheet the basis or adjusted basis of noncurrent property.

Capital Investment Accounts

Because of the data needed to develop and service the balance sheet it is important to have a record of capital purchases and sales. All the purchases farmers make to operate their businesses could properly be called capital, including feed, seed, fuel, repair items, livestock, machinery, and buildings. In this limited definition, investment capital refers only to items that are not purchased for resale purposes, that give services to other productive business activities, and that have a service life of more than 1 year. Included in this definition are breeding livestock, machinery and equipment, and buildings and improvements.

Often a trade is involved in the purchase of a new capital asset. The purchaser turns in a used item and pays a cash difference. The asset traded may have two values—the allowance given by the seller of the new asset and the value the asset had in the accounts of the purchaser. Thus, the total purchase price of the new item is the cash difference plus the trade-in allowance. However, from the standpoint of the purchaser the amount of total investment in the new item is the dollars of trade difference plus the value the traded item had in the purchaser's accounts. This amount is termed the tax basis or adjusted tax basis.

Thus, the capital asset investment purchase account should provide for recording the following:

- Date of purchase
- Description of item purchased
- Description of item traded
- Total market purchase price of item purchased
- Cash difference between the market price and the allowance given for the item traded
- Cash paid down when dealer financing is involved

- Value of the item traded in the accounts of the purchaser
- Tax basis of the item purchased

If the dealer finances part or the entire sale, a liability has been created, as already mentioned, and the cash difference and the cash paid down are not the same. The cash difference is the difference between the purchase (market) price and the trade-in allowance, whereas the cash paid down is the cash payment made at the time of purchase. If the dealer is not financing the purchase, the cash difference and the cash paid down are the same. This does not mean that some other source has not financed the purchase. Dealer-financed purchases should be recorded as loans or accounts payable in the liability accounts. Neither the purchase price nor the cash difference reflects the cash flow out of the business at the time of purchase. The cash paid-down column indicates the total cash drain on the business as a result of the new purchase. This figure is useful when tabulating the balance of cash coming in and going out of the business.

Capital asset sales also should be recorded. Assets disposed of through trade for a like item were discussed previously in this chapter. Direct sales of investment capital and credit received through trade when the item traded was unlike the item purchased should be recorded in this account. The items normally recorded in the capital-asset sales account are depreciable properties.

If the value received in sale or trade for unlike items is different from the undepreciated balance or adjusted basis for depreciable properties, a gain or loss has been sustained. If the sale price is larger than the account book value (usually the undepreciated balance) there is a gain, and if it is less there is a loss. For depreciable properties this adjustment could be considered a depreciation adjustment; that is, if the depreciation schedule truly reflects market values, the amount received in sale is the same as the undepreciated balance. Although neither depreciation methods nor market prices are that accurate or reliable, there is a bit of logic in the comparison. Regardless, it represents an income adjustment and the difference should be recorded in the income statement.

The following items should be recorded in the capital asset sales account:

- Date of sale
- Description of item sold
- Amount received through sale
- Cash received from sale if it is different from the sales amount
- Value that the item sold had in accounts of the seller at time of sale
- Gain or loss through sale
- Value of the item traded in the accounts of the purchaser
- Tax basis of the item purchased

Equity Enhancements

Equity (net worth) was defined as the residual difference between the assets owned and the corresponding liabilities. This definition is correct, but equity can be more meaningful than a single figure; it is not only interesting but also highly useful to know the sources of the equity position. In the beginning phases of a new farm business, as well as in later periods, money and other capital may flow into the business from outside sources (i.e., savings, gifts, inheritances, and other business activities). These sources could be accounted for in one or more categories of equity. Also, during the life of a business, assets may change in value due to general economic conditions outside the control of management. These capital gains or losses could be shown in a separate line item. Retained earnings—the sum of net income flows over time from business operations—is a major source of business growth and typically is shown in a separate line. This item is typically tabulated in double-entry commercial accounting systems. The current net income is usually separated from retained earnings and added at the beginning of the following accounting period. The following divisions of equity for the balance-sheet statement summarize the previous discussion:

1. Current net farm income
2. Equity balance forward:
  a. From retained earnings
  b. From other businesses
  c. From gifts and inheritances
  d. From capital gains
3. Current transfers in:
  a. From other business activities
  b. From gifts and inheritance
4. Current transfers out:
  a. To other business activities
  b. To family living or other uses
5. Change in capital values due to the general price level

This equity enhancement of the balance sheet is very similar to the "Statement of Owner Equity" recommended by the FFSTF. It could be formulated as a separate statement but it is better to make it part of the double-entry accounting system as developed in Chapter 3.

At the beginning of each new accounting period, the amounts in 1, 3, 4, and 5 are zeroed out and added to the items in 2. Net income (1), less family uses (4b), becomes part of the next year's retained earnings (2a), 4a is subtracted from 3a and added to 2b, 3b is added to 2c, and 5 is added to 2d. The growth that takes place as a result of the managerial ability of the owner is captured in 1 and retained in 2a. Note that the amount of income that can be retained in the business is the residual of the amount that is transferred out in support of personal and family uses. The sum of the changes in the business caused by outside influences is shown in 2b and 2c. Those changes resulting from general price-level increases and decreases are shown in 2d. The amounts recorded in 2d and 5 are related to the difference between the cash- and market-basis assets, less any contingency taxes recorded as liabilities. The amount in 5 is the difference from the previous year. Thus, changes in equity from increases and decreases in the general price level are only recorded in the cash-basis column. The growth in assets and equity is affected greatly by the amount removed from the business for family support (4b).

Beginning and Ending Balance Sheet Comparisons

The balance sheet for the FABA farm at the end of the current YEAR is shown in Table 2.4. In this table, both the beginning and ending dollar amounts are shown, which allows comparisons to determine where changes have occurred. Longer periods of comparison are even more revealing.

TABLE 2.4

Balance Sheet for the FABA Farm, YRXX

Assets	Beginning	Ending	Liabilities	Beginning	Ending
Current:			*Current:*		
Cash	$ 12,200	$ 10,819	Feed bill	$ 1,635	$ 885
Life insurance (cash value)	4,500	4,700	Cattle note	30,000	40,000
Savings and investments	10,050	10,840	Production	18,000	16,000
Feed	58,738	67,234	Interest, accrued	13,811	15,238
Nonfeed crops	8,086	8,978	Principal payments due:		
Feed hogs	41,093	29,016	Intermediate	2,130	11,061
Feeder cattle	36,518	51,600	Long term	22,524	24,141
Total current assets	$ 171,185	$ 183,187	*Total current liability*	$ 88,100	$107,325
Intermediate:			*Intermediate:*		
Machinery and equipment	$ 72,962	$ 115,864	Auto loan (farm share)	$ 2,279	$ 0
Gilts		2,400	Crop-harvesting machinery		16,487
Sows and boars	19,660	20,225	Tractor		24,731
Portable buildings	8,192	8,880	*Total intermediate*		
Total intermediate assets	$ 100,814	$ 147,369	*liability*	$ 2,279	$ 41,218
Fixed:			*Long term:*		
Farmland	$ 832,800	$ 832,800	Silo	$ 17,000	$ 12,680
Buildings and improvements	80,556	73,968	Farm mortgage	318,674	298,853
Total fixed assets	$ 913,356	$ 906,768	*Total long-term liabilities*	$335,674	$311,533
Total assets owned	**$1,185,355**	**$1,237,324**	**Total liabilities**	**$426,053**	**$460,076**
Landlord assets rented	180,000	180,000	**Equity**	**$759,302**	**$777,248**
			Change in equity		$ 18,016
Total assets managed	**$1,365,355**	**$1,417,324**	**Capital ratios:**	*Beginning*	*Ending*
Change in total assets managed		$ 51,969	Net capital ratio	2.78	2.69
			Current asset ratio	1.94	1.71

Contingent liabilities were purposely left out of this statement. Because the historical background was not known, it was not possible to designate the sources of equity.

The balance sheet for the FABA farm as shown in Table 2.4 is used in other parts of the discussion on farm accounting and business analysis that follow.

Notes

1. A copy can be obtained by writing Farm Financial Standards Council, 1163 E. Ogden Ave., Suite 103-051, Naperville, IL 60563–8529.
2. Agricultural Committee, *Farm Credit Analysis Handbook,* American Bankers Association.
3. For a discussion of IRS depreciation regulations see the current edition of *The Farmer's Tax Guide*, Bulletin 25.
4. *Farmer's Tax Guide* , Internal Revenue Service, Publ. 225.

CHAPTER 3

Financial Accounts—Income Statement

Introduction

All farmers keep track of their receipts and expenditures and compute an income statement, even though it be no more than required to meet Internal Revenue Service (IRS) regulations. However, tax reporting of income is not enough to take advantage of the management information available through a well-defined and executed set of farm records and accounts. Of course, they must be analyzed to glean information useful for improving business and production efficiency. The time required to keep a good set of records and accounts is not excessive and may not be greatly different from the time needed to meet IRS requirements. This chapter describes methods and procedures for determining business income and profitability, not only for the total business but also for cost and profit centers. The income statement gives meaning and adds clarity to the balance sheet discussed in Chapter 2. The income statement can be kept separately or integrated into the balance sheet. Procedures for combining them are presented in this chapter. The analysis of the balance sheet and income statement is presented in Chapter 4.

The income statement is a dynamic account of the production activities of the business. Receipt accounts include product sales and related income. Expense accounts include the financial requirements of production (e.g., costs for labor, machinery, feed, seed, fertilizer, utilities, insurance, property tax, and buildings).

ACCOUNTING SYSTEMS

Most farmers still use the single-entry system of accounting, although many have adopted double-entry procedures, particularly those with computer accounting programs. Both methods are acceptable and if accurately maintained give the same income statement. With single-entry accounting, income and expense transactions are

The content of this chapter is a revision of material in James, Sydney C. and Everett Stoneberg, *Farm Accounting and Business Analysis,* 3rd ed., Iowa State University Press, 1986.

recorded only once. The double-entry system rests on the principle that every transaction has both a source and a destination, and thus at least two entries must be made. Single-entry accounting is discussed first, and the general form of the income statement is developed.

Single-Entry Accounting

There are two general systems of recording data in the single-entry account, the cash method and the accrual method. With the cash method, entries are made in the accounts only after the cash or money has changed hands. A purchase is recorded when it is paid for, not when the transaction takes place (if at a different time); a receipt is recorded when the money is received and not when the sale was transacted (if at a different time). For example, fuel purchased in December YRXX but not paid for until January YRXX + 1 is recorded as a YRXX + 1 expense. Inventory changes reflecting crop and livestock increases or decreases are not included because they have not been converted to cash. Depreciation, a noncash item, is counted as an expense. In the long run, most of these deferred purchases and the sales and inventory increases and decreases show up as cash, and net income over time is reasonably accurate.

The accrual method follows the idea that income should be accurately described each year. Thus, transactions are recorded when binding, whether money changed hands or not; changes in inventory are included. The fuel purchase described above would be recorded as an expense for YRXX when delivery was received, not in YRXX + 1 when the payment was made. A beginning inventory that was later sold for cash is recorded not only as a cash sale but also as an inventory decrease. A crop that is raised but not sold is likewise shown as an increase in inventory value.

A farmer interested in analyzing the farm business for management purposes cannot tolerate the transfers of production and expenses, or the fluctuations of income inherent in the cash method. However, there are advantages and flexibility in the cash method compared with the accrual method when calculating taxable income. Thus, some farmers allow for both methods in their accounting systems. It is logical that cash income can be derived more easily from the accrual method than accrued income from the cash method; hence, the accrual method is emphasized.

Income Statement Formats

The statement of cash flows is described first to help understand the concept of an income statement, the nature of single-entry accounting, and the differences between the cash and accrual methods of accounting. Almost everyone is familiar with the monthly bank statement issued to each depositor. This cash-flow account is used to inform depositors of their cash balance at the close of the monthly accounting period. Included are all of the transactions that have affected the beginning cash balance to arrive at the ending cash balance. Every check (bank draft) written on the account is shown and every deposit, from whatever source, is shown. The depositor can compare this statement with personal records, such as checkbook stubs and business receipt and expense accounts, to determine if there are recording errors or omissions.

In addition to being useful for reporting balances and checking errors, the statement gives some information about the business. The size of the business can be estimated from the size of the individual transactions, and the volume of business can be estimated from the number of monthly transactions. The minimum monthly balance may give some indication of business stability, and the difference between the beginning and ending balance may tell something about business growth. But, it would be very risky to draw many conclusions about the business from 1 month's statement, or even a whole year of statements.

Although useful for reconciling with the checkbook, bank statements are not very helpful in analyzing productive activities of the business. Neither are they substitutes for farm financial accounts. Farmers should duplicate every bank transaction someplace in their financial accounts. A withdrawal made to purchase a capital asset should show up in the capital purchase account, one made to pay off a loan should be recorded in the credit account, and one made to pay for operating expenses such as labor or fertilizer should be recorded in the business-operating expense account. Similarly, a deposit from the sale of livestock should be recorded in the livestock sales account, and one from a bank production loan should be added to the credit account.

A cash journal or transaction list can be developed to keep the same financial data as the bank statement, but also to identify the sources of deposits and the destinations of withdrawals. Columns are labeled to record the date, check, or deposit number to whom the check was written or from whom the deposit was received; the dollar amount; and the running bank balance. Data are recorded chronologically. This account, which is very similar to the one many people keep with their checkbooks, adds considerably to the bare figures shown on bank statements. With this account, the sources of deposits are shown, as are the destinations of expenditures. Business items can be separated from those belonging to the family.

However, there are still difficulties in separating the activities of the business into meaningful categories. To do so requires sorting through many transactions and making lists of the categories wanted (e.g., crop sales, livestock sales, labor payments, feed purchases, machinery costs, and loan payments). This process is tedious but not unlike what some people do. A single check written to cover the purchase of unrelated items, such as gasoline and a candy bar or bolts and fertilizer, may not give this transaction detail in the journal. Another complication is that checks often are written to obtain pocket cash, which is then used to purchase various business or personal items that go unrecorded.

Assuming that it is possible to categorize and accumulate the many cash transactions over one accounting period (usually 1 year), a statement of cash flows could be summarized as shown in Table 3.1. Note that cash expenditures may be for labor, fuel, seed, fertilizer, and other farm operations; livestock and feed purchases; ownership expenses; capital purchases; and loan payments, including principal and interest. Receipts include the sale of farm products such as grain, vegetables, and livestock; program payments from the government; refunds and dividends and other miscellaneous farm income; sales of capital assets; and deposits from new cash loans. Because this account is for farm cash flow, payments for nonfarm items should not be shown on the farm business account. The net cash-flow amount reflects cash that has been available for family living and nonfarm investments, in addition to changes in the cash bank balance.

A multicolumn journal can be used to make the cash entries more functional. Each column would be for a different category of receipt or expense. Thus, livestock expenses, crop expenses, labor hired, feed purchases, etc., can be recorded in separate columns and summarized at the end of each month, year, both. It is desirable to have a column for recording family cash transfers if a joint farm-family bank account is maintained. Some farm account books have been organized this way. Others list the categories on different pages with a summary page to bring the accounts together.

TABLE 3.1

Statement of Cash Flows, FABA Farm, YRXX

Cash Expenditures		**Cash Receipts**	
Operating expenditures:		Farm production:	
Hired labor	$ 12,946	Livestock sales	$172,233
Livestock expenses	5,484	Crop sales	107,073
Crop expenses	43,448		$279,306
Fuel and oil	6,013		
Machinery repair	4,013	Government payments	$ 9,500
Building repair	1,925		
Machine hire	10,425	Other:	
Farm utilities	3,410	Interest and dividends	990
Miscellaneous	1,606	Miscellaneous	3,774
	$ 89,270		$ 4,764
Feed purchases	$ 27,301	Sale of capital assets	$ 400
Livestock purchases	$ 41,870		$293,970
		New farm loans:	
Ownership expenditures:		Capital purchase	50,000
Taxes	4,584	Livestock purchases	40,000
Insurance	3,562	Farm operation	16,000
			$106,000
Interest on loans	30,353	Total receipts	$399,970
Cash rent	12,500		
	$ 50,999	Cash-flow summary:	
Capital purchases:		Cash receipts	$399,970
Cash difference	$ 72,831	Cash expenditures	354,925
		Net cash flow	$ 45,045
Principal payments on loans:			
Real property	22,524		
Machinery and equipment	2,130		
Livestock purchases	30,000		
Farm production	18,000		
	$ 72,654		
Total expenditures	$354,925		

Net farm income—cash method. The statement of cash flows is a very helpful document in controlling cash, but it does not measure the net income produced by the business. An income statement in common use among farmers is the net income statement tabulated using the cash method. It resembles the statement (Schedule F, cash method) used by most farmers to report their taxable income to the IRS.

The major differences between the statement of cash flows and the income statement tabulated by the cash method are in how new loans and loan payments, and purchases and sales of capital assets are handled. First, consider a deposit from a new loan and the recording of principal payments to retire the debt. The money received from a new operating loan is typically deposited in the bank and used to finance farm production (i.e., buy feed, buy fertilizer, pay labor, and make insurance payments). Similarly, farm sales may be used to make principal payments on loans. But loan receipts are not product income and principal payments on debts are not business expenses. Borrowing money creates a liability that must be paid back. To include both the loan payment and the purchase of the goods for which the loan was contracted as

TABLE 3.2

Cash Method Income Statement for the FABA Farm, YRXX

Cash Expenses		Cash Receipts	
Operating expenses:		Farm production:	
Hired labor	$ 12,946	Livestock sales	$172,233
Livestock expense	5,484	Crop sales	107,073
Crop expense	43,448		$279,306
Fuels and lubricants	6,013		
Machinery repair	4,013	Government payments	$ 9,500
Building repair	1,925		
Machine hire	10,425	Capital asset sales gain	$ (210)
Utilities (farm)	3,410		
Miscellaneous	1,606	Miscellaneous	$ 4,764
	$ 89,270	Total receipts	$293,360
Feed purchases	$ 27,301		
Livestock purchases	$ 41,870		
Ownership expenses:		Farm cash summary:	
Property taxes	4,584	Cash receipts	$293,360
Insurance	3,562	Cash expenses	241,384
Interest	30,353	Net cash income	$ 51,976
Cash rent	12,500		
	$ 50,999		
Depreciation:			
Machinery	25,356		
Improvements	6,588		
	$ 31,944		
Total expenses	$241,384		

expenses would be double counting. A test of how to record a transaction is to consider how the balance sheet is affected. New loans increase liabilities and are offset by an increase in cash. Likewise, when a debt payment is made cash goes down, as do liabilities. In both cases equity is not affected. And if equity is not affected then there is no receipt or expense.

Second, the purchase of a farm capital asset with a life longer than 1 year is not totally a business expense in the year of purchase because the asset still has recoverable value at the end of the accounting period. The change in value that takes place from the date of purchase (or the beginning of the accounting period) to the end of the accounting period is the production expense. Hence, capital purchases are replaced with depreciation expenses. Similarly, the sale of a capital asset does not constitute a farm product sale. A capital asset sale merely converts into cash the market value of the asset. This form of wealth is more liquid, but the owner is no wealthier. However, the gain or loss from sale needs to be recorded in the income statement.

The value of products used in the home should be recorded as income only if a charge has been added to expenses for their production. This is an offsetting entry.

An income statement with these modifications is shown in Table 3.2. Note the following three changes from Table 3.1: (1) Capital purchases of $72,831 and capital sales of $400 were dropped. (2) Principal payments on loans of $72,654 and new loans of $106,000 were dropped. (3) Depreciation of $31,944 and capital sales loss of $210 were added.

Net farm income—accrual method. Whereas the cash farm income statement reports long-term average productive income accurately it cannot be relied on to give a true measure of net income produced in any one period. The cash method of tabulating farm income has been for convenience and not accuracy in reporting true farm income. Accrual-method accounting solves the problems of fluctuating inventories and annual changes in accounts payable and receivable.

The net income statement tabulated by the accrual method includes cash and non-cash income and expenses and accounts for inventory changes of productive enterprises. An income statement tabulated by the accrual method is shown in Table 3.3. Note how it differs from the cash income statement shown in Table 3.2.

- Inventories were included, and the increases exceeded decreases by $12,958.
- Feed purchases were less than feed account payments by $750 ($27,301 – $26,551).
- Interest payments were $1,427 less ($31,780 – $30,353) than had accrued.
- Net income was larger under accrual accounting by $12,281 ($64,257 – $51,976).

It is assumed that all other purchases and sales not otherwise identified were for cash, or the deferred payment purchases or sales were paid for or receipted within the accounting period.

TABLE 3.3

Accrual Income Statement for the FABA Farm, YRXX

Accrual Expenses		Accrual Income	
Operating expenses:		Receipts:	
Hired labor	$ 12,946	Livestock sales	$172,233
Livestock	5,484	Crop sales	107,073
Crops	43,448		$279,306
Fuels and lubricants	6,013	Government payments	$ 9,500
Machinery repairs	4,013	Other receipts	$ 4,764
Building repairs	1,195	Capital asset sales gain (loss)	$ (210)
Machine hire	10,425	Total receipts	$293,360
Utilities (farm)	3,410		
Miscellaneous	1,606		
Total operating	$ 89,270		
Feed purchases	$ 26,551		
Livestock purchases	$ 41,870	Inventory adjustments(+ or –):	
		Feeder livestock	3,005
Ownership expenses:		Breeding livestock	565
Property taxes	4,584	Crops and supplies	9,388
Insurance	3,562	Total inventory adjustments	$ 12,958
Interest, accrued	31,780	Total income	$306,318
Rent	12,500		
Total ownership	$ 52,426		
Depreciation:			
Machinery and Equipment	25,356		
Buildings and Improvements	6,588	Farm accrual income summary:	
		Income	$306,318
Total depreciation	$ 31,944	Expenses	242,061
Total expenses	$242,061	Net accrual income	$ 64,257

Some accounting systems leave the cash amounts for deferred-payment purchases the same as in the cash-method income statement and then add an adjustment to show the differences between purchases and payments (e.g., $750 for feed and $1,427 for interest in the previous example). Although this method accurately reflects business income, it does not accurately represent the categories in the income statement. Inaccuracy in reporting expense and income categories distorts business analyses and does not show corresponding changes in current assets and liabilities. The appropriate place to record accounts payable and receivable is the balance sheet.

The accrual income statement could be referred to as the production income statement because it reflects the total value of production and the costs of obtaining that production. Each statement reflects the total financial changes for the year in which it is tabulated; thus, one year can be compared with another to measure production, production requirements, prices, and income. This statement is the only true income statement of all those discussed, and thus it is the most useful for business analysis and decision making.

There is no common income statement format, and many businesses arrange their data specifically to fit their needs. However, the Farm Financial Standards Council (FFSC) has issued a recommended format containing most elements in Table 3.3, but it is organized slightly different (see note 1, chapter 2). We selected the format illustrated for clarity and because it is easily understood and adaptable to reporting income by responsibility and profit centers, as illustrated later in this chapter. The format selected ought to fit double-entry accounting procedures and be easily adaptable to computerization.

Table 3.4 repeats the data in 3.3 in a little different format, makes additional tabulations, calculates cash income from accrual income, and illustrates how a rented farm may be added to an owned farm. Inventory figures can be obtained from the net worth statement shown in Table 2.4. Note that Table 3.4 is divided into three columns—operator, landlord, and total farm. The "total farm" column reflects the situation that would exist if the farm were completely owner operated. Thus, the crop-share rent payment would be available for sale, and there would be no cash rent transfers. The leasing arrangement illustrated is for cash rent. A crop-share lease would have more shared expenses.

The income statement formats used in Tables 3.3 and 3.4 are organized around operating and ownership expenses, and around cash and noncash income and expenses. Cash as used here does not mean that cash is always exchanged, but rather that a current cash-payment obligation has been created. The terms *debits* and *credits* are used with respect to double-entry accounting. For now debits mean expenses and credits mean receipts in the income statement. A more strict definition of debits and credits is given in the double-entry accounting section of this chapter. Debits and credits may be for either cash or noncash purchases and receipts. A credit purchase is an example of a noncash debit, and a production to inventory is an example of a noncash credit. Operating or variable expenses include hired labor, machinery operating costs, irrigation water, fertilizer and other cropping costs, health and other livestock costs, and farm utilities. Purchases of feed and livestock are listed separately to facilitate the tabulation of income and efficiency measures discussed later. Ownership or fixed expenses include taxes, rent, farm insurance, and interest. Depreciation is an ownership expense but it is separated out because it is not a cash-type item. Ownership expenses tend to remain the same whether the farm is operated or not, and these expenses do not change with the selection of enterprises or with the level of production intensity.

TABLE 3.4

Income Statement for the FABA Farm, YRXX

Item	Operator	Landlord	Total Farm
INCOME MEASURES			
Cash Income			
Hog sales	$109,796	$ —	$109,796
Cattle sales	62,437	—	62,437
Feed crop sales	42,922	—	42,922
Nonfeed crop sales	64,151	—	64,151
Cash rent	—	12,500	—
Government payments	9,500	0	9,500
Capital asset sales gain (loss)	(210)	0	(210)
Miscellaneous receipts	4,764	0	4,764
Total cash	$293,360	$12,500	$293,360
Noncash Income			
Inventory adjustments:			
Livestock	3,570	—	3,570
Feed crops	$ 11,496	$ —	$ 11,496
Purchased feed	(3,000)	—	(3,000)
Nonfeed crops	892	—	892
Total noncash	$ 12,958	$ —	$ 12,958
Total Business Credits (Gross Revenues):			
Cash Income	$293,360	$12,500	$293,360
plus noncash income	12,958	—	12,958
	$306,318	$12,500	$306,318
EXPENSE MEASURES			
Cash Expenses			
Operating:			
Hired labor	$ 12,946	$ 200	$ 13,146
Livestock expenses	5,484	—	5,484
Crop expense	43,448	—	43,448
Machinery operating expenses	10,026	95	10,121
Machine hire	10,425	—	10,425
Buildings and improvements repair	1,925	240	2,165
Utilities (farm share)	3,410	—	3,410
Miscellaneous	1,606	235	1,841
Total operating	$ 89,270	$ 770	$ 90,040
Feed purchases	$ 26,551	$ —	$ 26,551
Livestock purchases	$ 41,870	$ —	$ 41,870
Ownership:			
Property taxes	$ 4,584	$ 950	$ 5,534
Insurance	3,562	320	3,882
Interest	31,780	—	31,780
Rent	12,500	—	—
Total ownership	$ 52,426	$ 1,270	$ 39,926
Total cash expenses	$210,117	$ 2,040	$198,387

TABLE 3.4

Income Statement for the FABA Farm, YRXX—*continued*

Item	Operator	Landlord	Total Farm
EXPENSE MEASURES *(cont.)*			
Noncash Expenses			
Depreciation:			
Machinery and equipment	$ 25,356	$ —	$ 25,356
Buildings and improvements	6,588	335	6,923
Total noncash	$ 31,944	$ 335	$ 32,279
Total Business Debits			
Cash expenses	$210,117	$ 2,040	$198,387
plus noncash expenses	31,944	335	32,279
	$242,061	$ 2,375	$230,666
GROSS RETURNS MEASURES			
Value of Farm Production			
Total business credits	$306,318	$12,500	$306,318
minus:			
Livestock purchases	41,870	—	41,870
Feed purchases	26,551	—	26,551
	$237,897	$12,500	$237,897
NET RETURN MEASURES			
Net Farm Income			
Total business credits	$306,318	$12,500	$306,318
minus total business debits	242,061	2,375	230,666
	$ 64,257	$10,125	$ 75,652
Net Farm Cash-Flow Income			
Total cash income	$293,360		
minus total cash expenses	210,117		
	$ 83,243		
Plus:			
New loans	106,000		
Unpaid accounts adjustment	(750)		
Cash interest adjustment	1,427		
Capital sales *minus* capital gains	610		
	$107,287		
Minus:			
Purchases (cash difference):			
Machinery and equipment	70,000		
Buildings and improvements	2,831		
Principal payments on loans	72,654		
	$145,485		
Net cash-flow	$ 45,045		

Noncash expenses are primarily depreciation. Recall that depreciation accounts for the change in value from using machinery, buildings, equipment, and other depreciable items in the business. Because the sum of depreciation charged over an asset's life seldom equals its total change in value (purchase price minus sale value), it may be necessary to adjust the income statement by the difference between the undepreciated balance and the sale value. The location of this adjustment in the income statement is arbitrary. It is shown in Tables 3.3 and 3.4 under receipts, "Capital asset sales gain (loss)." However, because it is generally a depreciation adjustment it logically could be shown next to depreciation under expenses. It was placed under receipts because with the fast write-off methods of depreciation typically in use, the amount received has generally been larger than the undepreciated value resulting in a cash gain. Also, some sales of capital assets may not be on the depreciation schedule.

The total of cash and noncash expenses equals the total business debits. The total of cash and noncash income equals the total business credits. Total business credits minus total business debits equals the accrual net farm income. Net income measures the following:

- The return to unpaid labor (operator and family)
- The return to unpaid management
- The return to unpaid capital (equity)

All other items have been paid a return for their use in the business. Market prices have been charged for operating and ownership expenses, depreciation has been charged for machinery and buildings, and interest has been charged for liability capital used in the business. Accrual net farm income as defined previously is very similar to income tax reporting under the accrual method as defined in the *Farmer's Tax Guide* and is discussed in Chapter 13.

Net farm income reflects the cash and noncash monetary value available for family living, debit retirement, increasing equity in the business, or other investments and savings. Some of this income may be in the form of inventory changes. With an inventory decrease, the cash-flow income would be greater than the accrual net income because more of the inventory items were sold than replaced.

Income tax is not usually included as a business expense. Regular corporations pay income tax, but tax-option corporations, partnerships, and single proprietorships do not. Income taxes are the responsibility of the owners and are paid out of net income.

Net cash-flow income was discussed previously and is illustrated in Table 3.1. It is possible to tabulate net cash-flow income, or net cash income (Table 3.2), from the accrual net income statement (Tables 3.3 and 3.4). Only a few adjustments are necessary as can be seen by comparing Tables 3.1–3.3. These adjustments were specifically noted previously, following the presentation of each statement. The differences in accounts receivable and accounts payable also can be found in the balance sheet. Table 2.4 shows that the feed credit account balance was $750 more in the beginning of the year than at the end, and the unpaid interest liability was $1,427 more at the end of the year than at the beginning. Cash-flow income includes the total proceeds from the sale of capital assets. Recall that the profit or loss amount from sale of capital assets was tabulated as a difference between the total sales receipts and the undepreciated balance (i.e., adjusted basis). Thus, to show the total proceeds from sale only the undepreciated balance needs to be added.

Gross revenues are often calculated from the income statement to measure gross volume of business and value of all items produced on the farm. Gross revenues is synonymous with total business debits as here defined. Note that gross revenues equals the sum of product sales, after adjusting for inventory changes and miscellaneous income. Gross revenues is a good measure of farm productivity and business volume, especially for nonlivestock farms. But when livestock is a major enterprise, or farm-raised grains are purchased for livestock feed, gross income is not a reliable statistic for comparing different farms, or even the same farm over time. Value of farm production (VFP) adjusts for these two items.

VFP is tabulated by subtracting from gross revenues the purchase costs of feeder livestock and livestock feeds. The contribution of livestock to VFP is tabulated by the formula [(closing inventory + sales) – (beginning inventory + purchases)] less feed costs. For example, consider two farmers with the same livestock sales, one selling farm-raised livestock and the other selling livestock that had been purchased. If the cost of purchasing the feeders were not subtracted, each farmer would have the same gross revenue from livestock production even though the purchased feeders were raised on some other person's farm.

The subtraction of purchased livestock feeds from total business credits is not so clear. Feed purchases are removed to adjust for difference among farm feeders and to give consistency to VFP. Consider two farmers with identical farms and operations except that one feeds home-grown grain to his or her livestock and the other farmer sells the grain and purchases a complete feed. If these two farmers are to have similar VFP figures (as they should), an adjustment must be made for the different source of feed for their livestock because the second farmer has larger grain sales and hence a larger gross income. If livestock feed purchases are subtracted, the VFP figures for the two farmers should be very similar. The farmer with the larger grain sales also would have the larger feed purchases. In summary, VFP is calculated by subtracting from total business credits livestock purchases and feed purchases. VFP also can be defined as the total value of crops produced plus the increase in value of livestock minus the total value of feeds fed plus miscellaneous income. This identity is illustrated in Chapter 4.

Note that the tabulation of net cash-flow income is the reverse of tabulating accrual net income from cash-flow income as illustrated in Tables 3.1–3.3. The formula for doing so is shown in the tabulation in Table 3.4.

The differences between the several income tabulations discussed in this section are illustrated in Table 3.5. The format deviates from those used previously to emphasize the differences between the income statements. Notice the way deferred payment purchases are handled as illustrated in feed purchases, how interest expense tabulations differ between accrual and cash accounting, how capital asset purchases and sales are accounted for, and what inventory information is used. Observe that the ending value of machinery inventory is equal to the beginning inventory plus purchases minus the undepreciated balance of sales minus depreciation.

Converting cash (single-entry) income to accrual income. The conversions of cash and accrual systems of accounting have been specifically applied in this section. But because so many farmers prepare only the Schedule F cash income statement for the IRS, a general conversion format to an accrual income seems useful. The following adjustments are required:

TABLE 3.5

Comparative Net Income Tabulations for Cash-Flow, Cash and Accrual Methods

Item			Amount ($)	Net Cash-flow Income ($)	Net Farm Income: Cash Method ($)	Net Farm Income: Accrual Method ($)	Value Farm Product ($)
Receipts:							
Hog sales			54,000	54,000	54,000	54,000	54,000
Breeding swine sales			9,000	9,000	9,000	9,000	9,000
Soybean sales			25,000	25,000	25,000	25,000	25,000
Corn sales			10,000	10,000	10,000	10,000	10,000
Miscellaneous farm income			3,000	3,000	3,000	3,000	3,000
Value of farm products used in home[a]			400		400	400	400
Capital asset sales: gain			3,000	3,000	3,000	3,000	3,000
Undepreciated balance			6,000	6,000			
Bank loan for farm operation			25,000	25,000			
Total receipts				135,000	104,400	104,400	104,400
Expenses or expenditures:							
Cash operating expenses			28,000	28,000	28,000	28,000	
Feed purchases, all on account			18,500			18,500	18,500
Feed account payments			18,000	18,000	18,000		
Swine purchases (breeding)			2,800	2,800	2,800	2,800	2,800
Taxes and insurance			9,500	9,500	9,500	9,500	
Interest on loans: accrual			7,100			7,100	
cash			6,500	6,500	6,500		
Loan principal payments			26,000	26,000			
Depreciation, machinery and buildings			17,500		17,500	17,500	
New machine purchase ($5,000 down, $15,000 credit)			20,000	5,000			
Total payments				95,800	82,300	83,400	21,300
Inventory:	***Beginning***	***Ending***	***Change***				
Feeder pigs	$21,000	$23,000	$2,000			2,000	2,000
Breeding swine	7,500	8,000	500			500	500
Purchased feed	3,600	3,000	(600)			(600)	(600)
Corn	10,000	12,000	2,000			2,000	2,000
Seed and supplies	1,900	2,100	200			200	
Machinery	86,000	87,000	1,000				
Improvements	22,000	17,500	(4,500)				
Land	340,000	340,000	0				
Total inventory change			600			4,100	3,900
Value of grain production			$34,000				
Net amounts				39,200	22,100	25,100	87,000

[a]Value of products for which a cost of production was charged to the farm.

Cash Basis Income Statement

plus:

- Increases in inventory (A)
- Increases in accounts receivable (A)
- Decreases in accounts payable (L)
- Increases in prepaid expenses (A)
- Decreases in accrued unpaid expenses (L)

minus:

- Decreases in inventory (A)
- Decreases in accounts receivable (A)
- Increases in accounts payable (L)
- Decreases in prepaid expenses (A)
- Increases in accrued unpaid expenses (L)

equals:

Accrual Income Statement
(usually synonymous with double-entry)

The adjustments are differences between the ending and beginning balance-sheet accounts. The letter in parentheses following the items listed refers to assets (A) and liabilities (L).

For many farmers in all years, and for most during some years, the changes in accounts receivable, accounts payable, prepaid expenses, and accrued unpaid expenses are negligible. Thus, these adjustments to cash income can be ignored without greatly distorting the accrual income tabulation. Changes in inventory are more serious items to consider when estimating accrual income from the cash-income statement. Consider for example the following inventory adjustments that would be ignored in the cash-income statement:

- A farmer with 100 mature beef cows in the beginning of the year increased the herd size to 120 from on-farm production. The increase in inventory value from $12,000 to $16,000 would be missed.
- A cattle feeder purchases 500 head of feeders. Assume these feeders gained 300 pounds from the time of purchase until the end of year. The purchase cost and the costs of this weight gain have been deducted as business expenses, but there is no offsetting income unless the inventory increase of approximately $100,000 is added.
- A grain farmer stores this year's crop for sale the following year. There would be no inventory increase to offset the many costs of production under the cash method and net income would be distorted.

Difficult Bookkeeping Transactions

Credit and debit transactions that may create problems for the bookkeeper are discussed in this section. Typical transactions for sales of crops, livestock, and livestock products; standard purchases of operating supplies, feed, and livestock; and ownership expenses for taxes and interest are not given special attention. However, these transactions should be recorded in enough detail to allow for business analyses. Quantities and weights are important for tabulating many efficiency measures and should be recorded along with the amounts paid or received.

TABLE 3.6

Net Income Aspects of the Inventory Account Relative to Depreciation, Purchases, and Sales

Transaction[a]	Beginning Inventory ($)	Purchases ($)	Sales ($)	Closing Inventory ($)	Depreciation ($)	Loss from Sales ($)	Gain from Sales ($)
Depreciable asset:							
Carried over	(30,000)			27,000	(3,000)		
Purchased at end of year		(15,000)		15,000	0		
Purchased mid-year		(66,000)		60,000	(6,000)		
Sold at a loss	(9,000)		7,500			(1,500)	
Sold at a gain	(3,000)		6,000				3,000
Purchased with trade[b]	(15,000)	(69,000)	21,000	60,000	(9,000)		6,000
Total	(57,000)	(150,000)	34,500	162,000	(18,000)	(1,500)	9,000
Debit–credit totals		(207,000)		196,500		(19,500)	9,000
Net debit or credit		(10,500)				(10,500)	

[a]Debits are shown in parentheses ( ) as the income statement would be affected.
[b]The item purchased with cash plus trade had a retail value of $75,000. The farmer paid $54,000 cash and received $21,000 trade allowance for an item with a undepreciated balance of $15,000. The basis of the new purchase is the cash difference plus the adjusted basis of the item traded (54,000 + $15,000).

Changes in inventory, and purchases and sales of capital assets are not always handled alike in various accounting systems; neither are different items within the same system. Sometimes, total inventories are shown instead of changes in inventory, with beginning inventories on the debits side and closing inventories on the credits side. This system does not change net income but does give different subtotals, such as total business credits and total business debits. Showing the change in inventory is the preferred practice.

Rather than using value changes, and purchases and sales for depreciable property, as is done with inventories, depreciation and gains or losses from sales are used. This relationship is illustrated in Table 3.6 where the changes in value are traced. Note that the net debit amount is the same on both sides of the solid line. These tabulations are a continuation of those in Table 3.5.

Purchased livestock subject to depreciation should be valued separately from farm-raised livestock. This separation is accomplished by using the depreciation schedule as a source for the beginning and ending inventory values. These values can be added to those for livestock not on depreciation or they can be listed separately.

Capital gains or losses create a problem that is directly related to the valuation methods used in the balance sheet, particularly the inventory. Even though price-level changes could be included as income, it was recommended in Chapter 2 that they be confined to the balance sheet as adjustments to assets and equity. However, price-level changes affecting inventories are not so easily handled in the balance sheet, and they become problems in the income statement if market-based prices are used, particularly for raised breeding livestock. Price cycles are common for many livestock, causing prices to repeatedly change over time. Because breeding livestock may be kept for several years and not sold, any changes in value due to price-level changes are paper gains or losses. For example, the farmer with breeding cows who values them in the inventory by the market-price method would have inventory gains while prices were increasing and losses when they were decreasing, but the farmer's bank account would not reflect any change as a result of the values placed on the breeding herd. Thus, there are good reasons for holding breeding livestock prices at near constant levels over time and for adjusting them only when the cost of their production changes considerably.

Repair versus investment is another item that may be difficult to determine. For example, should the total cost of overhauling a tractor or reroofing the barn be included as expense for the year in which the expenditure occurred? Guidelines for this decision were listed in Chapter 2. However, a farmer should be cautious about including farm-supplied labor as a cost when making repairs or adding new facilities that later will be added to the depreciation schedule. If the farm operator's labor is included as part of the development cost of an asset then, in effect, a wage has been paid to the operator. The same is true for hired labor paid by the month. If hired labor is included in the cost of a capital improvement, then the payment charged to other activities needs to be reduced accordingly.

Perquisites furnished to hired labor should only reflect their costs. Such benefits include living quarters, food, clothing, insurance, etc. If they are valued above their actual cost, there is a margin of profit to be reported. Suppose the farmer furnishes board, room, and laundry to the hired hand. If a wage for this service is included when calculating the salary adjustment or charge, it should be recorded as labor income. When calculating the cost of living expenses, it is appropriate to include an allowance for the wear and tear on the home, cooking utensils, bedding, etc., if these items are not accounted for elsewhere.

Farm products, such as beef and vegetables, used in the home should be included in income only when business expenses have been charged for their production. They should be valued at no more than has been charged to the business for their production. A preferable alternative is not to include their cost as a business expense.

Work performed off the farm should be counted with farm income unless it is considered a major separate business activity. This includes custom-machinery work as well as labor. Neighborly exchanges usually do not justify separate accounts.

Crop commodity loans can be handled in two ways. First, the loan can be counted as a sale. When the crop is later sold, any differences between the loan value and sale amount are recorded. Second, the loan can be counted as a money loan until the crop is actually sold. In the latter case, the amount of crop involved is included in the closing inventory if it ended between the time of the loan and sale of the crop. Either method is allowed by the IRS, providing the method selected is used in all future accounting periods. Treating as income the amount received as a money loan tends to even out year-to-year income variations.

Farm–family shared expenses should be divided according to the proportion used by each, such as the telephone, electricity, gas, family auto, and newspaper. There is nothing magical about 50 percent or 25 percent except that they are easy to apply. The point is that some cost items are jointly used and a division must be made. Be as realistic and objective as possible in making this division. Do not overlook the fact that the farm home often furnishes a farm business office, and the garage may shelter the farm truck, farm-shared auto, or even farm supplies.

Accrued expenses that have not been paid should be recorded in the year incurred rather than in the year paid. Examples are accrued interest on a loan for which payment does not fall due at the end of the accounting year, taxes that are paid the following year, and wages earned by an employee but not paid. These expenses take the form of an increased liability (credit) and a decreased owner's equity.

Prepaid expenses are the opposite of accrued expenses, in that a payment has been made covering an expense that has not yet occurred. Examples are interest paid at the beginning of a loan, insurance premiums paid in advance, wages paid in advance to employees, and fertilizer purchased in advance of application. Such expense should be prorated on the basis of the accounting year in which the service is rendered.

Principal payments on liabilities are not business expenses. An expense is incurred when borrowed money is used for business purposes and not when the loan is paid back.

Interest payments are considered a business expense in most accounting systems. Certainly, they are an expense to the farm operator or owner. However, there is some logic in considering interest payments as part of the returns to assets managed. When tabulating investment efficiency in Chapter 4, interest is added back into the income statement to tabulate an adjusted net income as if the business were free from debt.

There are probably areas not covered by this discussion that are or will be troublesome to some farm accountants or other readers. It is hoped that the reasoning used for the various items discussed carries over to other problems.

Cash-Flow Statement

Farm-business financial transactions can be considered a series of cash flows. These transactions were introduced earlier in this chapter, and an annual summary statement was illustrated (Table 3.1). Even though this summary is useful, some managers like a monthly cash-flow tabulation to gain greater financial control. A cash-flow statement is illustrated in Table 3.7, in which only bimonthly totals are shown.

Typical divisions for the cash-flow statement are shown here. Note particularly that nonfarm business transfers into and out of the farm business are included.

The sources of the data are the farm and family account books. Note in Table 3.7 that for any one month the income and expenditures from farm-production activities often do not balance. The manager must meet monthly cash-flow deficits from carryover cash balances of the previous month, or bring in money from some other source—usually by negotiating a new loan. In the Jan.–Feb. period it was necessary to borrow $4,000 in addition to the $12,200 beginning cash balance for operation expenses. Again, in the Mar.–Apr. and May–June periods it was necessary to borrow, but in the July–Aug. period there was a monthly cash surplus and most of the new loan money was repaid. In Sept.–Oct. the balance of the new production loans was repaid. Note that the cash balance at the end of Nov.–Dec. matches the cash balance in the balance sheet (Table 2.4).

Note the entries under the subheadings "nonfarm income" and "nonfarm expenditures." These nonfarm money flows often add to, or take away from, the cash available to the farm. They directly affect new farm investments and debt retirement. For this and other obvious reasons, the cash-flow income statement should not be considered a meaningful measure of profitability. Investments and sales of capital assets can affect the cash balance at any time. Thus, the cash-flow statement is more a measure of feasibility than profitability.

Farmers who generate the monthly cash-flow income statement gain the greatest benefit if they compare it with their budgeted cash-flow statement. This comparison provides a means of control over the finances of the business and is very useful when obtaining credit and planning repayment. The monthly cash-flow statement is easily generated on the computer and is a part of most electronic farm record-keeping systems.

Double-Entry Accounting

The use of double-entry accounting on farms has increased significantly since the advent of home computers. For most transactions the second entry is built into computer programs and the user is mostly unaware of it. Thus, keeping a double-entry set of ac-

TABLE 3.7

Bimonthly cash-flow statement for the FABA farm, YRXX

Months Description	Jan–Feb $	Mar–Apr $	May–Jun $	Jul–Aug $	Sep–Oct $	Nov–Dec $	Total $
CASH INFLOW							
Livestock sales	0	6,587	72,458	75,751	9,881	0	164,677
Crop sales	0	5,354	6,424	0	37,476	57,819	107,073
Other farm income	0	0	3,774	0	9,500	0	13,274
Sale of capital assets:							
Breeding livestock	378	2,267	382	659	3,015	855	7,556
Depreciation property	0	400	0	0	0	0	400
Nonfarm income:							
Off-farm wages	2,000	2,000	2,000	2,000	2,000	2,000	12,000
Interest and dividends	96	109	118	207	132	328	990
Total inflow	**2,474**	**16,717**	**85,156**	**78,617**	**62,004**	**61,002**	**305,970**
CASH OUTFLOW							
Operating expenses:							
Labor	2,212	2,315	2,334	2,118	2,343	1,624	12,946
Livestock expense	958	857	609	1,116	926	1,018	5,484
Crop expense	0	17,462	15,764	5,941	3,458	823	43,448
Machinery expense	1,189	1,237	1,828	2,214	2,075	1,483	10,026
Building repair	367	282	179	483	384	230	1,925
Machine hire	0	0	225	439	4,636	5,125	10,425
Farm utilities	615	601	616	528	467	583	3,410
Miscellaneous	367	284	167	348	54	386	1,606
Total	5,708	23,038	21,722	13,187	14,343	11,272	89,270
Feed purchases	6,894	3,384	3,680	3,760	4,205	5,378	27,301
Livestock purchases	0	1,295	0	0	39,375	1,200	41,870
Fixed expenses:							
Property taxes	0	0	0	0	0	4,584	4,584
Insurance	0	600	2,362	0	600	0	3,562
Interest expense	448	1,680	1,750	23,604	1,230	1,641	30,353
Cash rent	0	0	6,250	0	0	6,250	12,500
Total	448	2,280	10,362	23,604	1,830	12,475	50,999
Capital purchases:							
Machinery and equipment	0	46,000	24,000	0	0	0	70,000
Buildings and improvements	0	0	0	2,831	0	0	2,831
Nonfarm expenditures:							
Family living transfers	4,917	4,763	4,653	5,274	4,515	4,833	28,955
Income tax and social security	0	19,471	0	0	0	0	19,471
Nonfarm businesses	0	0	0	0	0	10,000	10,000
Total	4,917	24,234	4,653	5,274	4,515	14,833	58,426
Total outflow	**17,967**	**100,231**	**64,417**	**48,656**	**64,268**	**45,158**	**340,697**
SUMMARY:							
(Beginning Balance)	**12,200**						
Net cash flow (+ or –)	**(15,493)**	**(83,514)**	**20,739**	**29,961**	**(2,264)**	**15,844**	**(34,727)**
New loans:							
Operations (+)	4,000	55,000	0	0	0	0	59,000
Production (+)	0	0	16,000	0	40,000	0	56,000
Capital purchases (+)	0	30,000	20,000	0	0	0	50,000
Principal payments:							
New loans (–)	0	0	55,000	4,000	0	0	59,000
Old loans (–)	0	0	2,130	22,524	30,000	18,000	72,654
Monthly balance	**707**	**2,193**	**1,802**	**5,239**	**12,975**	**10,819**	**10,819**

counts is not much more difficult than keeping a single-entry set. And there are some important benefits to having the accuracy checks generated by double-entry listings. In addition, the keeping of enterprise accounts is facilitated. Professional accountants use double-entry accounting.

Actually, keeping a double-entry set of accounts is not greatly different from keeping the complete set of single-entry financial accounts previously described. If an accurate accounting is kept of cash; capital purchases and sales; credit, including payables; and accrual income and expenses, all the elements for a double-entry set of accounts are in place. In fact, it is possible (but not recommended) to have as few as three financial accounts under a double-entry system, each in the form of a journal organized according to the general accounting equation assets = liability + equity. The point was made earlier that every financial transaction and every productive farm activity affecting income, if recorded, influences this equation.

Double-entry accounting simply involves keeping track of all changes in assets and the claims upon those assets each time a transaction is made. This tracking is accomplished through a system of credits and debits. Each credit transaction must be balanced by a debit transaction and vice versa. Credits are defined as follows:

- Increase in owner's equity (net worth)
- Increase in a liability
- Decrease in an asset

Debits are defined as follows:

- Decrease in owner's equity (net worth)
- Decrease in a liability
- Increase in an asset

Debits and credits are illustrated in Figure 3.1.

This system as defined may seem unreasonable if the objectives of farm ownership are not kept clearly in mind. Ownership control of the business is exhibited as equity. Any increase in the owner's equity is a credit and any decrease is a debit. The pattern of debits and credits fits reasonably well together (Figure 3.2) if it is remembered that the value of assets must be equal to the claims upon them at all times, and the sum of credit entries must be equal to the sum of debit entries for every transaction. Thus, there is a built-in accuracy check of all accounting activities. Actually, credits and debits are merely terms used to keep the system in balance. Suppose an asset is added or increased; the following may occur:

1. If it comes from production on the farm,
   a. the asset account receiving the new product is increased (debited) and
   b. the owner's equity account is increased (credited).

FIGURE 3.1

Definition of Credits and Debits

	Assets	=	Liabilities	+	Owner's Equity
Credits	–		+		+
Debits	+		–		–

2. If it comes from outside the business, and is:
   a. a cash purchase, (1) the cash asset account is decreased (credited) and (2) the capital asset account receiving the item is increased (debited).
   b. a deferred payment purchase, (1) the liability account is increased (credited) and (2) the capital asset account receiving the item is increased (debited).

Double-entry journals are organized as illustrated in Figure 3.3. The debits are always recorded in the left column, and the credits in the right column. Transactions that normally are posted as single-entry expenses are posted as double-entry expenses (debited), and single-entry receipts are posted as double-entry receipts (credited). The marked difference is that every debit posting must have an opposite and equal credit posting, usually in the cash-asset account. Cash expenses reduce cash so the cash-asset account is credited; cash receipts increase cash so the cash-asset account is debited. Each has its opposite and a balance is maintained. Perhaps 90 percent of all farm business transactions are postings to the asset-checking account and income-expenses account. This approach does give a little tighter control over the checkbook.

Single-entry accounting is mostly concerned with the income statement—a categorized summary of credit (receipts) and debit (expenses) entries. It is noteworthy that

FIGURE 3.2

Relationship Between Credits and Debits and the Balance Sheet

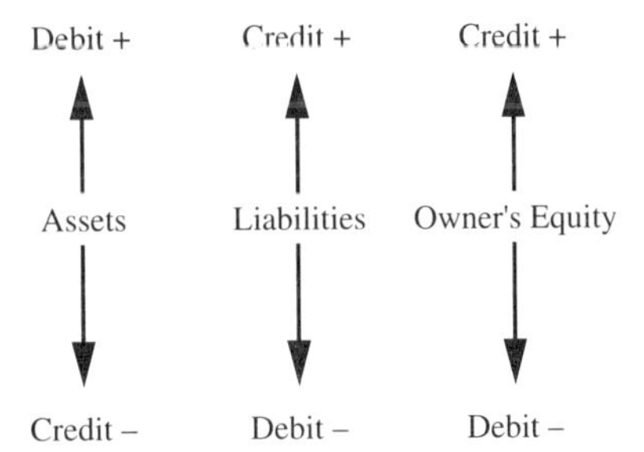

FIGURE 3.3

Organization of a Double-Entry Balance Sheet Ledger

Any Asset Account

Balance + Debits	– Credits

Any Liability Account

– Debits	Balance + Credits

Any Equity Account

– Debits	Balance + Credits

the definitions for credits and debits are the same in the income statement as for equity; that is, an increase is a credit and a decrease is a debit. Thus, it is not a large transition to recognize that the income statement is an extension of equity and of the balance sheet. This relationship is illustrated in Figure 3.4.

In double-entry accounting there are five accounts receiving transaction detail: assets, liabilities, equity, receipts, and expenses. Receipts and expenses still form the income statement, but now the income statement is an extension of equity and expenses can be credited in the case of a returned purchase. The income statement still shows the transaction detail of the firm's profit-making activities. Periodically (monthly or annually as the accountant desires), the income statement is summarized and the difference between the receipts and expenses is posted to equity. At the end of the accounting period the balances in the income accounts are zeroed out. Balance-sheet accounts retain their values and quantities. Transaction listings provide the detail of the changes in all of the balance-sheet and income accounts. Debits and credits are maintained in the expense and receipt accounts, in much the same way as in the asset, liability, and equity accounts (Figure 3.5).

A list of business financial events is presented here to illustrate double-entry postings. Each event is first described and then illustrated with an example (Examples 3.1–20). The transaction is shown in terms of the particular accounts affected. These examples cover most of the transaction types a farm accountant will encounter. The detail and logic of each example should help the reader to deal with business events not illustrated. There are different ways to handle some transactions so the

FIGURE 3.4

Relationship of Income Statement to Equity

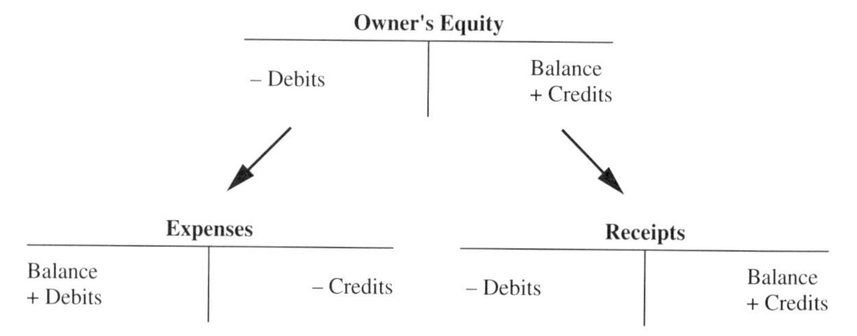

FIGURE 3.5

Organization of a Double-Entry Income Ledger

Any Expense Account

Balance + Debits	– Credits

Any Receipt Account

– Debits	Balance + Credits

method illustrated may not be appropriate for all situations. But the basics and rules shown do apply in all situations. Some deviations from the events illustrated are discussed later in the chapter. Following the examples is a combined balance sheet and income statement that summarizes all the transactions illustrated in the detailed accounts (Table 3.8). Much can be learned by studying this statement because it illustrates the applications of the five major accounts previously discussed (asset, liability, equity, receipt, and expense).

1. Hired labor is paid.
   a. The cash account is decreased (credited), and the expense account is increased (debited) (Example 3.1).

EXAMPLE 3.1

The owner or operator pays the hired hand $800 for labor.

Asset account - cash

	Balance	Debit (+)	Credit (–)
Before amount	$30,000		
Check to Todd Marks			$800
After amount	$29,200		

Expense account

	Debit (+)	Credit (–)
Labor hire	$800	

2. Fertilizer is purchased and applied.
   a. If this is a cash purchase, the cash asset account is decreased (credited) by the purchase amount and the expense account is increased (debited) in like amount (Example 3.2).

EXAMPLE 3.2

The owner or operator has fertilizer applied and pays $4,500 cash.

Asset account - cash

	Balance	Debit (+)	Credit (–)
Before amount	$29,200		
Check to Ace Fertilizer			$4,500
After amount	$24,700		

Expense account-crop fertilizer

	Debit (+)	Credit (–)
Crop-direct fertilizer	$4,500	

3. Farm-raised grain is sold.
   a. If this is new production and the grain has not been previously recorded (inventoried), and the operator does not want to post it into and out of inventory, the cash-asset account is increased (debited) and the receipt account increased (credited) (Example 3.3).

EXAMPLE 3.3

The owner or operator harvests oats and sells the entire crop for $9,000. The crop did not pass through inventory.

Asset account - cash

	Balance	Debit (+)	Credit (–)
Before amount	$24,700		
Check from Grain Coop deposited		$9,000	
After amount	$33,700		

Receipt account - grain sales

	Debit (–)	Credit (+)
Check from Grain Coop		$9,000

   b. If grain has been inventoried (last year's crop) the cash asset account is increased (debited) and the receipt account is credited with the amount of the sale. The grain-asset account is decreased (credited) and there is a debit entry under receipts, showing the value of inventory sold. The net effect on the owner's equity account is only the difference between the inventory value and the sale value (Example 3.4).

EXAMPLE 3.4

The owner or operator sells some of last year's corn crop for $6,000 that was held in inventory at $5,400.

Asset account - cash

	Balance	Debit (+)	Credit (–)
Before amount	$33,700		
Check from Grain Coop deposited		$6,000	
After amount	$39,700		

Receipt account - grain sales

	Debit (–)	Credit (+)
Check from Grain Coop		$6,000

Asset account - grain inventory

	Balance	Debit (+)	Credit (–)
Before amount	$40,000		
Grain sale out of inventory			$5,400
After amount	$34,600		

Receipt account - inventory adjustment

	Debit (–)	Credit (+)
Inventory value of goods sold	$5,400	

4. Fuel is purchased on credit account and placed in inventory. The operator wishes to charge its cost to the farm as it is used. Payment for the fuel is made when the bill is received.

   a. When the fuel is delivered, the asset-fuel inventory account is increased (debited) and the liability account for accounts payable is increased (credited) (Example 3.5).

EXAMPLE 3.5

The owner or operator has fuel delivered on account from Petro Products. The delivery person leaves a bill for $600. An inventory account is maintained.

Asset account - fuel inventory

	Balance	Debit (+)	Credit (–)
Before amount	$100		
Fuel delivered into tank		$600	
After amount	$700		

Liability account - accounts payable

	Balance	Debit (–)	Credit (+)
Before amount	0		
Fuel delivery voucher posted			$600
After amount	$600		

   b. When the fuel is used on the farm, the asset-fuel inventory account is decreased (credited) and the expense account is increased (debited) (Example 3.6).

EXAMPLE 3.6

The hired hand fills up the fuel tank of the big truck and records 50 gallons. The inventory price is $1.20.

Asset account - fuel inventory

	Balance	Debit (+)	Credit (–)
Before amount	$700		
Fuel out of inventory			$60
After amount	$640		

Expense account - machinery operations, fuel

	Debit (+)	Credit (–)
Farm use of fuel out of inventory	$60	

c. When the bill is paid, the cash-asset account is decreased (credited) and the liability account for accounts payable is decreased (debited) (Example 3.7).

EXAMPLE 3.7

The bill from Petro Products arrives for the fuel, and the owner or operator makes the payment of $600.

Asset account - cash

	Balance	Debit (+)	Credit (–)
Before amount	$39,700		
Check written to Petro Products			$600
After amount	$39,100		

Liability account - accounts payable

	Balance	Debit (–)	Credit (+)
Before amount	$600		
Petro Products bill paid		$600	
After amount	0		

5. Feeder livestock are purchased and sold.
   a. Feeder livestock were purchased from accumulated cash (Example 3.8). The cash account is decreased (credited) and the inventory account is increased (debited). There is no expense.

EXAMPLE 3.8

The owner or operator purchases feeder cattle for $15,000 and pays cash.

Asset account - cash

	Balance	**Debit (+)**	**Credit (–)**
Before amount	$39,100		
Check written to Pat Cattleowner			$15,000
After amount	$24,100		

Asset account - livestock inventory

	Balance	**Debit (+)**	**Credit (–)**
Before amount	$43,000		
Cattle inventory account increased		$15,000	
After amount	$58,000		

b. Finished feeder cattle are sold. There are four postings: (1) the cash account is increased (debited) by the total amount of the sale, (2) the receipt account is increased (credited) by the amount of sale, (3) the livestock asset account is decreased (credited) by the inventory value of the cattle being sold, and (4) the value of inventory, or cost of goods sold, is posted to the receipt account (debited) (Example 3.9).

EXAMPLE 3.9

The owner or operator sells cattle for $30,000 that had an inventory value of $18,000.

Asset account - cash

	Balance	**Debit (+)**	**Credit (–)**
Before amount	$24,100		
Check deposited in bank		$30,000	
After amount	$54,100		

Receipt account - cattle sales

	Debit (–)	**Credit (+)**
Cattle sale recorded		$30,000

Asset account - livestock inventory

	Balance	**Debit (+)**	**Credit (–)**
Before amount	$58,000		
Inventory account reduced			$18,000
After amount	$40,000		

Receipt account - value of inventory sold

	Debit (–)	Credit (+)
Value of inventory sold posted	$18,000	

c. Feeder livestock are purchased with a loan. (1) First the loan is posted—the cash-asset account is increased (debited) and the liability account is increased (credited) (Example 3.10). (2) Next the purchase is made and posted as illustrated in Example 3.11. The cash-asset account is decreased (credited) and the receipts account is increased (credited) (Example 3.11).

EXAMPLE 3.10

The owner or operator takes out a loan for $21,000 to purchase feeder cattle.

Asset account - cash

	Balance	Debit (+)	Credit (–)
Before amount	$54,100		
Check from ABC bank deposited		$21,000	
After amount	$75,100		

Liability account - short-term cattle loan

	Balance	Debit (–)	Credit (+)
Before amount	$35,100		
Loan check from ABC bank			$21,000
After amount	$56,100		

EXAMPLE 3.11

The owner or operator purchases feeder cattle for $22,000.

Asset account - cash

	Balance	Debit (+)	Credit (–)
Before amount	$75,100		
Check written to Pat Cattleowner			$22,000
After amount	$53,100		

Asset account - livestock inventory

	Balance	Debit (+)	Credit (–)
Before amount	$40,000		
Livestock received into inventory		$22,000	
After amount	$62,000		

d. Feeder cattle purchased with the loan are sold. (1) The cash-asset account is debited with the amount of the sale. (2) The receipt account is credited with the value of the sale. (3) The livestock inventory-asset account is credited with the value the livestock had when purchased (if sold in the year of purchase, or with the inventory value if sold in the following accounting period). (4) The receipt account is also debited with the amount of value the animals sold had in inventory before sale (Example 3.12). When the loan is retired, (1) the cash-asset account is credited with the combined amount of principal and interest payment, (2) the liability account is debited with the amount of principal payment, and (3) the expense account is debited with the interest payment (Example 3.13).

EXAMPLE 3.12

The owner or operator sells the cattle, which were purchased for $22,000, for $39,000.

Asset account - cash

	Balance	Debit (+)	Credit (–)
Before amount	$53,100		
Check from cattle buyer deposited		$39,000	
After amount	$92,100		

Receipt account - livestock sales

	Debit (–)	Credit (+)
Sale of cattle is recorded in income account		$39,000

Asset account - livestock inventory

	Balance	Debit (+)	Credit (–)
Before amount	$62,000		
Livestock inventory reduced			$22,000
After amount	$40,000		

Receipt account - value of goods sold out of inventory

	Debit (–)	Credit (+)
Value of inventory sold posted	$22,000	

EXAMPLE 3.13

The owner or operator pays off the cattle loan—$21,000 principal and $1,200 interest.

Asset account - cash

	Balance	Debit (+)	Credit (–)
Before amount	$92,100		
Check written to ABC bank to pay off loan			$22,200
After amount	$69,900		

Liability account - short-term cattle loan

	Balance	Debit (–)	Credit (+)
Before amount	$56,000		
Cattle loan paid off		$21,000	
After amount	$35,000		

Expense account - interest

	Debit (+)	Credit (–)
Interest expense posted	$1,200	

6. A new machine is purchased.
   a. If the purchase is from accumulated savings, the cash-asset account is decreased (credited) and the machinery-asset account is increased (debited). Owner's equity is not affected (Example 3.14).

EXAMPLE 3.14

The owner or operator purchases a new tractor for $34,000 and pays cash.

Asset account - cash

	Balance	Debit (+)	Credit (–)
Before amount	$69,900		
Check written to U.B. Machinery			$34,000
After amount	$35,900		

Asset account - machinery

	Balance	Debit (+)	Credit (–)
Before amount	$89,000		
Tractor to machinery capital		$34,000	
After amount	$123,000		

   b. If the purchase is with a new loan, the machinery-asset account is increased (debited) and the liability account is increased (credited). Again, owner's equity is not affected. The cash-asset account is debited with the amount of cash paid to secure the purchase (Example 3.15).

EXAMPLE 3.15

The owner or operator purchases a new plow for $4,000, financing it with dealer credit; $500 is paid down, and there is no trade-in.

Asset account - cash

	Balance	Debit (+)	Credit (–)
Before amount	$35,900		
Check written to U.B. Machinery			$500
After amount	$35,400		

Liability account - intermediate-term loan

	Balance	Debit (–)	Credit (+)
Before amount	$31,500		
New loan posted			$3,500
After amount	$35,000		

Asset account - machinery

	Balance	Debit (+)	Credit (–)
Before amount	$123,000		
Plow added to machinery capital		$4,000	
After amount	$127,000		

c. Only when the new machine depreciates is the owner's equity account affected. The machinery-asset account is decreased (credited), and the expense account is decreased (debited) (Example 3.16).

EXAMPLE 3.16

The owner or operator charges a depreciation expense of $10,200 against the farm machinery, including the new plow and tractor.

Asset account - machinery

	Balance	Debit (+)	Credit (–)
Before amount	$127,000		
Depreciation tabulated and posted			$10,200
After amount	$116,800		

Expense account - depreciation

	Debit (+)	Credit (–)
Depreciation posted to expense account	$10,200	

7. A farm crop is harvested and placed in inventory.
   a. If the operator wishes to give credit to the crop enterprise, the crop-inventory account is increased (debited) and the receipt account is credited with value that went into inventory (Example 3.17).

EXAMPLE 3.17

Hay is harvested and placed in inventory. Owner or operator measured 50 tons and valued it at $60 per ton ($3,000).

Receipt account - farm-raised feed inventory

	Debit (–)	Credit (+)
Value of farm-raised feed to inventory		$3,000

Asset account - forage inventory

	Balance	Debit (+)	Credit (–)
Before amount	$12,000		
Farm-raised feed posted to inventory		$3,000	
After amount	$15,000		

8. A farm crop is fed to the livestock on the farm. The operator wants to charge the receiving enterprise for its cost.
   a. The crop-inventory account is decreased (credited), and a debit entry is posted to the expense account under feed (Example 3.18).

EXAMPLE 3.18

Hay is fed out of the inventory cattle. Thirty (30) tons are recorded at a value of $60 for $1,800.

Asset account - forage inventory

	Balance	Debit (+)	Credit (–)
Before amount	$15,000		
Feed fed out of inventory			$1,800
After amount	$13,200		

Expense account - livestock feed

	Debit (+)	Credit (–)
Feed fed out of inventory	$1,800	

9. Money is transferred from the farm for family living expenses.
   a. Because this is an out-transfer from the farm to the family, the farm expense account is not involved. The two accounts that are affected are cash (credited) and equity (debited) (Example 3.19).

EXAMPLE 3.19

A check is written out at the grocery store for $100 for the owner or operator's family.

Asset account - cash

	Balance	Debit (+)	Credit (–)
Before amount	$35,400		
Check written to XY Grocery			$100
After amount	$35,300		

Equity account - family transfer

	Balance	Debit (–)	Credit (+)
Before amount	$300,000		
Family living transfer account		$100	
After amount	$299,900		

10. Once the books are ready to close for an accounting period, it is necessary to close out the receipt and expense accounts. Closure is done by summing the totals for each of these accounts and posting the difference to the equity account. The receipt and expense accounts start over at zero, but asset, liability, and equity accounts retain their balances (Example 3.20).

EXAMPLE 3.20

Receipt and expense transactions posted in Examples 3.1–3.19 are summarized and posted to the equity account. Receipt and expense accounts are zeroed out.

Receipt account

	Debit (–)	Credit (+)
Summary of all entries	$45,400	$87,000
Posting to equity account	$41,600	

Expense account

	Debit (+)	Credit (–)
Summary of all entries	$18,560	0
Posting to equity account		$18,560

Equity account - net farm income posted

	Balance	Debit (–)	Credit (+)
Before amount	$299,900		
Net farm income posted			$23,040
After amount	$330,340		

These illustrated transactions are summarized in Table 3.8. The postings in this table are devoid of any specific asset, liability, or equity account names, making it easier to observe the debit–credit balance. Also note the closing out of the receipt and expense records at the end of the period and of the posting of net income to equity. Other methods of posting these same transactions could be defined to fit the needs of management. Some of these methods are discussed later.

Chart of Accounts

The accounts established for a double-entry accounting system need not be greatly different from those for a single-entry system. However, an important difference may be with how responsibility or enterprise accounting takes place. In single-entry accounting additional accounts are added to identify particular costs or returns important to management (e.g., separate feed accounts would be defined for dairy cows, beef cows, beef feeders, hogs, etc). In double-entry accounting this can be more easily accomplished with cost and profit centers. A single feed account can be used to allocate feed to several livestock enterprises or profit centers. This organization of accounts can be thought of as a two-way table (Figure 3.6). The responsibility centers shown in Form 3.1 are suggestive of how this can be accomplished.

FIGURE 3.6

Organization of Balance Sheet and Income Accounts with a Cost-Center Designator

	Cost Centers:			Profit Centers:		
	Overhead	Labor	Machinery	Crop 1	Crop 2	Livestock
Balance	XXX.XX	XXX.XX	XXX.XX	XXX.XX	XXX.XX	XXX.XX
sheet	XXX.XX	XXX.XX	XXX.XX	XXX.XX	XXX.XX	XXX.XX
and	XXX.XX	XXX.XX	XXX.XX	XXX.XX	XXX.XX	XXX.XX
income	XXX.XX	XXX.XX	XXX.XX	XXX.XX	XXX.XX	XXX.XX
accounts	XXX.XX	XXX.XX	XXX.XX	XXX.XX	XXX.XX	XXX.XX

FORM 3.1

Master Enterprise and Cost-Center Codes

Enterprise/Cost Center	Codes
Overhead	1
Service Centers:	2–9
Labor	
Machinery	
Irrigation	
Production Centers:	10–89
Crops	10–49
Livestock	50–89
Control Centers:	90–99
Education and Training	
Travel	

TABLE 3.8

Effect of Business Events Illustrated in Examples 1–19 Upon the Balance Sheet and Income Statement

Ref. No.	Ex. No.	Description	Asset Accounts: Debit (+) $	Asset Accounts: Credit (–) $	Balance $	Liability Accounts: Debit (–) $	Liability Accounts: Credit (+) $	Balance $	Equity Accounts: Debit (–) $	Equity Accounts: Credit (+) $	Balance $	Expense Accounts: Debit (+) $	Expense Accounts: Credit (–) $	Receipt Accounts: Debit (–) $	Receipt Accounts: Credit (+) $
		Beginning balance			600,000			300,000			300,000				
1	1	Labor paid		800	599,200			300,000				800			
2	2	Fertilizer purchased		4,500	594,700			300,000				4,500			
3	3	Current crop sold	9,000		603,700			300,000							9,000
3	4	Crop inventory sold	6,000	5,400	604,300			300,000						5,400	6,000
4	5	Fuel purchase account	600		604,900		600	300,600							
4	6	Fuel dispensed		60	604,840			300,600				60			
4	7	Fuel bill paid		600	604,240	600		300,000							
5	8	Cattle purchased	15,000	15,000	604,240			300,000							
	9	Cattle sold	30,000	18,000	616,240			300,000						18,000	30,000
5	10	Cattle loan	21,000		637,240		21,000	321,000							
5	11	Purchased cattle	22,000	22,000	637,240			321,000							
5	12	Cattle sold	39,000	22,000	654,240			321,000						22,000	39,000
5	13	Loan paid		22,200	632,040	21,000		300,000				1,200			
6	14	Tractor purchased	34,000	34,000	632,040			300,000							
6	15	Plow purchased	4,000	500	635,540		3,500	303,500							
6	16	Depreciation		10,200	625,340			303,500				10,200			
7	17	Hay harvested	3,000		628,340			303,500							3,000
8	18	Hay fed		1,800	626,540			303,500				1,800			
9	19	Groceries purchased		100	626,440			303,500	100		299,900				
10a		Income totals										18,560	0	45,400	87,000
10b		Equity posting								23,040	322,940	0	18,560	87,000	45,400
10c		Ending balances			626,440			303,500			322,940	0	0	0	0

FORM 3.2

Master Chart of Accounts

A. BALANCE SHEET ACCOUNTS

	Account Range		*Account Range*
1. Asset Account Codes	**1000–2999**	**2. Liability Account Codes**	**3000–3999**
Current Assets	1000–1999	Current Debts	3000–3199
Liquid Assets	1000–1499	Intermediate Debts	3200–3399
Inventory	1500–1799	Long-Term Debts	3400–3599
Production in Progress	1800–1999	Other Debts	3300–3499
Intermediate Assets	2000–2599		
Depreciable Assets	2000–2299	**3. Equity Account Codes**	**4000–4999**
Depreciation	2300–2499	Net Income Forward	4000
		Equity Balance Forward	4100–4199
Long-term Assets	2600–2999	Transfers In	4200–4499
Depreciable Assets	2600–2699	Transfers Out	4500–4799
Depreciation	2700–2799	Adjustments	4800–4999
Land Assets	2900–2999		
Other Assets	2900–2999		

B. INCOME STATEMENT ACCOUNTS

1. Income Account Codes	**5000–5999**	**3. Expense Account Codes**	**7000–8999**
Crop sales	5000–5799	Operating cost accounts	7000–7999
Livestock sales	5300–5599	Labor	7000–7099
Miscellaneous income	5600–5699	Machinery	7100–7199
Capital gains (losses)	5800–5999	Irrigation	7200–7299
		Crops	7300–7399
2. Inventory Adjustment Codes	**6000–6999**	Livestock	7400–7499
Inventory increases (+)	6000–6400	Utilities	7500–7599
from production	6000–6199	Business office	7600–7699
from transfers	6200–6399	Other	7700–7999
from other	6400–6499		
Inventory decreases(–)	6500–6899	Ownership cost accounts	8000–8999
from sales	6500–6699	Rent, repair, taxes	8000–8499
from transfers	6700–6799	Depreciation	8500–8899
Death and loss	6800–6899	Other	8900–8999
Purchases to inventory(–)	6900–6998		
Purchases offset	6999	**4. Information**	**9000–9999**
		Family living	9800–9999

With responsibility-center accounting in mind the accounts shown in Form 3.2 were developed. They are suggestive of an organizational pattern and are not expected to fulfill individual needs. Charts of accounts can be organized so that summaries can be made by specific digits in the account code to give different levels of detail in printed reports. However, program developers do not all use the same coding system. In fact, some programs use alpha codes instead of numeric.

Look at the inventory adjustment accounts (item 2 of the income statement account codes in Part B). Two options are available for recording inventory adjustments: the perpetual basis and the periodic basis. Either or both of these two methods may be used, so long as the user is careful to define how each item of inventory is to be handled and then is consistent in application. A discussion of these two methods follows.

Perpetual Inventory Accounting

An income statement patterned after the chart of accounts shown above and using the figures developed for the FABA farm is illustrated in Table 3.9. This table should be compared with Tables 3.3 and 3.4, prepared using single-entry accounting. Most important is the partitioning of the income statement by cost and profit centers, made possible by double-entry accounting. Also note that there are no direct inventory adjustment postings and that feed costs are the value of transfers out of inventory and not purchases. Purchased feed could have been posted directly as an expense but if there were a change in inventory, the amount charged to the animals would not be accurate.

The production-adjustment amount may not be the value of harvested production. In this illustration the total value of production is shown as a single posting to inventory for crops but not for livestock. The explanation is that all production may not be inventoried, with some part being directly sold without going through inventory. Inventory adjustments are a summary of increases and decreases throughout the accounting period. The inventory reconciliation at the bottom is not an income-statement tabulation but was added to show the balancing of inventory accounts for each profit and cost center. Note that any one element of inventory can be tabulated if all of the other elements are in place. Explanations for some of the categories are discussed later in this chapter.

Under the single-entry system of accrued income accounting, the periodic system of inventory adjustment was used to adjust for inventory changes. The values of farm-raised products going into inventory were not recorded as receipts, nor were the values of farm-raised feeds fed recorded as expenses. The system accurately reflected total farm income by posting periodic (annual) changes in inventory to the income statement and counting purchases to inventory as expenses. This method is illustrated again in Table 3.10 under "Accrual Periodic." A comparison is made with other forms of the income statement discussed earlier in this chapter. Note the similarities and differences. The periodic system of inventory accounting also can be used in double-entry accounting, but the user loses a major benefit by doing so.

With perpetual inventory accounting, it is necessary to account for the value of items leaving the inventory (items with a new use on the farm or items sold out of the farm business), as well as those entering. Items are not held in the inventory for a profit. Profits belong to productive activities, which include storage and market growth. Thus, the value of items leaving the inventory should closely resemble the value entering the inventory. Three systems are used to value the items going out of inventory:

- First in, first out (FIFO)
- Last in, first out (LIFO)
- Value out = average value of items in inventory

We prefer the last method because it does not require keeping track of particular additions. A running-quantity dollar balance is maintained and anytime an item leaves the inventory a new price is computed by dividing the total dollar amount by the total quantity. There is not a business expense for purchases entering the inventory; dollar assets are exchanged for inventory value. But there is an expense when an inventoried item is used in the business. An expense account is debited with the value this item had in inventory.

When inventoried items are sold off the farm and the total amount of the sale is recorded as a receipt, it becomes necessary to show a debit in the receipt account or

TABLE 3.9

Summary Income Statement for the FABA Farm by Profit and Cost Centers, YRXX

Category	Farm Total ($)	Overhead ($)	Corn ($)	Soybeans ($)	Forage Crops ($)	Hogs ($)	Cattle ($)
I. Receipts:							
Crop sales	107,073		42,422	64,151	500		
Livestock sales	172,233					109,796	62,437
Miscellaneous	14,264	4,764	5,500	4,000	0	0	0
Capital gain/loss	(210)	(210)	0	0	0	0	0
Total receipts	293,360	4,554	47,922	68,151	500	109,796	62,437
II. Inventory adjustment:							
Production adjustment (+)	215,397	0	93,000	65,973	7,550	35,842	13,032
Sales from inventory (–)	(192,240)	0	(42,422)	(64,151)	(500)	(48,649)	(36,518)
Losses (–)	(4,103)	0	(811)	(930)	(355)	(1,200)	(807)
Subtotal	19,054	0	49,767	892	6,695	(14,007)	(24,293)
III. Gross revenues (I + II)	312,414	4,554	97,689	69,043	7,195	95,789	38,144
IV. Feeds fed:							
Farm raised	44,966					26,733	18,233
Purchased	29,551					26,775	2,776
Subtotal	74,517	0	0	0	0	53,508	21,009
V. Value of Production (III – IV)	237,897	4,554	97,689	69,043	7,195	42,281	17,135
VI. Expenses:							
Operating costs:							
Labor	12,946	1,865	2,876	2,150	480	4,900	675
Equipment operating	10,026	640	2,345	1,473	560	4,587	421
Machine hire	10,425	450	5,250	4,500	225	0	0
Cropping costs	43,448		27,348	15,025	1,075		
Livestock costs	5,484					3,347	2,137
Building repair	1,925	125	900	0	0	500	400
Utilities	3,410	900	400	350	200	1,200	360
Miscellaneous	1,606	1,606	0	0	0	0	0
Subtotal	89,270	5,586	39,119	23,498	2,540	14,534	3,993
Fixed costs:							
Interest	31,780	443	12,130	10,650	1,544	3,240	3,773
Taxes	4,584	200	2,055	1,801	148	300	80
Insurance	3,562	800	1,125	1,237	0	300	100
Cash rent	12,500	0	6,500	5,000	1,000	0	0
Subtotal	52,426	1,443	21,810	18,688	2,692	3,840	3,953
Depreciation	31,944	1,000	10,000	8,750	800	8,588	2,806
Total Expenses	173,640	8,029	70,929	50,936	6,032	26,962	10,752
VII. Net Income (V – VI)	64,257	(3,475)	26,760	18,107	1,163	15,319	6,383
VIII. Inventory Reconciliation:		Concentrates					
Beginning inventory (+)	164,095	9,000	43,838	8,086	5,900	60,753	36,518
Purchases (+)	68,421	26,551	0	0	0	2,495	39,375
Production (+)	302,463	0	93,000	65,973	7,550	96,989	38,951
Sales (–)	(279,306)	0	(42,422)	(64,151)	(500)	(109,796)	(62,437)
Farm use (–)	(74,517)	(29,551)	(38,888)	0	(6,078)	0	0
Losses (–)	(4,103)	0	(811)	(930)	(355)	(1,200)	(807)
Ending inventory (–)	(177,053)	(6,000)	(54,717)	(8,978)	(6,517)	(49,241)	(51,600)
Balance	0	0	0	0	0	0	0

TABLE 3.10

Perpetual Versus Periodic Inventory Accounting on a Dairy Farm

Income from Operations		Accrual Perpetual	Accrual Periodic	Cash Periodic
A. Receipts:				
Sales of crops	(+)	$ 12,000	$ 12,000	$ 12,000
Sales of livestock	(+)	85,000	85,000	85,000
Milk sales[a]	(+)	530,000	530,000	510,000
Miscellaneous	(+)	4,000	4,000	4,000
Subtotal	(+)	$631,000	$631,000	$611,000
B. Inventory adjustments:				
Production to inventory:				
Crops	(+)	$ 52,000	—	—
Cows (net)	(+)	84,000	—	—
Calves	(+)	26,000	—	—
Inventory sold:				
Crops	(–)	12,000	—	—
Cows	(–)	127,400	—	—
Calves	(–)	24,000	—	—
Loss and death adjustment	(–)	8,900	—	—
Purchases of breeding livestock	(–)	—	6,500	—
Purchases of feeder stock	(–)	—	5,000	5,000
Beginning inventory, crops	(–)	—	55,000	—
Beginning inventory, livestock	(–)	—	417,200	—
Ending inventory, crops	(+)	—	110,000	—
Ending inventory, livestock	(+)	—	378,400	—
Subtotal	(±)	$–10,300	$ 4,700	$ –5,000
C. Expenses:				
Labor	(–)	$134,000	$134,000	$134,000
Equipment operations	(–)	36,000	36,000	36,000
Crops, direct[b]	(–)	8,000	8,000	7,500
Livestock, feed[c]	(–)	305,000	320,000	320,000
Livestock, direct	(–)	18,000	18,000	18,000
Utilities	(–)	14,000	14,000	14,000
Real property costs	(–)	26,000	26,000	26,000
Business and overhead	(–)	2,000	2,000	2,000
Depreciation	(–)	29,000	29,000	29,000
Subtotal	(–)	$572,000	$587,000	$586,500
Net income (*A* + *B* + *C*)		$ 48,700	$ 48,700	$ 19,500

[a]Milk deliveries exceeded receipts by $20,000.
[b]Five-hundred-dollars worth of crop expenses were outstanding in accounts payable.
[c]Feed expenses under the perpetual account were transfers out of inventory, under the periodic account they are total purchases.

expense account for the inventory value of the item sold. The difference between the sale and the loss in inventory value is the gain in value the item sold had since it was last inventoried. Thus, at least one "value of inventory sold" account must be included in a perpetual system of inventory accounting. This account is similar to a "cost of goods sold," which is discussed later in this chapter. Not shown in Table 3.10 are the corresponding entries necessary in the balance sheet. The cash account goes up in like amount to the sale and the inventory account of the item sold goes down by a like amount to the value of inventory sold.

The periodic validation (audit) of the balance sheet takes on a different meaning under the perpetual inventory. A running inventory balance is maintained under the perpetual system of accounting; a beginning balance is entered and after that there is an accounting of all amounts going in and coming out. Hence, the purpose of a physical count is not to develop a value for the balance sheet or a change in value for the income statement, but rather to see if the quantities and values inventoried are the same as those shown in the accounts. If they are not, an adjustment must be made.

Because adjustments are necessary to correct inaccuracies in quantities and value, it is necessary to have an account to adjust for inaccuracies caused by measurement errors, price swings, and death and waste. To maintain an accounting balance it is necessary to post to other accounts (normally receipt or expense) the offsetting amount. This posting can be thought of as an adjustment of an amount already entered at an earlier period.

Cash-Flow Accounting

Cash-flow accounting was illustrated in Table 3.7, and a conversion of accrual income to cash-flow income was illustrated in Table 3.4. Cash-flow statements also can be generated with double-entry accounting. It is a matter identifying all transactions that impact the cash-asset account. Computerized accounting programs usually have this feature built in; it is one of the report selections. The same chart of accounts can be used as for the accrual account. For this purpose it is useful to maintain a record of purchases into inventory, either in the inventory-adjustment section of the income statement or in the expense section. In either case, there needs to be an offsetting credit entry.

A cash-flow account also can be tabulated from the double-entry accrual income statement in the same manner as in single-entry accounting. The procedure is illustrated in a little more detail in Table 3.11. Note the following adjustments to accrual net income:

Account adjustments

- Accounts payable and receivable
- Inventory increases and decreases
- Depreciation
- Capital gains

Accounts added

- Principal payments on loans
- Cash paid out for capital purchases
- Cash received from capital sales

Accounts Payable and Receivable

Accounts payable and receivable tend to cause problems for farm accountants, not because of the difficulty of collection but because double posting is a tedious process. Farmers are not accustomed to keeping track of this much detail and it is not required to file a valid tax report. The reversing entry system makes this easier and does not invalidate the accrual account that is built into the double-entry system. Under the reversing entry system only cash payments and receipts are regularly posted. All known accounts payable or receivable are not entered and posted until an accrual statement is

TABLE 3.11

Cash-Flow Balances Tabulated from Accrual Income, FABA Farm, YRXX

Accrued net farm income		$64,257
Cash flow from operations:		
Additions:		
Depreciation	$ 31,944	
Accounts payable, ending	885	
Accrued interest, ending	15,238	
Inventory decreases	—	
Capital losses	210	
Total	$ 48,277	
Subtractions:		
Accounts payable, beginning	$ 1,635	
Accrued interest, beginning	13,881	
Inventory increases	12,958	
Capital gains	—	
Home-used products	—	
Total	$ 28,404	
Net to operations		$84,130
Cash flow after investments:		
Capital sales (+)	$ 400	
Capital purchases (–)	72,831	
Net after investments		$11,699
Cash flow after debts:		
Loan payments: (–)		
Old loans	$ 24,654	
New loans	48,000	
Total	$72,654	
New loans: (+)		
Production	$ 56,000	
Capital purchase	50,000	
Total	$106,000	
Net after loans		$45,045

Note: This table is based on the information in Tables 2.4 and 3.4.

wanted. These same entries are then reversed at the very beginning of the next accounting period. This reversal allows posting only cash payments and receipts except when an accrual account is desired. When these payable and receivable accounts are of an inconsequential nature or do not fluctuate greatly from month to month, the balance sheet and income statement are not greatly in error, and over time will reflect the same average income. This feature can be built into accounting packages. Income statements that are not greatly in error and over time will reflect the same average income.

Accounting for Production in Progress

There are accounting systems, particularly in nonfarm manufacturing industries, in which operating expenses, sometimes referred to as gross margin expenses, are considered asset accounts instead of business expenses. This conclusion is logical! Growing crops and feeder livestock have value, and fertilizer applied to fields and feeds fed add to value; building up assets rather than taking away from them. Thus, some

FORM 3.3

Production and Construction in Progress Accounts (See Table 3.9)

Fall-planted crops:
- Seed
- Fertilizer and chemicals
- Tillage
- Water

Fall-purchased livestock:
- Feed
- Health services
- Transportation

Construction:
- Labor
- Materials
- Contract services

"expenses" are really adding a store of value. When this system is applied, accounts must be defined in the asset ledger to receive such "costs of production." If the transaction is a factor of production, such as feed or fertilizer, and is purchased for cash, the cash-asset account is credited and the "production-in-progress" account receiving the item is debited. These accounts have some similarity to inventory accounts. When the production is complete and the item is ready for sale or inventory as finished goods, a receipt is posted, reflecting the value of the finished product. Also, the costs of production are zeroed out of the asset account(s) and a "cost of goods produced or sold" is posted in the expense-income account. Asset accounts such as those in Form 3.3 could be defined. Production-in-progress accounts may be particularly useful for carrying over costs from one season to the next to bring together in 1 year all of the costs of production as in the cases of winter wheat and fall-purchased feeder livestock. During the year in which the finished product is added to inventory or sold, the costs in the balance sheet can be transferred to the expense account. Thus, the profitability of the finished product, whether retained in inventory or sold, are shown in the income statement and profit center.

We would be more enthusiastic about this system if production were not so unstable in agriculture, and if farmers and their accountants were accustomed to recording things in this manner. The farm tax system, a major reason for keeping farm accounts, is not built around this method. However, for special situations it may be a very useful procedure.

Keeping track of production in progress with a series of asset accounts, as many nonfarm industries do, may become a major system in the future, but a double-entry system constructed like the single-entry conventional farm system is a necessary first step.

Allocations of Ownership, Overhead, and Service Expenses

In enterprise analyses, whether the accounting system is single or double entry, there are always a group of costs that do not relate directly to any particular enterprise or activity but rather belong to the total business or group of productive enterprises and activities. These costs are discussed here because the income statement shown in

Table 3.9 illustrates how allocations are reported. These accounts usually include general business and overhead expenses, depreciation, interest and taxes on land, liability insurance, and other ownership costs. Each can be allocated separately or grouped as illustrated. Labor is sometimes difficult to allocate, particularly when it is used for general farmwork, such as fence and building maintenance. The allocation of these overhead costs does not affect profitability and may even be of little benefit in evaluating alternative production activities. But sometimes it is good to know the approximate total enterprise returns and costs, particularly when considering farm policy, farm rental agreements, and other total concept matters.

Service-center costs comprise those activities that are directly related to the profit-making enterprises but are not associated with an enterprise at the time of their occurrence. Machine services are a good example. A farmer may not keep track of machine or labor time during the season but still may want to charge the specific crop or livestock enterprise with their use. There are no magic formulas or methods that are painless to apply and that will always produce highly accurate results. Methods that we have experienced, read about, or heard of include the following:

- Allocating on the basis of use records
- Proportioning ownership costs on the same percentage basis as are the relevant operating costs to the total operating costs
- Allocating on the basis of gross income or net income less allocated costs
- Allocating on the basis of particular related costs
- Charging a standard rate at the time of use and allocating residuals at the end of the accounting period

A combination of these methods is probably the most accurate but not as easy to apply. The easiest is to sum all of the unallocated costs and allocate them to the enterprises by using one set of percentage figures, but this approach is often not very accurate. Overhead labor costs can be allocated on the basis of operating labor; machinery ownership costs can be allocated on the same basis as machinery operating costs, and so forth. Ownership land cost can be allocated on a land-use-value basis. For some costs an educated estimate system may be the most accurate. Telephone expenses are an example—for which crops and livestock is the telephone used the most?

For assets such as machinery and labor, the charging of a standard use rate is an acceptable accounting practice. The enterprise is charged a fee based on the quantity of use times the established dollar rate per unit of use. This rate can be estimated historically or from standard rates established by others. The quantity of use requires keeping track of the time each machine or other item is used for each enterprise. The product enterprise is debited with this cost, and the machinery service center is credited with a receipt for the same amount. At the end of the accounting period, the balance of debits and credits in the service center indicates how close the per-unit service costs were to reality. Any net or residual amount (plus or minus) in the service center can be allocated to the receiving enterprises or cost centers on the basis of the amounts previously charged.

Net incomes over all costs should be interpreted on the basis of the allocation process. It is always good to have two net income figures, one before allocations and one after. When greater accuracy is needed, more effort should be used to obtain better data on which to base the allocation. This process may require activity records for labor or machinery and for other accounts and activities. Particular needs dictate the amount of energy and time to spend on making more accurate allocations.

CHAPTER 4

Financial Accounts—Business Analysis

Introduction

Analyzing the financial and business records and accounts is the focal point of all farm accounting activities. Only through this phase can a farm manager gain insights into the business. Without analyses, the record-keeping function is little more than accounting for income tax purposes. Even though monitoring income tax savings is a useful endeavor, keeping accounts to minimize taxes seems less purposeful than keeping accounts to maximize after-tax income. Records and accounts must be accurate and reliable to be useful in business management. Accurate records demand consistency and attention to detail that at times may seem like drudgery. The paycheck at the end of the work period keeps most laborers on the job, and this paycheck also helps farmers maintain their records and accounts. The size of the paycheck is dependant on the use made of records and accounts to manage the farm as a profitable business. This chapter provides guidelines for analyzing farm records and accounts to make their use more beneficial. The desired information may direct which records to keep and the amount of detail to include.

The analyses need to include the same set of measurements and elements from year to year so that meaningful comparisons can be made over time. Because of the variable nature of farm prices and production, any one year's results may not indicate efficiency, or the lack of it. This shortcoming, however, does not limit the scope of the analyses that are desirable in any one year. The analyses selected should include those measures available for comparison from agricultural cooperative extension services and farm record-keeping associations. The same tabulations should be made each year to follow trends and check improvements. Tabulation procedures need to be consistent. The comparison of results internally and externally could be an enjoyable activity and certainly can suggest items for business improvement.

The content of this chapter is a revision of material in James, Sydney C., and Everett Stoneberg, *Farm Accounting and Business Analysis,* 3rd ed., Iowa State University Press, 1986.

The discussion that follows does not cover all the many analyses that are possible; it concentrates on measurements meaningful to the majority of farmers. Even though some detail is required, an elaborate set of records and accounts is not necessary to make meaningful analyses. A concerned farm manager may be able to do a better job than a trained accountant. It is surprising how much information can be gleaned from a regular set of farm records and accounts, even about particular enterprises. Data that are not 100 percent complete may be good enough to guide a business out of failure and into success.

To repeat, if farm records and accounts are to be useful in identifying farm management problems, then they must be summarized, analyzed, evaluated, and compared. This process needs to take place not only for the total farm but also with profit and responsibility centers. The balance sheet, income statement, and production records have been discussed and procedures for adding enterprise accounting were illustrated. There are analyses that pertain to each of these procedures, and there are analyses that tie them all together. The discussion in this chapter is confined to those measures commonly used. It is more illustrative than exhaustive; each problem may dictate a different analysis.

Methods and Standards of Comparison

Several steps are necessary to analyze or monitor any particular aspect of the business:

- Determine what is to be monitored.
- Select criteria that can serve as indicators of strengths or weaknesses.
- Set standards or norms for these criteria.
- Set acceptable deviations for these norms.

These steps are implicit in the following discussion.

There are three major levels of comparison for any farm analysis:

- A comparison with the current farm plan
- A comparison over time within the farm
- A comparison with other farms

These comparisons should be applied to the various analyses that follow.

The end-of-year results measure the effectiveness of the farm plan or budget. A disappointing outcome may not mean the plan was not well conceived; there may be too many uncertainties on the side of the budget and too many variables on the side of records. Nevertheless, a comparison of the budget with the results can help future planning and data gathering. It may be as important to see what went right as to determine what went wrong.

A comparison of efficiency criteria over time measures the progress of the farm toward the goals and objectives of the farmer. These goals might include a larger farm, more pigs weaned per litter, the highest corn yield in the county, or a larger net farm income. Regardless of the financial goal, if it can be quantified, records are a useful tool to measure its accomplishment.

Analysis of the Balance Sheet

The makeup and tabulation of the balance sheet were discussed in Chapter 2. The analyses made to illustrate the measures discussed in this section are based on the farm accounting and business analysis (FABA) used throughout that chapter. Three major categories are monitored in the balance sheet:

- Asset levels and combinations
- Liabilities and debt structure
- Equity and business growth

Most business persons would like to increase their assets, reduce their liabilities, and increase their equity. All analyses of the balance sheet relate to these goals in some way. The following measures commonly are tabulated for the balance sheet:

- Liquidity
- Solvency
- Leverage
- Growth
- Balance

Tabulation procedures are applied here to determine efficiency for each of these measures.

Liquidity

Liquidity refers to the anticipated ability of the business to have sufficient cash available to meet financial commitments as they become due without disrupting the ongoing operation of the business. This ability is at least partially shown in a comparison of current assets with short-term liabilities. It is out of the proceeds received from the sale of current assets that debt payments are made. This process is not always direct because new income is being generated, but so also are new expenses. Off-farm dollar transfers to meet family living costs and other business obligations could be important to consider. It was for the purpose of measuring business liquidity that principal payments, accrued interest and unpaid taxes were moved to the current section of the balance sheet. If there are insufficient liquid reserves available to meet current obligations, it may be necessary to obtain new loans by using intermediate and fixed assets as security. This requirement may not be a healthy condition. Thus, the liquidity ratio becomes an important tabulation to the business, and a necessary one when obtaining new production credit. The current ratio, also referred to as the liquidity ratio, is illustrated using the FABA farm for YRXX as an example (see Table 2.4, Chapter 2):

$$\text{Current asset ratio} = \frac{\text{Current assets}}{\text{Current liabilities}} = \frac{\$183{,}187}{\$107{,}325} = 1.71 \text{ or } 1.71{:}1$$

This ratio indicates there is \$1.71 of current assets on hand for each \$1.00 of current liabilities. This ratio is lower than that from the previous accounting period when the ratio was 1.94:1. Such lowering does not necessarily mean that management is losing control and tightening is necessary. A large ratio may mean that the business could use more borrowed capital to increase efficiency and income.

What is a safe ratio? Unfortunately there are no exact answers. Ratios on the high side are never explicitly specified. Ratios on the low side have been provided by lending institutions, but this group has not been consistent. For most businesses a ratio of 2:1 is thought to be relatively safe, and some loan agencies loan at 1.5:1 or less. The safety of the ratio is dictated somewhat by the amount of uncertainty associated with income—the greater the variability and uncertainty, the higher the ratio must be. Some productive enterprises are more stable than others. The historical records of a business may be the best guide for measuring a safe ratio.

Given that there are two or more methods of valuing assets in a balance sheet, the question might be asked about which method is best for making the current ratio. The answer is that the market-based method is best, but the cost-based method probably uses the same values so it doesn't make any difference. Being that it is a current ratio, current conditions should be reflected.

The ratio is a static or "stock" concept of the financial resources available to meet specified obligations at a point in time. It does not measure or predict the timing of future fund flows, nor does it measure the adequacy of future inflows in relation to outflows. Committed lines of credit to finance production are ignored.

The amount of working capital is tabulated by subtracting current liabilities from current assets. Theoretically, the difference is available for operations, the purchase of inventory, and making new investments. For the FABA farm the working capital is illustrated in the following equation: working capital = current assets – current liabilities = $183,187 – $107,325 = $75,862.

Solvency

Solvency relates to the strength of the equity position. It is concerned with the ability to retire all debts if all assets were liquidated (i.e., sold). It measures the amount of borrowed capital (or debt) used by a business relative to the amount of owner's equity investment in the business. Because debt extracts interest and maintains a schedule of exacting payments, solvency provides a measure of the ability to continue operations as a viable business after a financial adversity. For this measure to be accurate, assets need to be valued using a current market and contingent liabilities should be included. Contingent liabilities are those costs generated by the sale of assets, including capital gains taxes, income taxes, and selling fees (see Table 2.3). For the example FABA farm for YRXX, the ratio is as follows:

$$\text{Net capital ratio} = \frac{\text{Total assets}}{\text{Total liabilities}} = \frac{\$1,321,187}{\$523,076} = 2.53 \text{ or } 2.53{:}1$$

The net capital ratio (NCR) is tabulated by institutions before making all major loans, whether used to finance fixed assets or current farm production. It is the credit industry's "security blanket." Thus, lending institutions have been more vocal in specifying what they consider to be a safe ratio. It is very difficult to obtain loans to finance real property at less that 25 percent equity (NCR = 1.33). Having one-third equity (NCR = 1.5) or higher is thought to be much safer. Private loans below this ratio can sometimes be negotiated, but safety may not be the issue in these situations. For the FABA farm the 2.5:1 ratio seems safe, and furthermore, cash flow for YRXX was adequate to make payments.

The debt-to-asset ratio is often tabulated to show the relationship of assets and liabilities. It is the reciprocal of the net capital ratio (total liabilities/total assets). Liabilities are reported as a percentage of total assets. For the FABA farm this value would be 39 percent [($523,076/$1,321,187)100]. Ratios under 40 percent generally are considered safe, whereas those over 60 percent can cause extreme stress. The higher the ratio the greater portion of income goes to pay interest and service debt. (See Figure 8.6.)

The nature of the business being financed dictates to a large degree the safety of the ratio required. Some farms have enterprises and activities that provide a more stable and reliable income than others, and these farms can support lower ratios than farms with unstable income flows and less secure markets. The ability of the farm to generate income to meet debt payments is also important in assessing debt loads.

A working capital ratio is sometimes tabulated. It is used less but has some value when machinery and breeding livestock loans are being evaluated. For the FABA farm it is as follows:

$$\text{Working asset ratio} = \frac{\text{Current + intermediate assets}}{\text{Current + intermediate liabilities}} = \frac{\$351,187}{\$148,543} = 2.36 \text{ or } \$2.36:1$$

The safety of this ratio is similar to that of the net capital ratio previously discussed.

The equity-to-asset ratio is another solvency measure often tabulated. This ratio measures the proportion of assets financed by owner equity. For the FABA farm it is as follows: equity-to-asset ratio = equity/total assets = $798,111 / $1,321,181 = 0.60 or 60%.

Leverage

Leverage measures the proportion to which debt is used, as related to equity capital, to finance the total business. Most businesses borrow with the expectation and hope that the net returns to the total investment, expressed as a percentage, are greater than the interest cost required to finance borrowed capital. Any excess of returns over costs accrues to equity on the basis of the leverage ratio. The debt-to-equity ratio measures leverage and is tabulated for the FABA farm (Table 2.3) as follows:

$$\text{Debt-to-equity ratio} = \frac{\text{Total liabilities}}{\text{Equity}} = \frac{\$523,076}{\$798,111} = 0.66 \text{ or } \$0.66:1$$

This means there are $0.66 of debts for each $1 of equity. The larger the ratio, the more the business is leveraged. A business having two-thirds of its assets secured by outstanding debt would have a ratio of 2:1.

How much individual farmers should leverage their business is conditional. Sure investments can be leveraged more than those with highly variable returns. Highly leveraged investments that return an amount that is below the interest rate being paid on debt give a very low-to-negative return to equity. Higher leverage implies increasing risk with variable returns. Compare the effect of a 1:1 debt-to-equity ratio and a 2:1 debt-to-equity ratio on the return to equity (ROE) for an investment of $30,000 with a 15 percent return on investment or assets (ROA) and the interest on debt (i) is 10 percent. For a 1:1 ratio, the balance sheet would show assets at $30,000, debt at $15,000, and equity at $15,000. For a 2:1 ratio, the balance sheet would show assets at $30,000, debt at $20,000, and equity at $10,000. The return to equity is 25 percent for the higher leverage compared with 20 percent for the lower-leverage ratio. If the return on investment is 5 percent then the lower leverage results in a 0 percent return to equity and the higher leverage has a negative 5 percent return. This is known as the increasing risk principle as illustrated in Table 4.1.

The relationship between returns to equity and the debt-to-equity ratio is as follows: $ROE = ROA + (ROA - i)D/E$. It is apparent from this relationship that higher leverage results in higher returns to equity when return on assets is greater than interest on debt, and returns to equity are lower when return on assets is less than interest on debt.

Farm rental arrangements are a form of leverage. The landlord's capital is added to that of the tenant with the hope that the tenant's limited capital can be ex-

TABLE 4.1

Illustration of the Increasing Risk Principle with Higher Leverage

	Assets	Debt	Equity
		Return on Investment 15%	
D/E 1:1	$30,000	$15,000	$15,000
Returns ($)	$ 4,500	$ 1,500	$ 3,000
Returns (%)	15%	10%	20%
D/E 2:1	$30,000	$20,000	$10,000
Returns ($)	$ 4,500	$ 2,000	$ 2,500
Returns (%)	15%	10%	25%
		Return on Investment 5%	
D/E 1:1	$30,000	$15,000	$15,000
Returns ($)	$ 1,500	$ 1,500	$ 0
Returns (%)	5%	10%	0%
D/E 2:1	$30,000	$20,000	$10,000
Returns ($)	$ 1,500	$ 2,000	–$ 500
Returns (%)	5%	10%	–5%

Note: D/E = debt-to-equity ratio.

panded to return a higher-than-market rate of interest. This situation is illustrated later in this chapter.

Growth

Growth measures the progress being made in acquiring assets and increasing equity. Thus, it is a historical perspective. For the FABA farm for YRXX (see Table 2.4), assets owned changed from $1,185,355 to $1,237,324, for an increase of $51,969. This increase is significant, but at the same time liabilities also increased by $34,023 ($460,076 – $426,053). Equity increased $18,016. A savings of this amount is no small sum for some businesses. The longer the period of comparison, the more this measure reflects management expertise. Normal business fluctuation can cause this measure to be sporadic.

Growth can be measured using either the market basis or the cost basis of asset valuation; but if it is to show progress made by management, the prices used need to be kept relatively constant. Otherwise, it is difficult to tell whether the change was from business savings and retained earnings or reflects price-level changes. Financial growth is also important for purposes of future retirement and estate planning, as well as for business expansion. A market-based comparison may be more useful for retirement planning. It is important to keep in mind cost-of-living increases when comparing business growth and retirement needs. Assuming a 5 percent land inflation rate, the FABA farm would have increased by $41,640 ($832,800 × 0.05) in YRXX. This increase is considerably above the $18,016 increase in equity achieved through retained earnings.

Balance

This measure considers the distribution of investments in the asset column. The FABA farm had 15 percent of its owned assets (13 percent of managed assets) in current assets, 12 percent of its owned assets (10 percent managed) in intermediate assets, and 73 percent of its owned assets (77 percent managed) in fixed assets. Is this balance

TABLE 4.2

Relationship of Capital Investment to Farm Size, Farm Type, and Income Efficiency

	Farm Size[a]			Farm Type[a]			Efficiency		
	309 acres	525 acres	888 acres	Grain farms	Hog farms	Beef-feeding farms	High-profit farms[b]	Low-profit farms[b]	FABA
Current assets	24	26	29	27	28	45	31	20	15
Intermediate assets	19	19	19	17	20	13	17	36	12
Land and improvements	57	55	52	56	52	42	52	45	73
Total	100	100	100	100	100	100	100	100	100

Sources: [a]*1996 Iowa Farm Costs and Returns,* FM-1789. Iowa State University Extension, Ames; [b]*1997 Annual Report, Southeastern Minnesota Farm Business Management Association,* University of Minnesota, 1998.

good or bad? There are no standards for comparison, but comparisons with similar farms may be enlightening. Investment percentages for some farms in Iowa are shown in Table 4.2. The FABA farm is more like the grain farms than the livestock farms. Even though an exact answer is elusive, and may even change over time, comparison with similar businesses raises questions worth investigating.

Over- or undercapitalization are problems of many farms. This can be measured partially when looking at investments per unit of production, such as per acre and per head, and fixed costs per bushel and per hundredweight. These will be illustrated later in this section under "Machinery Efficiency."

Cattle and sheep livestock enterprises may depend on fixed-pasture investments for their profitability. This relationship may lead to a higher percentage of investment in livestock and feed. The balance of investments is important to profitability even if the exact relationship cannot be established. There is a best investment strategy for each farm.

Analysis of the Income Statement

Net income is probably the best overall measure of farm efficiency, whether considering the total business or one of it's enterprise parts. Lack of money to buy supplies and pay bills is where the pinch is felt first when a significant problem exists. Figure 4.1 illustrates how income problems relate to other parts of the business. Lack of income could be a problem of efficiency or size of business. Farm-size efficiency is discussed later in this chapter. Efficiency relates to the use of resources and general business management.

The fact that cash flow is not a problem does not mean there is no inefficiency in the business. An entrepreneur can be content operating an inefficient business if there is adequate cash to meet farm and family needs. A business with few debt obligations can have a sizable cash flow even when investment returns are low.

The income efficiency measures evaluated in this section include the following:

- Allocation of net farm income
- Sources of income
- Rate of asset turnover
- Revenue per dollar of expense
- Operational ratios

FIGURE 4.1

Location of the Income Problem

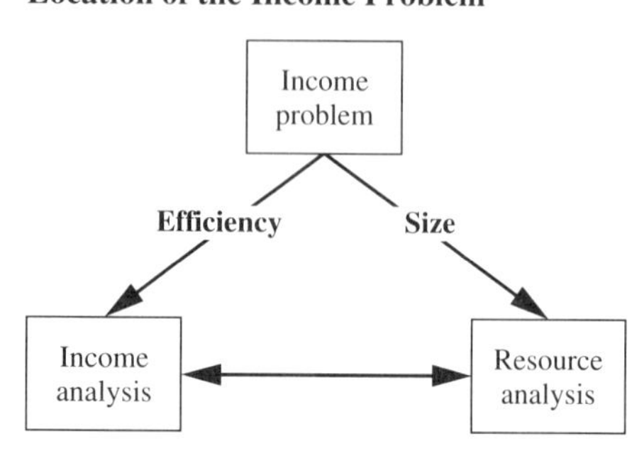

The above measures help to determine if there is an income problem , and if so, the general direction of its location (Figure 4.1). The analyses of enterprises discussed later are more specific in locating solutions to problems. Some income-efficiency measures link the balance sheet to the income statement.

When analyzing income efficiency it may be helpful to place the net income figure on a nonbiased basis by subtracting out the value of unpaid family labor used in the business and adding back interest payments included with expenses. This approach places net income on a uniform basis for comparison with other farms.

Allocation of Net Farm Income

Net farm income was previously defined to measure the following:

- A return to equity
- A return to unpaid labor
- A return to unpaid management

If opportunity returns are imputed in any two of these factors, the remaining residual of net income can be said to be a return to the third factor. There is no easy and accurate way of determining the contribution of intermediate and fixed capital, or operator labor and management, to net farm income. Any allocation is arbitrary. However, to obtain some relative measure of efficiency in the use of capital, labor, and management, an allocation process is useful. For labor and capital there are well-established market prices that can be imputed as a cost to the business for their use (this also could be viewed as a return to these two input factors). This is not true of management. From an owner–operator standpoint the management return is similar to profit, which is the residual left after all other factors have been paid a market price.

Figure 4.2 illustrates how accrual net farm income can be allocated to pay for the services provided by the unpaid factors previously named. Use it as a guide when following the tabulations for the FABA farm shown in Table 4.3. The data used come from balance sheets and incomes statements previously developed in Chapters 2 and 3. (See Tables 2.4, 3.4, and 3.9.) The distributions to management and capital are the most meaningful. Note first that this allocation process can be for the operator or landlord alone, or combined for the total farm. The "total farm" column represents the situation in which the operator was the sole owner. If comparisons are made with other farms, the total-farm approach is the one with the best comparable meaning.

The asset values used are averages of the beginning and ending of year. Some accounting systems use only the end-of-year values. This approach is permissible if ap-

FIGURE 4.2

Tabulation Diagram for Calculating Distributive Payments to the Unpaid Factors of Net Income

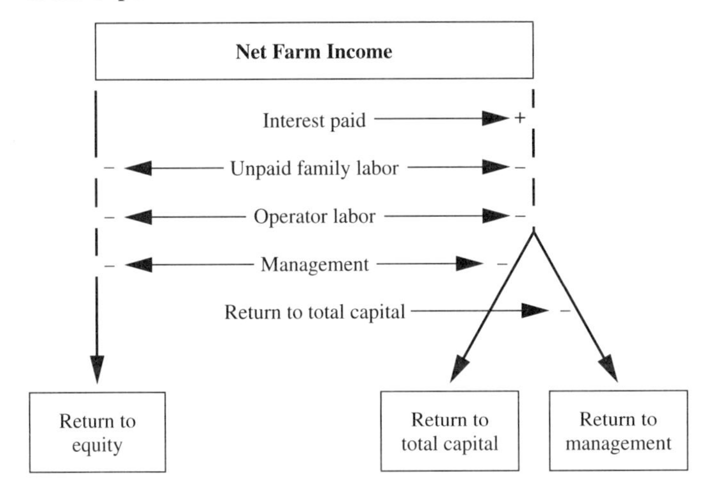

plied consistently, but it gives a less-representative estimate of the business assets used during any one accounting period.

To make meaningful comparisons between farms, income tabulations need to be applied uniformly. Because all farms do not have the same liabilities or furnish the same amount of unpaid labor it is necessary to make unifying adjustments. These adjustments add interest costs that had been subtracted as an expense and subtract a value for unpaid family labor. Adjusted net income now measures as follows:

- A return to total asset capital
- A return to operator labor
- A return to management

Returns to management and total capital are tabulated in Table 4.3. Note the tabulation of adjusted net farm income as explained previously. A return to the operator's labor and management is calculated by subtracting a return for the asset capital. In the example a 5 percent composite return was applied to the total capital asset value. This rate represents individual return rates for operating, intermediate, and fixed capital. The rates used should be in line with competing uses for this capital. Farmers need to ask themselves what they could expect to receive if they loaned this money or invested it (e.g., government bonds or corporate stock). The operator's labor was valued at $20,000. The return-to-operator labor is based on farm-labor wage rates and alternative nonfarm employment opportunities. The returns to management is $10,470 under the operator column and $11,595 for the total farm.

The return-to-farm asset capital (ROA) is calculated in a similar manner. The operator's labor and management is subtracted from adjusted net farm income. Surprisingly, the same figure ($20,000) is used to represent both the value of the operator's labor and the operator's labor and management. This implies that management has zero value, which may or may not be true because the opportunity value of management is not known. However, it is interesting and useful to compare the residual management return with the fee charged by some professional farm managers. Using a rate of 5 percent of the value of farm production (VFP), the fee would be $11,895 on this

TABLE 4.3

Distribution of Net Farm Income for the FABA Farm, YRXX

Item	Operator	Landlord	Total Farm
Total asset value:			
Operating capital	$ 177,186	—	$ 177,186
Working capital	124,091	—	124,091
Fixed capital	910,063	180,000	1,090,063
	$1,211,340	$180,000	$1,391,340
Adjusted net farm income:			
Net farm income	$ 64,257	10,125	$ 74,382
Plus interest expense	31,780	—	31,780
Minus value of unpaid family labor	5,000	—	5,000
	$ 91,037	$ 10,125	$ 101,162
Return to operator labor and management:			
Adjusted net farm income	$ 91,037	$ 10,125	$ 101,162
Minus 5% interest return assets	60,567	9,000	69,567
	$ 30,470	$ 1,125	$ 41,595
Minus value of operator labor	20,000	0	20,000
Return to management	$ 10,470	$ 1,125	$ 11,595
Return to capital investment:			
Adjusted net farm income	$ 91,037	$ 10,125	$ 101,162
Minus value of operator's labor and management	20,000	—	20,000
	$ 71,037	$ 10,125	$ 81,162
Percent return to capital:			
(return to capital ÷ total asset value) × 100	5.9	5.6	5.8

farm, which is $4,398 less than the residual return to management of $16,293 for the total farm.

Because the return to capital is generally expressed as the rate of return on farm assets, it is necessary to divide the dollar return to capital by total assets. Percentages thus obtained are comparable to interest rates and expected returns from alternative investments and among farms. For the FABA farm this return is 6.7 percent on operator capital and 6.5 percent on total capital.

One useful calculation not illustrated in Table 4.3 is the return to operator's equity capital. Many investors make their living by expanding their limited capital through credit to make relatively large investments possible. Leverage was tabulated before and the effects of returns above and below the cost of interest were illustrated. The return to equity capital is calculated for the FABA farm as follows (see Table 2.4, chapter 2, for equity):

1.	Net farm income	$64,257
	Minus value of unpaid family labor	–10,000
	Minus value of operator's labor and management	–25,000
2.	Return to equity capital	$29,257
3.	Percent return to average equity capital:	

$$\$29{,}257/\frac{(\$759{,}302 + \$777{,}248)}{2} \times 100 = 4.1\%$$

TABLE 4.4

Value of Feed Fed on the FABA Farm, YRXX

Add		Subtract	
Beginning feed inventory	$ 58,738	Closing feed inventory	$ 67,234
Value of crops raised	100,550	Feed crop sales	42,922
Feed purchased	26,551	Value of crops used for seed	0
		Storage losses	1,166
		Government payments	Exclude
		Crop-share rent	0
	$185,839		$111,322
		Value of feed fed	$ 74,517

Farmers who cannot experience an acceptable return for the use of their resources (equity capital) and labor and management over a period of years should perhaps consider other employment. However, if returns are high, farmers should consider expanding their businesses. The return to equity and total assets, regardless of their size, are indicative of the efficiency with which farm resources are used.

Sources of Income

Gross revenues (GR) and value of farm production (VFP) were defined and illustrated previously in Tables 3.4 and 3.9. In Table 3.9 these measures were tabulated for each profit center. Also shown in Table 3.9 was a reconciliation of inventories showing the productive value for each crop and livestock enterprise, including the transfers of on-farm feeds. These elements are used when looking at the sources of income. This value is a gross volume of business-activity evaluation without dealing with the details of expenses, except for livestock feeds. The following formula is very useful for tabulating missing elements, including values, numbers and weights for all crops raised, livestock produced and feeds fed: beginning inventory + production + purchases = sales + farm use + losses + ending inventory. This formula is illustrated in Table 4.4 to tabulate the value of feeds fed.

The contributions of crops and livestock to gross income are shown below:

Contributions to:	Gross revenues		Value of production	
Corn	$ 97,689	31%	$ 97,689	41%
Soybeans	69,043	22%	69,043	29%
Forage crops	7,195	2%	7,195	3%
Total	$173,927	55%	$173,927	73%
Cattle	38,144	12%	17,135	7%
Hogs	95,789	31%	42,281	18%
Total	$133,933	43%	$ 59,416	25%
Miscellaneous	$ 4,554	2%	4,554	2%
Total	$312,414	100%	$237,897	100%

For Iowa farmers in 1996 crops contributed 72 percent of gross revenues, livestock 16 percent, and other 12 percent as measured by the middle third of farmers shown in Table 4.13. This table compares high- and low-profit farms by using several efficiency measures.

Rate of Asset Turnover

Asset turnover ratios are common measures of business efficiency. Profit margins of many companies are very low per unit of production; but when the volume of business (turnover) is large, the returns to assets and management could be large. There are two methods for measuring turnover rates on farms. The first uses gross revenues and the other value of farm production. Both are expressed as percentages of total assets managed.

(Gross revenues/Assets managed) × 100
($306,318/$1,391,340) × 100 = 22.0%

(VFP/Assets managed) × 100
($237,897/$1,391,340) × 100 = 17.1%

The method chosen depends on the comparisons being selected and their use and farm type. Gross revenues as a percentage of total assets managed is the tabulation generally used.

The rate of capital turnover in most farming activities is low compared with many other businesses. The 17 to 23 percent for the FABA farm illustrates this fact. In comparison, a grain marketing cooperative in Iowa showed more than 500 percent turnover, a regional farm sales cooperative more than 300 percent, and a national meat-packing plant more than 700 percent.

Within the farming industry the rate of turnover varies with the enterprises selected and the capital position of the operator. The turnover rates are higher for livestock than for crops, higher for dairy cows than for beef cows, higher for broilers than for layers, etc. Consider beef cows and feeder cattle, for example. Using some average figures for the Midwest, turnover rates were determined as shown in Table 4.5. Farmers short on capital may wish to emphasize enterprises that have a relatively high turnover rate, providing there is also an adequate profit potential. This aspect is particularly important to beginning farmers.

Farmers that rent have a higher turnover rate than do owner-operators. Renting is a means of spreading limited equity capital over a larger total investment. The rate of turnover on equity capital is increased considerably by this means. For the FABA farm this rate would be as follows:

(Operator GR/Owned assets) × 100
($306,318/$1,211,340) × 100 = 25.3% (Inverse = $3.95)

(Operator VFP/Owned assets) × 100
($237,897/$1,211,340) × 100 = 19.6% (Inverse = $5.10)

The turnover rates for owned assets are about 3 percent higher than for the total assests managed, which was previously calculated. The turnover rate increases as the proportion of rented assets goes higher. These rates are lower than for Iowa farms in 1996 that showed a rate of 42 percent on owned assets. (See Table 4.6.) The typical Iowa farm has more rented assets than does the FABA farm.

Looking at equity the turnover rates for the FABA farm are considerably higher than when total assets are used as the divisor.

(Operator GR/Equity) × 100
($306,318/$759,232) × 100 = 40.3% (Inverse = $2.48)

(Operator VFP/Equity) × 100
($237,897/$759,232) × 100 = 31.3% (Inverse = $3.19)

TABLE 4.5

Average Rate of Turnover for Livestock

Item	Returns	Investment	Turnover (%)
Using GR:			
Beef cows	$ 199	$ 525	38
Beef feeders (yearlings)	490	413	118
Dairy cows	1,234	1,505	82
Using VFP:			
Beef cows	67	525	13
Beef feeders (yearlings)	69	413	17
Dairy cows	712	1,505	47

Reciprocals (inverses) of the above tabulations are shown in parentheses. They measure the dollars of investments required to generate $1 of income.

Two other measures of input efficiency are the assets required to produce $1 of net income and the dollars of assets per farmworker.

Assets per $100 net income:

(Total assets managed/Adjusted total net income) × 100
($1,391,340/$101,162) × 100 = $1,375

Assets per person:

Total capital/Worker-years
$1,391,340/(28 months/12) = $596,289

Operating profit margin ratio. The operating profit margin ratio (OPMR) is the proportion of gross revenue that is a return to farm assets. The operating profit margin ratio for the FABA farm is as follows:

Operator OPMR = [(NFI + farm interest expense – unpaid labor and management)/operator GR] × 100
Operator OPMR = ($71,037/$306,318) × 100 = 23.19%

Total Farm OPMR = [(NFI + farm interest expense – unpaid labor and management)/total farm GR] × 100
Total Farm OPMR = ($81,162/$312,414) × 100 = 25.98%

Multiplying the rate of asset turnover times the operating profit margin ratio gives the rate of return to farm assets. This relationship suggests that to increase the rate of return to farm assets requires either a greater operating profit margin or a greater volume of business.

	Rate of asset turnover	×	OPMR	=	Rate of return to farm assets
Operator:	25.29%	×	23.19%	=	5.9%
Total Farm:	22.45%	×	25.98%	=	5.8%

The effect of renting land for the operator compared with the total farm is to increase the rate of asset turnover and reduce the operating profit margin, which results in a higher rate of return to assets than the total farm. This same analysis can be done by substituting value of farm production for gross revenue in both the asset turnover rate and operating profit margin.

Revenue per Dollar of Expense

This tabulation measures the capacity of expenses to generate income. Two tabulations are made. The first divides gross revenues (TBC) by total expenses (TBD). For the FABA farm the following is tabulated (See Table 3.4.): TBC/TBD = $306,318/$242,061 = $1.27 of revenues per $1 expense. This is a useful statistic but may not be as comparable as VFP per dollar of adjusted expense. Not only does VFP remove the differences between livestock farms as previously explained but also subtracting interest costs removes differences caused by business debt. TBD must be adjusted not only for feed and livestock purchases but also for interest expenses. These tabulations for the FABA farm are as follows: (See Table 3.4.)

Adjusted expenses = $242,061 – $26,551 – $41,870 – $31,780 = $141,860
VFP/Adjusted expenses = $237,897/$141,860 = $1.672

If another $1 is expended, neither $1.27 nor $1.67 more of production value will be created. Income per dollar of expense is an average concept and not a marginal one. It does relate, however, to the efficiency with which resources are used and products marketed.

Other meaningful income relationships can be calculated, such as net income per cultivated acre and per hour of operator labor. These two measures for the FABA farm are calculated as follows: (See Table 4.3 for a tabulation of adjusted net income.)

Net income per cultivated acre:

Adjusted net income/Cultivated acres
$91,037/393 = $231.65

Net income per hour of operator labor:

Adjusted net income/Total operator labor
$91,037/2,640 = $34.48

Operational Ratios

Operational ratios reflect the distribution of gross revenue or value of farm production to operating expense, depreciation, interest, and net farm income. This distribution is illustrated by the pie charts (Figure 4.3). The ratios differ depending on the use of gross revenue or value of farm production. The purchase of livestock and feed used to calculate value of farm production is part of operating expenses for the gross revenue chart. The following lists the various types of operational ratios.

Operating expense ratio—operating expenses divided by gross revenue expressed as a percentage. Operating expenses for this ratio includes all expenses necessary for the ongoing operation of the business, excluding depreciation and interest. This ratio indicates percentage of gross revenue used to cover operating expenses.

Depreciation expense ratio—depreciation expense divided by gross revenue expressed as a percentage. A high ratio may indicate underutilization of machinery, equipment, and buildings.

Interest expense ratio —interest expense divided by gross revenue expressed as a percentage. A high ratio may be a signal that the farm is too highly leveraged and may have liquidity problems.

Net farm income ratio—the residual or net farm income divided by gross farm returns. This ratio indicates the percentage of gross revenue going to owner's equity and unpaid labor and management. (The FFSC suggests the use of net farm income from operations as the appropriate measure because typically net farm income tabulations include capital gains and losses that do not occur regularly.)

FIGURE 4.3

Distribution of Gross Revenue or Value of Farm Production to Expenses and Net Farm Income

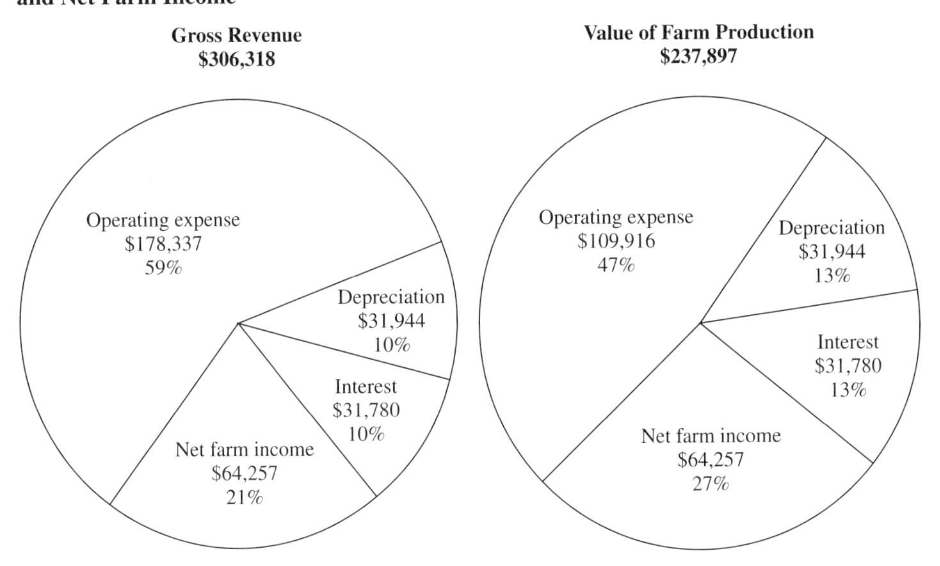

It is important to note that high or low ratios may suggest strengths or weaknesses in the business, but they also could be explained by differences in how resources are acquired. Leasing machinery versus owning would increase operating expenses and decrease depreciation expense. Cash renting land versus equity acquisition of land would increase operating expense and reduce net farm income. From year to year, the percentages also will change due to changes in market prices, which affects gross revenue.

Sources and Uses of Funds

Balance sheet, income, and cash-flow statements have all been discussed in detail in earlier sections. (See Tables 2.4, 3.4, 3.7, and 3.9.) Integrating these statements into a single statement showing the sources and uses of funds is commonly referred to as the statement of changes in financial position. Any money-earning person, including the farmer, has only about three options for the use of cash: consumptive buying, including gifts; investing in capital purchases; and saving. The sources of the cash may be earnings, sale of assets, savings, and gifts. Net farm income can be used for family living expenditures, income and social security (self-employment) taxes, new on-farm investments, savings, and off-farm investments. In lean years the farmer may even disinvest (reduce current asset or working asset balances) to obtain adequate family living funds. Credit is an equalizer, making many of these transfers of funds possible. As explained earlier, a new loan does not make the farmer richer or poorer. The farmer is made better off only by investing borrowed money and receiving a return higher than interest costs. Rather than sell assets to obtain funds for family living or investment, assets may be used as security for new loans from which the funds are obtained. Likewise, loans may be used as a source of funds to be combined with earned savings for making investments.

The relationships of the three statements discussed are illustrated in Figure 4.4. The income statement and statement of sources and uses of funds both show the same amounts of income transfer, but the latter provides more detail of the changes that affect net worth or business equity.

FIGURE 4.4

Relationship of the Balance Sheet, Income Statement, and Sources and Uses of Funds Statement

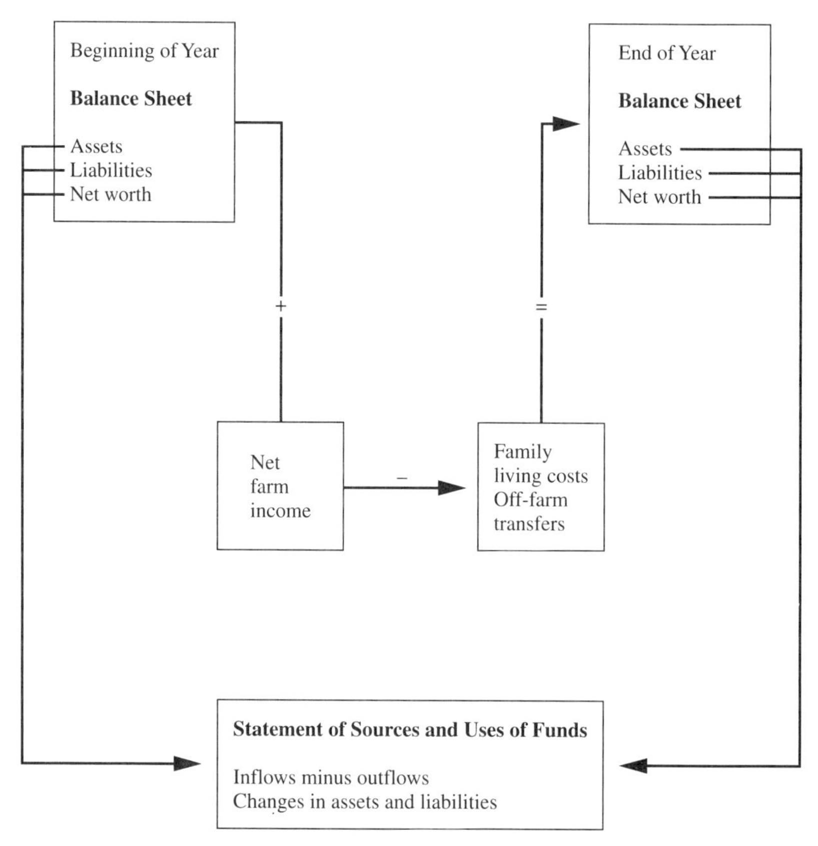

Table 4.6 has been generated to show the sources and uses of funds on the FABA farm in YRXX. First, note that growth funds (1c) are tabulated as a residual of net income adjusted for transfers in from off the farm minus family living expenditures, personal taxes, and off-farm transfers. Second, note that this not only is the difference between net investments and net indebtedness but also is the same as the increase in equity tabulated in Table 2.4. Thus, for farms with little or no outside employment or income investments, family living transfers plus the increase in equity equal net farm income.

The sources and uses of funds statement has three main uses:

1. It serves as a reminder of the source and destination of business funds and the balance with off-farm transfers. The necessity of a family commitment to the farm business can readily be seen.
2. It provides insight into investment and financing decisions. Lenders realize that their own earnings can increase only as their clients' investments increase and prosper. A large operation's income is necessary to provide the required support for new loans. Disinvestment, particularly of working and fixed assets, is alarming. Large off-farm transfers also may be alarming.
3. It is possible to visualize how growth funds are being used. They can be used for debt liquidation or new investment. This affects the size of the business and income-generating power of the operator. If growth funds are combined with new indebtedness, new investments may be comparatively large, whereas if growth

TABLE 4.6

Sources and Uses of Funds Statement, FABA Farm, YRXX

Sources of Funds	Dollars	Uses of Funds	Dollars	Net Change in Funds	Dollars
1a. Income funds:		1b. Nonfarm withdrawals:		1c. Equity increase (1a – 1b):	
Net farm income (NFI)	64,257	Family living	28,955	NFI – Family living	35,302
Labor income	12,000	Income and social security taxes	19,471	Labor – Taxes	(7,471)
Transfers in	0	Transfers out	10,000	Transfers	(10,000)
Other	0	Other	(115)	Other	115
Total	76,257	Total	58,311	Total	17,946
2a. Reductions of assets:		2b.Increases of assets:		2c. Net change in assets (2b – 2a):	
Current assets	0	Current assets	12,002	Current assets	12,002
Intermediate assets	0	Intermediate assets	46,555	Intermediate assets	46,555
Fixed assets	6,588	Fixed assets	0	Fixed assets	(6,588)
Total	6,588	Total	58,557	Total	51,969
3a. Increases of liabilities:		3b. Decreases of liabilities:		3c. Net change in liabilities (3a – 3b):	
Current liabilities	19,225	Current assets	0	Current liabilities	19,225
Intermediate liabilities	38,939	Intermediate liabilities	0	Intermediate liabilities	38,939
Fixed liabilities	0	Long-term liabilities	24,141	Long-term liabilities	(24,141)
Total	58,164	Total	24,141	Total	34,023
				(1c – 2c + 3c or)	0
4a. Total sources	141,009	4b. Total uses	141,009	4c. Balance (4a – 4b)	0

funds are largely used for debt retirement, the assets managed may remain static. Many factors must be combined to conclude the best use of growth funds. This statement aids the decision-making process.

Farm Size Efficiency

Economies of size are reflected in most other measures of efficiency. A common idea is that it is necessary to have a large volume of business to have a large net income. However, there are diseconomies of size as well as economies, so a farmer also must be efficient in using resources and spending money.

Farm size, like people, can be measured in a number of ways. Some persons are judged on the basis of height, some by weight, some by brawn, and others by the size of their investments or income; measurements differ for each individual or purpose. So too, farm size may be measured by the amount of investments; inputs of land, livestock, or labor; or volume of income produced.

Some size efficiencies for Iowa farms in 1996 are shown in Table 4.7. Several things can be observed by studying this table. First, note that acreage increased 3.2 times between the smallest farms with 251 acres and the largest farms with 801 acres, whereas labor increased only 1.6 times. One person at the largest size level took care of 200 more acres than one person at the smallest size level. Machinery costs per acre decreased by $5 while investments in machinery per acre increased by $44. Value of crops produced per acre and livestock returns per $100 feed fed were both higher for

TABLE 4.7

Size Efficiency Measures as Related to Investment, Acres, and Labor on Iowa Farms, 1996

Size of Farm (Sales [$1,000])	40–99	100–249	250+	FABA
Acres in farm	251	464	801	393
Months of labor	13	15	21	28
Investment ($1,000):				
Short-term assets	62	135	302	177
Intermediate-term assets	62	101	195	124
Long-term assets	150	287	528	910
Total	264	523	1,025	1,211
Value of farm production (VFP) ($100):				
Livestock income over feed	48	216	899	594
Crop production	598	1,904	2,668	1,739
Miscellaneous	47	118	221	140
Total	693	1,640	3,788	2,379
Expenses ($100):				
Operating	308	683	1,487	893
Fixed	216	517	1,197	1,817
Net farm income ($100)	180	445	1,127	668
Efficiency measures:				
Crop value per crop acre($)	299	313	331	442
Machinery cost per crop acre ($)	61	62	56	100
Livestock returns/$100 feed fed ($)	145	149	159	179
VFP/$100 expense ($)	142	143	145	139
% return to assets owned	3.3	8.7	12.4	5.9

Source: *1996 Iowa Farm Costs and Returns,* FM-1789. Iowa State University Extension, Ames.

the larger-sized farms. VFP per $100 expense, a measure of overall efficiency, was not greatly different for all size levels, but usually the larger farms have a clear advantage. Comparative tabulations for the FABA farm are shown in the last column. These tabulations were for a different year so they are not totally comparable. (Data are from Tables 2.7 and 3.9).

Acres farmed is the most common and easily understood unit for measuring farm size. It is particularly applicable where the major source of income is from crop production and similar crops are being produced. For the Midwest, where 70 to 80 percent of VFP is from crop production and the major cash crops are corn and soybeans, the acre is a good unit of measure. It is not a good unit of measure where land productivity differs greatly or the major enterprises are not the same. For some comparisons the number of acres in cultivation may be a better measure than total acres.

Assets managed includes rented assets so it is more broadly based as a measure of size. It takes into account land productivity differences as reflected in land values and is aggregative of different enterprise investments. However, it is less well understood and not easily derived. Also, it may not accurately describe size as it relates to income when the turnover of capital in the productive activities differs greatly.

Gross revenues and VFP measure size in relation to the volume of business and the income-generating power of the farm. They, like assets managed, can be broadly applied but are probably less well understood than either of the other two measures of size, although they may more accurately describe size. It would not be a good measure of size when business expenses, as related to production activities, use up greatly different percentages of VFP. Vegetable production, which has very high expenses compared with dryland wheat production, illustrates this point.

Worker-months of labor as a measure of size relates to one of the major input factors in many production situations. When labor accounts for a large portion of business expenses or inputs it is a better unit of measure than when capital investments replace much of the labor. However, when enterprises are similar and production practices uniform, months of labor is still a good unit for measuring size. As can be seen in Table 4.7, the worker-months of labor used do not increase in proportion to acres.

Units of livestock may be one of the best measures of size in areas where livestock contributes the major portion of income. The number of livestock breeding animals, number of feeders, number of layers, or number of poultry may be the most descriptive and meaningful unit for describing business size in many areas of the country. A 400-cow unit is descriptive of size to the cattle rancher in Wyoming, Nevada, or New Mexico.

Size is an important aspect in all expansion considerations on the farm. Whether an expansion or increase in size will reduce unit costs and increase income must be determined by budgeting or programming each individual situation. There is a different cost curve for each farm. A study of a farmer's records should give him or her some indication of position on the cost curve and provide some directional guides for decision making.

When size is determined to be insufficient to produce enough income to meet the needs of the farm family, it may be necessary to increase the size of the business by the following methods:

- Intensifying the production of crops
- Increasing livestock production
- Renting additional land
- Doing custom work
- Taking an off-farm job

Measures of Production Efficiency

This section discusses several aspects of enterprise efficiency. It is divided into the following parts:

- Crop efficiency
- Livestock efficiency
- Machinery efficiency
- Labor efficiency

Evaluation of crop and livestock efficiency can be viewed as in Figure 4.5. Note the branching lines in the figure. First, revenues are separated from expenses. Revenues are a combination of quantities produced and prices received. Expenses are a combination of quantities purchased and prices paid. It is recognized that there is a relationship between the inputs purchased and production. This diagram may be helpful in finding the cause of a felt difficulty or need. Many problems are obvious, whereas others are harder to locate. The analyses outlined may be useful in this process. The discussion is more illustrative than exhaustive of the analyses that can be applied. The

FIGURE 4.5

Diagram Showing the Elements That May Contribute to Crop or Livestock Income Efficiency

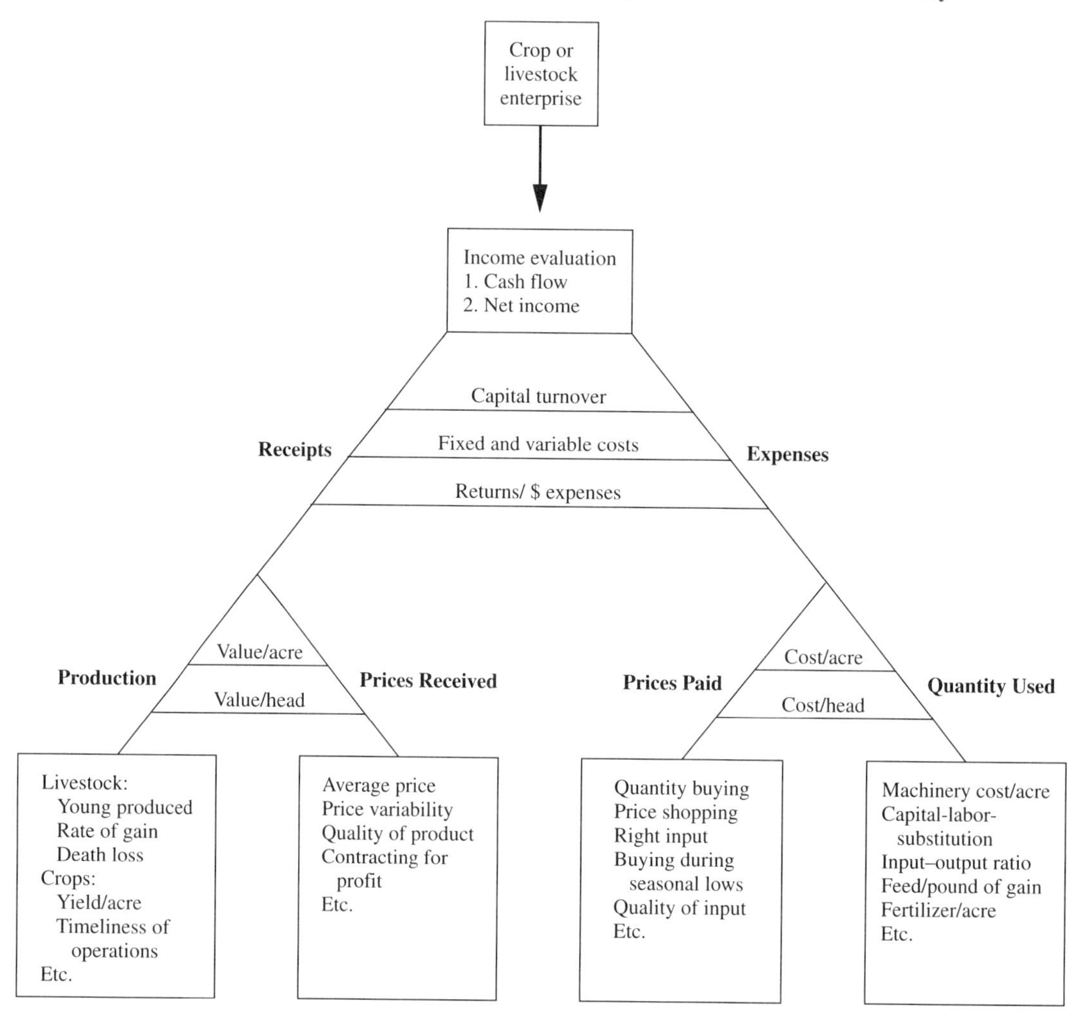

approach used in this section is that of the total farm records and accounts. Individual enterprises should be analyzed but procedures and elements are not discussed here.

Crop Efficiency

Measures of net income, investments, and production can be generated for each major enterprise on the farm as illustrated in Table 3.9. The importance of crops in generating gross revenue and the percentage of total acres in cultivation are both important concepts that have been discussed previously.

Yield is the most common, and perhaps the most useful, measure of crop efficiency, particularly when crops contribute the bulk of gross revenues. Yields are measures that farmers know and understand well and provide interesting conversations in many farmer meetings. Comparisons with neighbors, record-keeping groups, and county or township reports are helpful. Discussions and comparisons involving varieties, fertilizers, cultural practices, pests, etc., as related to yield can be very benefi-

TABLE 4.8

Summary of Yields for the FABA Farm, YRXX

Crop	Unit	Acres	Cultivated Acres (%)	Yield
Corn	bushel	200	51	145
Silage	ton	175	44	20
Soybeans	bushel	8	2	45
Hay	ton	10	3	3.5
		393	100	

TABLE 4.9

Crop Yield Index Tabulation

Crop	Unit	Production (FABA farm)	Community Yields Per Acre [a]	Acres Required at Community Yields
Corn	bushel	29,000	140	207
Silage	ton	160	21	8
Soybeans	bushel	7,000	38	184
Hay	ton	85	3.5	24
				423

Crop yield index = (423 acres ÷ 393 acres)100 = 107%

[a] Hypothetical figures.

cial, particularly to the less-efficient producer. It should be realized that yield can be too high as well as too low. Economic considerations dictate the most profitable level of production for each farm and they are discussed in Chapter 6.

Yields for the FABA farm are summarized in Table 4.8. A crop yield index that compares composite crop yields for a farm with yields from some norm is illustrated in Table 4.9. The method shown for calculating a crop yield index weighs the individual crop yield difference by the number of acres for each crop. This is accomplished by determining the number of acres required at community yield levels (423) (or some other norm) to produce the amount of crops on the reference farm. Dividing these acres by the number of acres on the reference farm (393 acres) and multiplying by 100 gives a percentage of comparison (107 percent).

The distribution of crops also may be important in relation to net farm income. Not all crops have the same income-producing potential. Table 4.10 shows some suggested income figures for major crops in Iowa. Of course, individual farmers do not always achieve the average. For example, some farmers may find oat production more profitable on parts of their farm than soybeans or corn. Also, some unmeasured costs and values are not included here, such as nitrogen from soybeans and alfalfa and disease control. These figures emphasize that the combination of crops selected can be an important consideration in maximizing net farm income. These percentages were shown for the FABA farm in Table 4.8.

Value of crops per cultivated acre relates to the distribution of crops just discussed. This figure is calculated by dividing the value of production from crops by the number of acres in rotation. For the FABA farm this is calculated as gross value of crops produced per cultivated acres = $183,427/393 = $466. This compares to $310 crop value per cultivated acre for Iowa in 1996 (Tables 4.6 and 4.13).

TABLE 4.10

Budgeted Profitability for Some Iowa Crops, 1998

Crop	Yield Per Acre	Price	Income Per Acre	Operating Cost Per Acre	Income Over Cost Per Acre
Corn	145 bushel	$2.50	$362	$182	$260
Soybeans	45 bushel	6.00	270	102	168
Corn silage	18 ton	25.00	450	206	244
Alfalfa hay	4 ton	70.00	280	143	137

Source: *Estimated Costs of Crop Production - 1998,* Fm 1712, Iowa State University Extension, Ames.

Fertilizer and lime costs per rotated acre is another important efficiency measure because fertilizer has become such an important crop input. For the FABA farm this cost is calculated as fertilizer and lime costs per crop acre = $18,348/393 = $47.

Crop efficiency measures and their importance may be summarized as follows:

- Farms with high crop value per tillable acre have higher net incomes than those with low crop values within the same size group.
- High crop yields contribute to high crop values per acre.
- Crop expenses (largely seed and fertilizer) average higher for farms with high crop values per acre, and these farms still have higher net incomes.

Livestock Efficiency

Production and feed determinations are the major efficiency measures that can be developed from the general set of farm accounts and records for livestock. Measures of production relate to output per livestock units, such as pigs farrowed or weaned per litter, percentage of calf crop or pounds of calf weaned per cow, milk per cow, eggs per hen, and weight gain per day. For the FABA farm the following measures were calculated:

Hogs:

Number of litters farrowed = 104
Pigs weaned per litter = 836 pigs weaned/104 = 8.0

Feeder cattle:

Number of feeders sold = 100
Rate of gain = weight gain/feeder days = 45,525/30,273 = 1.5 pounds/day

For feeder cattle, weight gain is determined by weighing and from purchase and sale records. Feeder days is a summation of the number of days each animal is on the farm. Feeder days can be tabulated easily by using purchase and sale dates. Extreme accuracy is not essential for useful tabulations. For the most part each measure is only a clue for detecting some suspected problem.

The importance of feed efficiency has already been discussed, as well as tabulation procedures for all livestock and individual enterprises. Feed efficiency can be a physical or financial measure. Because tabulation aspects are similar, only the financial measure is illustrated. However, keep in mind that the amount of feed required to produce a pound of meat, a dozen eggs, or 100 pounds of milk can be determined by using the same procedures. In most cases the only requirement is to keep a record of the weights of feed, livestock, and products purchased and sold.

TABLE 4.11

Livestock Income per $100 Feed Fed for Illinois

	Corn Price	Sheep	Hog Farms	Beef Feeding	Beef Raising	Dairy Farms
1985	$2.54	$130	$170	$121	$101	$202
1986	2.01	156	254	149	125	210
1987	1.61	141	232	196	168	237
1988	2.32	115	158	150	150	198
1989	2.48	96	167	145	144	209
1990	2.44	98	247	162	165	220
1991	2.41	64	199	109	129	188
1992	2.35	116	167	164	142	211
1993	2.28	95	197	143	133	191
1994	2.44	146	166	114	117	196
10-year average	2.29	116	196	145	137	206
High, 2-year average	2.51	151	250	180	166	228
Low, 2-year average	1.81	79	162	111	109	192

Source: *1995 Summary of Illinois Farm Business Records,* Illinois Cooperative Extension Service, Urbana.

Returns per $100 feed fed is a common measure of feed efficiency. This value can be calculated for a composite of all classes of livestock on a farm with very few additional records beyond those required for income tax purposes. Production of livestock and feeds fed tabulations have been previously illustrated for the FABA farm (see Tables 3.9 and 4.4). Returns per $100 feed fed are given below for cattle and hogs combined:

Livestock increase/Feed fed
($133,933/$74,517) × 100 = $180

When there is only one major livestock enterprise, the all-livestock returns per $100 feed fed figure can be very meaningful, but less meaningful if several livestock enterprises are represented. Some comparative returns per $100 feed fed for Illinois are shown in Table 4.11.

Machinery Efficiency

Investment and cost per cultivated acre are major efficiency measures that can be developed without rather detailed records of machine use. This detail can be justified only for individual machines when particular information is desired. This discussion treats the total investment without reference to types of machine or use.

Investment per cultivated acre is calculated by dividing the average of the beginning and ending machine values by the number of cultivated acres. For the FABA farm this value is calculated as investment per cultivated acre (avg.) = $94,413/393 acres = $240.

Cost per cultivated acre is calculated by adding operating and investment expenses for machinery and equipment. These figures can be found in the income statement. For the FABA farm, machinery costs were calculated as shown in Table 4.12. Use caution in interpreting these figures—figures that are too low as well as those that are too high may indicate machinery investment and cost problems. These figures must be interpreted with regard to farm size, crops and livestock produced, labor availability, and alternative returns on investment.

TABLE 4.12

Machinery and Equipment Cost Per Rotated Acre, FABA Farm

Variable Costs		Fixed Costs	
Machinery and equipment hire	$10,425	Depreciation	$31,944
Machinery and equipment repair	6,046	Taxes	—
Fuel and lubricants	5,020	Insurance	800
Farm share of auto expense[a]	—		$32,744
Farm share of utilities	3,410		
	$24,901	Total all costs	$57,645
$57,645 ÷ 393 rotated acres = $146/acre			

[a]Included with machinery.

Labor Efficiency

Even though labor may not be as important as it once was, it is still a major input on most farms. The input unit is usually measured in worker-months or worker-years of labor. Common measures relate to physical output or value. Typical of these tabulations are the measures illustrated for the FABA farm:

Labor efficiency:
- Worker-months of labor = 22
- Rotated acres per worker = 393/(22/12) = 214
- Value of crops produced per worker = $173,927/(22/12) = $95,042
- Value of livestock produced per worker = $133,933/(22/12) = $73,187
- Value of production per worker = $237,897/(22/12) = $129,998

Key Analysis Factors

When making an analysis of the total farm business, the following five factors are the most important and probably the minimum needed for a useful information system:

- Gross value of crops per rotated acres
- Livestock returns per $100 feed fed
- Value of farm production per worker
- Value of farm production per $100 expense
- Power and machine costs per rotated acres

An analysis of these five factors often accounts for 75 percent or more of the income difference between farms when a comparison is made. Farms with above-average performance in all five factors are high-profit farms a very high percentage of the time. Farms that are below average in all five factors (or most of them) are low-profit or low-income farms a high percentage of the time.

Gross value of crops per rotated acre indicates a combination of crops selected and yields. This value provides guidelines to the performance of the cropping program relative to the value of the land resource being used. Livestock returns per $100 feed fed indicate the margins realized on livestock over feed costs. This margin must pay for costs other than feed and provide for profit. Value of farm production per worker is the only efficiency factor that indicates the volume of production. It shows the volume being produced, and implies, or gives some indication of labor efficiency. Value of

TABLE 4.13

Comparison of High- and Low-Profit Farms in Iowa to FABA Farm, 1996

	High Third	Middle Third	Low Third	FABA Farm
Resource used:				
Acres in crops	751	530	460	393
Months of labor	20	16	16	22
Machinery investment ($)	133,678	96,460	97,980	102,911
Gross revenues:				
Crops ($)	225,791	144,521	123,485	173,281
Livestock ($)	214,881	92,255	82,897	133,933
Miscellaneous ($)	49,508	24,303	17,533	4,554
Total	490,180	261,079	223,915	312,414
Value of Farm Production:				
Feeds fed ($)	125,709	59,950	61,024	74,517
Total (GR-feeds fed) ($)	364,471	201,130	162,891	237,897
Net Farm Inome:				
Operating expenses ($)	132,507	83,011	77,998	89,270
Fixed expenses ($)	102,301	64,615	63,265	84,370
Subtotal ($)	234,808	147,626	141,263	173,640
Total (VFP-Expenses) ($)	132,142	53,940	22,742	64,257
Efficiency of operations:				
Corn yield/acre (bushel)	142	138	132	145
Crop value/rotated acre ($)	335	321	296	440
Machine and power cost/acre($)	53	56	68	133
Livestock increase/$100 feed ($)	168	151	139	180
Crop acres/worker	453	386	345	215
VFP/$100 expense ($)	165	145	119	139
Debt-to-asset ratio (%)	32	33	33	37

Source: *1996 Iowa Farm Costs and Returns*, FM-1789. Iowa State University Extension, Ames.

farm production per $100 expense provides some guidelines on the overall efficiency of the business. It indicates whether margins over costs are satisfactory to produce profits. Machine and power costs per rotated acre indicate whether the farm may have excessive costs in this area. These costs must be related to volume of business, yields, and livestock production to be meaningful.

Comparison with other farms has been mentioned throughout the discussion on efficiency. A comparison of the FABA farm with other farms in Iowa is shown in Table 4.13.

CHAPTER 5

Using Budgets in Farm Planning

Introduction

A farm budget is a definite financial plan of future operations. It is a summary of expected production, prices, receipts, expenses, and net income. Budgeting is planning on paper where changes can be made without costly mistakes. Budgeting formats are similar to those in farm accounting. (See Chapter 3.) The major purpose of budgeting is to test the profitability of alternative farming plans or courses of action. This test could be a comparison of present methods with proposed changes. Another use of budgets is to plan purchase requirements of seed, fertilizer, feed, labor, and other inputs. Budgets are a major planning tool for farm managers.

Budgets require two major kinds of data. One kind concerns physical relationships of transforming raw materials (i.e., land, labor, machinery service, feed, seed, fertilizer) into products (i.e., vegetables, fruits, grains, forages, livestock, livestock products). These conversions are often referred to as input–output relationships. The inputs are often called factors and the outputs products. The other major kind of data is prices of inputs and products. Prices are studied in Chapter 7 and are assumed known in this chapter.

Listed below are a few reasons for changing farm plans:

- Farm may not have been planned properly in the first place.
- Depletion of soil from erosion or overcropping.
- Adoption of higher capacity tractors and machines may call for larger fields and specialization.
- Product demand may have changed so that current products are no longer profitable.
- Marketing channels may have changed requiring greater attention to uniformity and quality.

Some text in this chapter was taken from Chapter 12, *Farm Accounting and Business Analysis* by James, Sydney, and Everett Stoneberg, Iowa State University Press, 1986.

- Technological developments affecting crops or livestock may call for changes in production practices and enterprises.
- Government regulations may restrict the production of some products or limit the use of certain inputs and practices.
- Leasing or renting agreements have changed.
- Death, injury, disease, and illness of the operator or family member.

Planning of most farms lags far behind the time when it should have been done.

The study of accounting procedures provide the best guidelines to follow when budgeting. The uses of farm records and accounts in budgeting are summarized below.

- Provide the formats into which data can be organized to produce reliable planning information.
- Provide the operator with a performance history, including past and current financial conditions, financial growth in assets and equity, and total business and enterprise profitability.
- Provide basic data needed in the budget, such as: crop yields, livestock performance, and production requirements.
- Provide the financial information useful in computing future taxes. Budgeted plans should be compared on an after-tax basis.
- Provide the basic information necessary for a sound credit policy. Implementing budgeted plans may need outside financing.
- Guide the implementation of farm plans.
- Provide the test of time of how well-implemented plans turn out.

Farm records and accounts and budgets need to be integrated to improve the usefulness of each.

Budgets need to have beginning and ending dates between which expected performance is evaluated. This planning period is not the same for all products or businesses. The planning period should fit the purpose of the plan and the production cycle of the products.

Planning period:

Beginning ———————————————————— End

months—years—lifetime

Consider for example the following time intervals for completing production cycles in farming. It takes about 3 months to finish a batch of broilers, 5 months to finish hogs, 10 months to finish beef calves, and 12 months for a dairy cow to freshen. A hen is usually kept only 1 year, a sow may give birth twice each year for about 4 years, and a cow is usually not kept past her 8th birthday. Most machinery is placed on a 5- to 10-year depreciation schedule, but it may last considerably longer, and many buildings are in use for 20–40 years. Soil seems never to wear out, but may erode or lose productivity. The plans of an operator may be for a single production season, over a leasing period of 3 to 5 years, until children finish school, until retirement, or even longer. The time period selected affects the input–output coefficients used as well as prices. The specifications for next year's cropping season should be considerably different from those required when making a decision about purchasing more land to farm. Budgeting begins with a specification of objectives, inventory of resources and listing of operator-imposed limitations. This approach presumes that the problem has been identified and specified. The objectives to be accomplished through farm planning may include one or more of the following considerations:

- A high standard of living for family members.
- A sound land use and cropping system.
- A livestock program, if any, adjusted to available feeds, the operator's knowledge and skill, available labor, buildings, equipment, and markets.
- A choice of enterprises that fit together well, integrating crop and livestock production activities, providing for good labor distribution, and giving proper balance in the use of other resources.
- Sufficient volume of business to permit the efficient use of labor, machinery, and buildings.
- Adequate labor-saving equipment, farmstead arrangement, and building design consistent with the type and volume of business.
- Adoption of practices that will give the most profitable crop yields and rates of livestock production.
- Enough flexibility in farm organization to facilitate adjustments to changing economic conditions.

Resource limitations might include the following:

- Land—including the productivity of soil, precipitation patterns, slope, erosion, weed and insect problems, and previous crops.
- Livestock on hand—including breed, genetics, productivity, and health.
- Machinery—including kind, usefulness, size, age, and condition. Kinds may include tractors, cultivating and harvesting machines, and special-purpose equipment.
- Buildings—including kind, usefulness, size, age, and condition. Kinds may include grain storage bins, machinery storage, livestock facilities, and special-purpose buildings.
- Improvements—including the condition and capacity of water wells, tiled land, terraces, fences, and water ponds.
- Labor—including age, education, and health of operator, availability of family labor, and sources and quality of hired labor.
- Sources of business finance—including assets and debt, credit availability, and available rental properties and equipment.
- Land and livestock tenure arrangements—including lease specifications that might limit the activities of the operator.
- Available markets—including distances, type of market such as open versus contractual, competition, and buyer specifications.
- Operator-imposed limitations that would limit farm plans—including education, training and experience, likes and dislikes, debt aversion, ability to withstand uncertainty of production and income, and willingness to work long hours and weekends and go without vacations.

Questions relating to business organization, product selection, production methods and procedures, yield levels, and markets were introduced in Chapter 1. The use of budgets in farm planning should be helpful when answering these and other types of questions. If the problem can be quantified and organized into a financial statement then budgeting should be an appropriate tool to use for planning the optimization of goals and objectives.

Types of Budgets

Budgets may include the total farm business or be confined to specific parts. The former are commonly called total farm budgets and the latter are generally referred to as partial budgets.

Total Farm Budgets

Total farm budgets, also called complete farm budgets, are more concerned with the profitability of the total farm than any one part. It is possible that a seemingly profitable enterprise or activity may be dropped from the farm because it does not fit in with other enterprises or activities. Only in the complete plan can the mix of enterprises and activities be tested and worked out.

Partial Budgets

Partial budgets are used to adjust production methods, change input levels, and test new techniques that do not affect appreciably other parts of the total farm plan. When developing the total farm budget the partial budget is very helpful in selecting products and in developing profitable input–output combinations. Partial budgets should maintain the same set of goals and objectives specified for the total farm.

Unit Budgets

Unit budgets, also called enterprise budgets, provide a useful format for both total and partial budgeting. The unit budget specifies the quantity and price amounts and relationships for 1 acre of a crop, one head of livestock, or other meaningful unit. The idea is that the unit budget can be expanded or multiplied in a partial or total farm budget. This requires that the size level be anticipated so that economies of size can be incorporated into the specification of input–output coefficients and prices used.

Cash-Flow Budgets

Cash-flow budgets have become significant in farm planning. Their popularity is associated with changes in the financial structure of the agricultural industry. Due to the magnitude of certain investments, the size of the business, small margins of profit, and the tightness of credit, farmers have had to tighten the control of their money and make advance arrangements for credit and its repayment. Cash-flow budgets follow the form of cash-flow income statements discussed in Chapter 3. The cash-flow budget as a planning device anticipates the flow of cash into and out of the business over specified periods of time. The difference between a cash-flow budget and the total farm budget is illustrated in the differences between the cash-flow income statement and the accrual income statement. Although cash-flow budgets are most often associated with the total business, it is possible to make cash partial-budget evaluations.

Cash-flow budgets test the feasibility of alternative farm plans. Profitability budgets for the total farm, or part thereof, may overlook financing requirements or other acquisitions. Furthermore, cash-flow budgets usually integrate farm and nonfarm financial flows. It may be possible to operate a farm with a negative net farm income in the short run, but it is impossible to operate with a negative cash balance. Also, a particular enterprise or activity may appear to be very profitable in the long run but may have impossible cash requirements in the short run. Cash-flow budgets do not identify profitability.

Procedures to Be Used in Budgeting

Steps in Budgeting

The following steps and procedures are part of the budgeting process. The detail and inclusion of each of these depends upon the nature of the problem being assessed.

1. *Identify and limit the problem.* Review farm records and accounts for problem areas. Observe technology to see if machinery or practices need to be updated. Make personal and family goal assessments. Develop a hypothesis about the solution to the problem selected.
2. *Inventory relevant resources and the current situation.* Determine the beginning point. Make an inventory and net worth statement. Specify conditions. Use units that are common and easy to manipulate.
3. *Specify operator-imposed limitations.* Consider physical, financial, emotional, and other restraints of the operator, family, and owner of rented resources. Some operators do not want to be tied to the farm with livestock chores, some have allergies against certain products, and still others have physical handicaps. Some families do not want hired labor around, and for others debt causes great distress. These limitations need to be recognized and accounted for. There is no benefit from developing a plan that cannot or will not be made operational.
4. *Gather production data and specify input–output relationships.* These relationships should represent production practices used for each of the products or systems under consideration. This step relates to specifying the amounts of labor, seed, fertilizer, water, and machinery services required to produce crops; and feed, health care, and housing required for livestock. Past performance rates and requirements may be a good guide to future planning.
5. *Choose the "most probable" prices for products marketed and costs for purchased inputs.* In some cases, opportunity costs may be important to identify. It is important to be realistic when selecting prices. Overly optimistic prices may cause an operator to select an activity that is doomed to failure. Prices that are too pessimistic may cause the operator to never invest in profitable activities. It may be a good practice to plan with three sets of prices—the most probable price and one level on either side.
6. *Tabulate net income and make comparisons among plans.* Place the physical and financial relationships in an income-statement format and tabulate profitability, and measure other goals.
7. *Check the feasibility of each plan developed.* It is possible that the best of plans cannot be implemented. Lack of finances is a big deterrent but not the only one. Sometimes the physical resources are not available, such as a farm to rent or labor to hire. Even obtaining a particular repair part for a machine can be a big problem at times. Perhaps the necessary market has not been developed for the product.
8. *Choose the best plan.* Consider the chances and consequences of being wrong and make contingent provisions.

Where to Begin

The place to begin in budgeting is often arbitrary and depends on the specific problem. In the total farm plan it may depend on the relative importance of crops or livestock to the total farm business. Logically, the crop and livestock program should be developed simultaneously. Traditionally, most budgeting specialists have selected the cropping system as the logical place to begin because the land is considered the basic farm resource. Livestock were thought of as a means for marketing farm crops. Exceptions might be specialized dairy, hog, chicken, or beef feed-lot programs. In these instances, the livestock program might control the selection and production methods of crops. But, even in these instances, alternative crop–pasture–livestock combinations should be evaluated in the search for the most profitable one. Some crops can only be harvested by grazing livestock.

Assuming there are both crops and livestock in the plan it is important to consider the integration of the two. Examine particularly the way in which the livestock enterprise blends with the cropping system in the use of crops produced. Are the livestock providing a better market for the home-grown feeds than if the feeds were sold for cash? Adjustments will probably be necessary in cropping plans as different livestock systems are tried.

Competitive relationships are important to identify. Although some crop or livestock activities are competitive, others do not compete for resources. It is possible that some crops or livestock enhance or contribute to the production of another product. For example, some crops reduce pest problems in other crops, and livestock may furnish fertilizer useful to crops. After crop and livestock enterprises are balanced, look for unused resources (i.e., surplus labor, machinery, buildings, and land). Thus, a series of partial budgets may be necessary in developing the complete farm plan that yields the highest net farm income. The plan having the greatest net income or other yield is the one that makes the fullest use of available resources and achieves the highest possible production efficiencies. Successive budgets test for improvements in resource use and efficiency.

Suggested Procedures for Developing a New Plan for the Total Farm

The following considerations may prove helpful when setting up a new farm program, or when completely revising or evaluating an existing farm program:

1. Develop a sound land-use program that places each acre in its most productive use.
2. Work out a profitable crop and pasture system consistent with conserving the soil and maintaining or increasing soil fertility.
   a. Develop a tentative cropping plan according to
      - suitability of land for different crops;
      - kinds of livestock and cash crops to be produced;
      - needs for pasture, roughage, and grain;
      - available labor, markets, equipment, and facilities; and
      - abilities and interest of the operator.

   b. Adjust tentative crop selections to support livestock feed requirements if needed.

   c. Crop plans should
      - contribute to maximizing net farm income over a long period of time,
      - maintain soil fertility, and
      - be efficient in the use of labor and machinery.

3. Determine if livestock enterprises fit into the revised plan.
   a. Select the kinds of livestock considering
   - livestock now on hand;
   - skills and preferences of the operator;
   - available labor, buildings, equipment, and facilities;
   - kinds of feed to be produced, including pasture, hay, and grains;
   - available markets and prices;
   - size of farm;
   - accessibility to a water supply; and
   - capital strength of operator, including borrowed money.

   b. Decide on number of livestock by checking
   - pasture balance,
   - grain and roughage balance,
   - amount of labor available, and
   - building facilities (most limiting in the short run).

4. Complete the details of the revised cropping system; specify soil treatments and develop field arrangements consistent with land use patterns, the cropping system, and the livestock program.

5. Prepare a livestock management calendar of breeding, feeding, and marketing plans. Make provisions for change.

6. Provide the necessary equipment and buildings consistent with the cropping system and the livestock program.

7. Prepare a revised farmstead layout map.

8. Summarize any planned farm improvements and specify credit needs. Work out a calendar on all proposed capital improvements.

9. Test the revised plan against the present plan or alternative plans from the standpoint of volume of business, labor distribution, diversity of the business, risk, efficiency of operation and a satisfactory income. Estimate probable income and expenses of each plan and compare net farm income.

10. Choose the plan that is consistent with the long-run objectives and interests of the operator and his family. Work out and show plans for changing from the present to the new organization. This plan involves annual or short-run budgets adjusted to current prices, costs, weather, and technological conditions.

Ownership and Operating Costs in Budgeting[1]

Only operating costs and returns affect decisions when budgeting. The planner can control and manipulate these costs. Operating costs and ownership costs are separated in the accrual income statement. Operating costs include hired labor, fuel and repair of machinery, seed and fertilizer for crops, feed and veterinary supplies for livestock, and electricity. Ownership costs include property taxes, farm owners insurance, interest, farm cash rent, and depreciation. These income-statement definitions are a good place to start when identifying the operating and ownership elements in farm budgets.

A less specific definition of ownership and operating costs is that operating costs are those that change as enterprises are selected and production takes place. Ownership costs are those that continue at the same level whether the business, or part thereof, is operated or not. For example, crop insurance for a new crop would be an operating cost, whereas crop insurance for a new variety or change in production practices of the same crop would be an ownership cost.

Another broad guideline when determining whether an expenditure is an operating or ownership cost is to consider the time period over which the budget pertains. That which may be fixed in one budget could be variable in another. For example, the depreciation expense of a tractor on hand would be an ownership cost, whereas depreciation on a tractor the operator is planning to buy would be a variable cost. What is defined as a fixed or variable cost is dependent on the amount of decision-making power the planner has over the cost and whether or not the specific element will be affected by the anticipated change. These issues all relate to the planning period. In the very long run all elements are variable.

On-farm produced inputs such as feed can be troubling when budgeting. If the budget is organized around profit centers (enterprises), it is important to credit the producing center (crop and forage) and charge the receiving center for the item (feed). Feed and other inputs produced on the farm may have an opportunity cost (i.e., they could be sold off the farm). This same kind of reasoning applies to operator and family labor used on the farm.

Suggested Computational Rules in Budgeting

The following rules mostly pertain to the total farm budget but have application to partial and unit budgets because the parts make up the whole.

1. Keep the budget as simple and free from unnecessary detail as possible. Use prepared budget forms or computerized programs when available if they fit the problem. It is generally easier to modify existing forms than make new ones. Minute details are not necessary for making many decisions.
2. Choose unit measures of production, expense, and income that are easy to calculate, readily understood, and a valid measure for comparing alternative farm plans for relative profitability.
3. Include only receipts and expenses that relate directly to the crop and livestock activities being considered. Itemize operating costs, aggregate ownership costs.
4. Recognize all income and expenses pertaining to the activities being considered. Income (TBC) represents the value of production for the budget year. Income will be from one or more of three sources: (1) products sold from the current year's production; (2) products produced during the year but used on the farm as seed, feed, or in the household; and (3) products produced into inventory and left as an inventory increase. The sum of these three sources of income must be adjusted to account for sales of inventory from a previous year. In the long-run budget representing the average year, inventory levels should be held constant. Only the normal depreciation and upkeep of real property and power machinery are treated as expense. Price changes of resources not being held for sale are not considered in the budget. Price changes for products being held for sale are considered because they influence production decisions.

5. Purchases and sales of capital assets such as land, buildings, machinery, and equipment are not entered as costs or income for the budget year—only the depreciation and repairs on such items are treated as costs. However, the purchase costs and sales receipts of capital assets are important in the cash-flow budget.
6. Summarize the budget by bringing variable income and expense items together as discussed previously. Operating costs that were not included in profit centers can be added as overhead costs in the summary. Also, ownership costs can be added to give a total perspective to income and farm profitability.

Budgeting Formats and Illustrations

Unit Budgets

Unit or enterprise budgets consider the physical input requirements with accompanying outputs and respective prices for one unit of production. For most crops, this unit is 1 acre and for most livestock it is one head. These are useful in and of themselves and can be part of the budgeting process for partial and total farm budgets. Unit budgets are necessary when building linear programming (LP) models. LP is discussed later in this chapter.

As mentioned, it is important to consider the level at which these activities will be produced to select the correct input coefficients. For example, the per-head requirement for labor would be greatly different in a 1,000-head cattle feedlot compared with a 100-head feedlot. If the unit budgets are not used at the level budgeted, it may be necessary to make adjustments so that the unit budgets are predictive of the activity being planned.

Two unit budgets are illustrated—one for a crop and one for livestock. The crop budget is typical of those available through agricultural extension services. Also, online budgets are available for a variety of products. The livestock budget was developed on a spreadsheet. Computer templates are available for many farm software programs. The procedure used is more important to observe than the particulars of inputs and prices.

EXAMPLE 5.1 Crop Budget for Corn

Even though the budget in Table 5.1 was developed for 1 acre it is important to recognize that other variables were considered as the title suggests (e.g., the cropping sequence, the type of tillage system used, soil type, and size of farm operation). This information is important to determine the appropriate quantities of inputs such as fertilizer, pesticides, labor, and machinery costs. Quantities, prices, unit of measure, and total value are shown. Operating expenses are separated from ownership and overhead costs. Costs per bushel are shown indicating the break-even prices to cover operating costs and all costs. Two net income figures are shown: gross margin, measuring income less operating costs, and net income, measuring income less total costs. A column for listing the amounts for a total field or farm is helpful sometimes.

TABLE 5.1

Continuous Corn Mulch Tillage System for 640-Acre Farm on Fayette Soil

Item	Unit[a]	Quantity	Price	Value/Acre
RECEIPTS	bu	125.00	$ 2.70	$337.50
OPERATING COSTS				
Preharvest Costs:				
Seed	1,000 ker	30.00	$ 0.85	$ 25.50
Fertilizer:				
N-anhydrous ammonia	lb	150.00	$ 0.20	$ 30.00
P_2O_5	lb	53.75	$ 0.24	$ 12.90
K_2O	lb	35.00	$ 0.12	$ 4.20
Lime	ton	0.30	$14.00	$ 4.20
	SUB-TOTAL: FERTILIZER			$ 51.30
Herbicides	pt	5.8	$ 4.31	$ 25.00
Insecticides	lb	8.7	$ 1.98	$ 17.23
	SUB-TOTAL: PESTICIDES			$ 42.22
Crop insurance			$ 4.80	$ 4.80
Machinery: fuel, oil, and repairs	acre		$ 8.37	$ 8.37
Labor	hr	0.81	$10.00	$ 8.10
Harvest Costs:				
Machinery: fuel, oil, repairs	acre		$ 4.81	$ 4.81
Labor	hr	1.27	$10.00	$ 12.70
Trucking	bu	125.00	$ 0.02	$ 2.50
Drying	pt/bu	7.00	$ 0.02	$ 0.14
Storage	mo/bu	0.00	$ 0.02	$ 0.00
Hired machinery			$ 0.00	$ 0.00
Miscellaneous			$ 0.00	$ 0.00
Interest (7 mo. preharvest costs)			9.00%	$ 6.56
	SUB-TOTAL: OTHER OPERATING COSTS			$ 65.34
TOTAL OPERATING COSTS	per acre			$184.37
	per bu			$ 1.47
GROSS MARGIN				$153.13
OWNERSHIP COSTS				
Machinery: depreciation, interest, taxes, housing, and insurance			$83.33	$ 83.33
Land rent			$80.00	$ 80.00
Real estate taxes			$ 0.00	$ 0.00
TOTAL OWNERSHIP COSTS				$163.33
TOTAL OPERATING AND OWNERSHIP COSTS	per acre			$347.70
	per bu			$ 2.78
NET INCOME OVER ALL COST				($ 10.26)

Source: Developed from "Crop and Livestock Budgets Examples for Illinois 1998-1999," Farm Laboratory, Department of Agricultural and Consumer Economics, University of Illinois, Urbana-Champaign.

[a]Abbreviations: bu, bushels; hr, hours; ker, kernels; lb, pounds; mo, months; pt, pints.

EXAMPLE 5.2 Livestock Budget for Feeder Cattle

This feeder cattle budget is presented in three parts. Table 5.2a is a summary table showing costs and returns per head and expanded to the total number on feed. Note that physical requirements and prices are shown. Even though this is the first table shown it is really the last one completed. Planning begins in Table 5.2b in which the variables are set forth (purchase weight and price, sales weight and price, months on feed, feed prices and nutrients, death loss, and other factors affecting profitability). Table 5.2c is an expanded feed requirements budget. Rates of gain and days on feed are first specified, followed by a listing of nutrient requirements. Rations are then built to meet these parameters. For each ration tested, a summary of nutrients provided is compared with the requirements. Experience and training are needed to build good rations and those elements listed do not cover all considerations. But the result is a tabulation of the feeds needed. Tables 5.2b and c are copied into Table 5.2a. Feeding efficiency as entered in Table 5.2b is built into Table 5.2a but is not part of Table 5.2c. Note that break-even purchase and selling prices are tabulated. Changes made in Table 5.2b allow the user to play "what if?" games.

Partial Budgets

Partial budgets are a practical application of marginal analysis discussed in Chapter 6. Partial budgets take the following form:

A. Added incomes
   1. Increased sales from new activity
   2. Decreased costs from replaced activity

B. Added costs
   1. Increased costs from new activity
   2. Decreased sales from replaced activity

C. Net income increase = A – B

The use of partial budgets is illustrated in Example 5.3 from material adapted from "Positioning Your Pork Operation for the 21st Century," ID-210, Purdue University Cooperative Extension, Indiana.

TABLE 5.2A

Spreadsheet Template for Budgeting Cattle Feedlot Profitability

Item	Unit[a]	Quantity	Price	Amount	One Head	Total Head
I. *Value of production:*						
A. Sales (1% loss)	cwt	12.58	75.00	934.07		
B. Sales or Inv. Inc.	cwt	12.58	0.00	0.00		
Total sales					$934.07	$93,407
II. *Cost of production:*						
A. Purchase or value	cwt	5.50	85.00	467.50		
Total feeder cost					$467.50	$46,750
B. Feeds fed:						
Pasture (dm)	aum	0.00	0.00	0.00		
Roughage 1 (dm)	ton	0.75	75.00	56.61		
Roughage 2 (dm)	ton	2.72	28.00	76.04		
Grain 1	cwt	34.44	4.50	154.97		
Grain 2	cwt	0.00	0.00	0.00		
Supplement	cwt	3.83	12.00	45.90		
Minerals	cwt	0.62	5.00	3.12		
Additives	cwt	0.00	20.00	0.00		
Other	cwt	0.00	0.00	0.00		
Total feed cost					$336.64	$33,664
C. Health care	hd	1.00	4.00	4.00		
D. Machinery and equipment	hr	1.00	10.00	10.00		
E. Yardage fee	$/day	330	0.15	49.50		
F. Marketing	hd	0.99	15.00	14.85		
Total other operating cost					$ 78.35	$ 7,835
G. Interest	%/mo	11.00	0.58	43.31		
H. Labor	hr	2.00	10.00	20.00		
Total costs (IIA-H)					$925.80	$92,580
III. *Net income:*					$ 8.26	$ 826
IV. *Analysis:*						
A. Necessary selling price to break even		($/cwt)			$ 74.34	
B. Maximum purchase price to break even		($/cwt)			$ 86.50	
V. *Summary of feed needs:*		*Quantity*	*$ Cost*			
Pasture	aum	0	0			
Roughage 1	ton	75	5,661			
Roughage 2	ton	272	7,604			
Grain 1	cwt	3,444	15,497			
Grain 2	cwt	0	0			
Supplement	cwt	383	4,590			
Minerals	cwt	62	312			
Additives	cwt	0	0			
Other	cwt	0	0			
Total feed cost			$33,664			

[a]Abbreviations: aum, animal unit month; cwt, hundredweight; dm, dry matter; hd, head; hr, hours; mo, months.

TABLE 5.2B

Feeder Cattle Budget Variables Specifications

Variable:	% dm	% TDN	% Protein	Unit[a]	Amount
Number of head				hd	100.00
Purchase month				Jan=1	(3.00)
Beginning weight				cwt	5.50
Beginning price				$/cwt	85.00
Selling weight A				cwt	12.50
Selling price A				$/cwt	75.00
Selling weight B				cwt	0.00
Selling price B				$/cwt	0.00
Months on feed				mo	11.00
Feed composition and prices:[b]					
Pasture[c]				$/aum	0.00
Roughage 1	90.0	55.0	18.0	$/ton	75.00
Roughage 2	30.0	70.0	8.4	$/ton	28.00
Grain 1	90.0	83.0	13.0	$/cwt	4.50
Grain 2				$/cwt	0.00
Supplement	90.0	85.0	48.5	$/cwt	12.00
Minerals	100.0	0.0	0.0	$/cwt	5.00
Additives	100.0	0.0	0.0	$/cwt	20.00
Other				$/cwt	0.00
Machinery cost				$/hr	10.00
Labor cost				$/hr	10.00
Interest cost[d]				%/yr	7.00
Yardage fee				$/day	0.15
Death loss A				% sold	99.00
Death loss B				% sold	0.00
Feed efficiency, roughage				% util	95.00
Feed efficiency, grain				% util	98.00

[a]Abbreviations: aum, animal unit month; cwt, hundred weight; hd, head; hr, hour; util, utilization, yr, year.
[b]Feed composition is on a dry matter basis; prices are on a wet-weight basis.
[c]One aum = 260 lb TDN, 430 lb dry matter.
[d]Interest cost = (rate/100) (mo/12)[feeder cost + 0.5(feed cost + variable costs)].

TABLE 5.2C

Feeder Cattle Budget Feed Requirements Table for One Head

Explanation:[a]	Unit	Period 1	Period 2	Period 3	Period 4	Period 5	Period 6	Summary
Beginning weight	cwt	5.50	6.85	7.90	9.22	10.78	12.58	5.50
Daily gain	lb	1.50	1.75	2.20	2.60	3.00	0.00	2.15
Days on feed	day	90	60	60	60	60	0	330
Months on feed	mo	3.00	2.00	2.00	2.00	2.00	0.00	11.00
Ending weight	cwt	6.85	7.90	9.22	10.78	12.58	12.58	12.58
Nutrient requirements/day:	(Obtained from feed requirements table to match rates of gain.)							
Dry matter	lb	15.00	17.00	19.00	21.00	23.00	0.00	
TDN	lb	9.50	12.00	14.00	16.00	17.50	0.00	
Protein	lb	1.50	1.70	1.90	2.10	2.30	0.00	
Roughage	%	80.00	65.00	50.00	40.00	25.00	0.00	
Feeds fed/day (wet weight):	(To be entered by feedlot operator.)							*Totals*
Pasture	aum	0.00	0.00	0.00	0.00	0.00	0.00	0
Roughage 1	lb	5.50	5.00	4.50	3.50	2.65	0.00	1,434
Roughage 2	lb	14.00	16.00	18.00	17.00	14.00	0.00	5,160
Grain 1	lb	5.00	8.00	9.75	14.00	17.00	0.00	3,375
Grain 2	lb	0.00	0.00	0.00	0.00	0.00	0.00	0
Supplement	lb	0.75	1.00	1.25	1.50	1.50	0.00	383
Minerals	lb	0.50	0.06	0.07	0.08	0.08	0.00	62
Additives	lb	0.00	0.00	0.00	0.00	0.00	0.00	0
Other	lb	0.00	0.00	0.00	0.00	0.00	0.00	0
Ration balancing:	(Tabulated internally from coefficients given.)							
Dry matter check:								
Pasture	lb	0.00	0.00	0.00	0.00	0.00	0.00	0
Roughage 1	lb	4.95	4.50	4.05	3.15	2.39	0.00	1,291
Roughage 2	lb	4.20	4.80	5.40	5.10	4.20	0.00	1,548
Grain 1	lb	4.50	7.20	8.78	12.60	15.30	0.00	3,038
Grain 2	lb	0.00	0.00	0.00	0.00	0.00	0.00	0
Supplement	lb	0.68	0.90	1.13	1.35	1.35	0.00	344
Minerals	lb	0.50	0.06	0.07	0.08	0.08	0.00	62
Additives	lb	0.00	0.00	0.00	0.00	0.00	0.00	0
Other	lb	0.00	0.00	0.00	0.00	0.00	0.00	0
Total	lb	14.83	17.46	19.42	22.28	23.32	0.00	6,283
TDN check:								
Roughages	lb	5.66	5.84	6.01	5.30	4.25	0.00	
Concentrates	lb	4.31	6.74	8.24	11.61	13.85	0.00	
Total	lb	9.97	12.58	14.25	16.91	18.10	0.00	
Protein check:								
Roughages	lb	1.24	1.21	1.18	1.00	0.78	0.00	
Concentrates	lb	0.91	1.37	1.69	2.29	2.64	0.00	
Total	lb	2.16	2.59	2.87	3.29	3.43	0.00	
Roughage check:	%	61.72	53.26	48.66	37.03	28.24		

[a]Items in bold type are entered by the operator.

EXAMPLE 5.3 Partial Budget for Split-Sex and Phase Feeding

Feed costs are a major component of any livestock production system. Split-sex feeding and phase feeding are two practices that improve feed efficiency for swine producers. Barrows require less protein than gilts. Barrows also overconsume energy and should be restricted in fat content as they approach market weight. The required amount of protein as a percentage of diet declines over the life of a pig for both barrows and gilts. Adoption of split-sex and phase feeding can improve feed efficiency from 3.5 pounds of feed per pound of gain to a 3.19 feed conversion ratio.

Consider a swine producer desiring to implement split-sex and phase-feeding technologies. The producer's current system is designed to feed only one diet. Thus, modification of the existing facilities is necessary. The producer has two modification options: the three-diet option and the four-diet option. The three-diet option would improve the feed conversion ratio to 3.25. The four-diet option would improve the feed conversion ratio to 3.19. The three-diet option requires a $1,750 investment in feeding equipment. The four-diet option requires a $5,140 investment. Two partial budgets are prepared to evaluate the changes from the present system to the three-diet system and from the three-diet system to the four-diet system. Table 5.3a illustrates the adoption of the three-diet system. Table 5.3b compares the four-diet option with the three-diet option. The reason for comparing the four-diet system to the three-diet system instead of the present system is to observe the incremental addition to income resulting from the additional investment needed for four diets. The beginning section of each table identifies the changes that will occur if the alternative diet systems are adopted.

Decision-making three-diet, four-diet, or present system: Both alternatives promise a higher net income than the present system. The three-diet system captures most of the gain in feed efficiency with the least amount of additional investment. Income increases by $7,609 for an investment of $1,750. The four-diet system increases income by $1,886 for an additional investment of $3,390. The swine producer must evaluate whether this smaller return on investment is better than other investment opportunities. The swine producer also must consider the risk of actually realizing the additional gain in feed efficiency as the optimum limit in feed efficiency is approached.

Total Farm Budgets

There are at least two very different approaches to tabulating a total farm budget. The first might be called the functional or activity approach. Functional budgeting proceeds by considering which crops to produce and the acres of each, and which livestock to produce, each with their respective productivity. Specific budget categories would include seed; fertilizer and chemical requirements for crops; feed and health requirements for livestock; machinery needs and costs for both crops and livestock; and labor requirements, charted against available on-farm labor to determine employment needs and costs. Individual enterprises and profit centers are not evaluated according to their profits. There are many good computer budgeting routines available. An example is *K-farm,* developed at Kansas State University.[2]

TABLE 5.3A

Comparison of the Present Hog System with a Three-Diet, Split-Sex, and Phase Feeding System

Item	Present System	Three-Diet System	Change
Size of operation	150 sows/259 litters	150 sows/259 litters	0
Hogs marketed	2,047 hogs/4,913 cwt	2,047 hogs/4,913 cwt	0
Corn/litter (bu)	98	91	–7
Soybean meal/litter (ton)	0.74	0.69	–0.05
Other feed/litter (cwt)	2.57	2.39	–0.18
Total equipment investment	$251,943	$253,693	+$1,750*
Equipment investment/litter	$973	$980	+$6.75

*Additional equipment consists of feed bins, bin boot, concrete pad, power unit, sort box, and auger.

Tabulations for Three-Diet, Split-Sex, and Phase Feeding System

	Per Litter	Total System
A. Added Income		
1. Reduced corn cost (7 × 2.25)	$15.25	$3,950
2. Reduced soybean meal cost (0.05 × $180)	$ 9.00	$2,331
3. Reduced other feed cost (0.18 × $36)	$ 6.48	$1,678
	$30.73	$7,959
B. Added Costs		
1. Equipment depreciation (8 years zero salvage)*	$ 0.84	$ 219
2. Equipment interest (9% on average investment)*	$ 0.30	$ 79
3. Equipment repairs and insurance (3% of investment)	$ 0.20	$ 52
	$ 1.35	$ 350
C. Net Income (A – B)	$29.38	$7,609

*An alternative method to calculate depreciation and interest is to amortize the investment or use the capital recovery method, (investment – salvage value) × amortization factor + (salvage value × interest rate). The formula for the amortization factor is,

$$\frac{i}{1-(1+i)^{-n}}$$

where i is the interest rate or required rate of return on the investment, and n is the life of the investment. This method indicates the amount required each year over the life of the investment to recover depreciation and provide a given rate of return on the investment. An amortization factor for 8 years at 9 percent is 0.1807. The alternative cost for depreciation and interest is $1.22 ($6.75 × 0.1807) per litter.

The second method approaches the total farm budget from individual profit centers, or enterprises. The idea is that the total business is the sum of its enterprise parts. The profitability of each enterprise is evaluated separately from the rest of the business. An overhead, or cost center, is necessary to receive all receipt and expense items not directly associated with a specific profit center. The profit center approach to budgeting the total farm is illustrated in Example 5.4. A long-range planning horizon is frequently used in developing total farm budgets. Rather than centering on particular year comparisons, the projected future (the next 5- or 10-year average) is compared to the average of the past several years. Thus, there is not a complete connection between the inventory and equity of the period before the planned changes with the inventory and equity for the period after the changes. This causes no problem because inventory increases and decreases are not shown. Average and projected prices are used rather than those for a particular year.

TABLE 5.3B

Summary Comparison of a Three-Diet with a Four-Diet System of Hog Production

Item	Three-Diet System	Four-Diet System	Change
Size of operation	150 sows/259 litters	150 sows/259 litters	0
Hogs marketed	2,047 hogs/4,913 cwt	2,047 hogs/4,913 cwt	0
Corn/litter (bu)	91	89	–2
Soybean meal/litter (ton)	0.69	0.67	–0.02
Other feed/litter (cwt)	2.39	2.34	–0.05
Total equipment investment	$253,693	$257,083	+$3,390*
Equipment investment/litter	$980	$993	+$13

*Additional equipment consists of two feed bins, two bin boots, two concrete pads, and two power units.

Tabulations for Four-Diet, Split-Sex, and Phase Feeding System

	Per Litter	Total System
A. Added Income		
1. Reduced corn cost (2 × 2.25)	$4.50	$1,166
2. Reduced soybean meal cost (0.02 × $180)	$3.60	$ 932
3. Reduced other feed cost (0.05 × $36)	$1.80	$ 466
	$9.90	$2,564
B. Added Costs		
1. Equipment depreciation (8 years zero salvage)	$1.63	$ 389
2. Equipment interest (9% on average investment)	$0.59	$ 153
3. Equipment repairs and insurance (3% of investment)	$0.39	$ 102
	$2.62	$ 678
C. Net Income (A – B)	$7.28	$1,886

EXAMPLE 5.4 Enterprise Approach to Planning the Total Farm

This is a computerized budget using the enterprise approach where the total farm is the sum of the profit centers. The set of budgets is integrated, consisting of a balance sheet, income statement, and cash-flow statement. Each template can be run independently and is developed from a series of smaller worksheet subbudgets.

The balance sheet is typically formatted as shown in Table 5.4a. Current, intermediate, and long-term assets and liabilities are shown. Values can be assessed on a cost or market basis. Unpaid or contingent tax liabilities are estimated using current tax laws. There are mini-worksheets that provide data for the line items in the balance sheet. The mini-worksheets include inventories, depreciation, land, and loans. These items are not illustrated to conserve space. At the bottom, a few useful efficiency measures are shown.

The income statement is the product of eight enterprise budget worksheets. There are four crop budgets and four livestock budgets. (In another version there are eight crop budgets and no livestock.) The enterprise worksheets feed into a detailed income statement that is patterned after the categories in the enterprise budgets. The income statement is summarized on one page and is shown in Table 5.4b. The enterprise worksheets and detailed income statement are not shown to conserve space.

This system allows products to enter the inventory from production or by purchase. Sales have an offsetting "value of inventory sold" entry. Another feature of this budget set is that farm-raised feeds can be shown on the livestock worksheets. The quantities fed of each crop are transferred automatically to the respective crop worksheet. The farm planner can then balance feed needs with crop production. If more crops are produced than are needed by the livestock some can be sold. If there is a shortage additional feed can be purchased, or the cropping plan can be modified to produce the additional feed needed.

Once the crop and livestock worksheets have been completed, the farm planner can return to the income statement and complete the "general" or overhead column. In this column, overlooked items and overhead costs are entered. Examples include management labor, utilities, farm liability insurance, office expense, property taxes, and depreciation.

At the bottom of the income statement, the overhead and ownership expenses can be allocated to the several profit centers by using a variety of procedures appropriate to the category of expense.

Cash flow considers feasibility and is illustrated in Table 5.4c. It, like the other budgets, can be independently tabulated. However, having the balance sheet and income statement available is of great help. It is developed in a two-step procedure. First, the dollar amounts for the total year for each budget item are entered. Next, these amounts are allocated to the months by using a system such that the total of the months must equal 100 percent. Those receipts and expenditures from the farm can be obtained from the income statement (Table 5.4b). Both detailed and summary cash-flow statements were tabulated but only the summary is shown. Only bimonthly figures are shown for illustration. A study of the titles of the rows reveals the following organizations:

1. Business receipts from sales and miscellaneous activities (I-A, I-B), and expenditures for business operations and ownership (II-A, II-B, II-C).
2. Sales of capital assets (IC) and purchases of depreciable property and real property (II-D).
3. Transfers into the business (I-D) from off-business sources, and transfers out of the business (II-E) for family living, income taxes, and other uses.

An important part of cash-flow budgeting is tabulating monthly cash surpluses and deficits. Surpluses can be used to support future deficits, but when these surpluses are insufficient new borrowing must take place. These tabulations are illustrated in Section III. This tabulation requires that principal and interest payments on prior loans and notes be separated as illustrated in Section II. The farm planner must keep the monthly cash balances positive by initiating new borrowing as shown in Section IV. If there is sufficient cash on hand at the end of any month, the surplus can be used to pay back some or all of the new loans, plus interest. In Section V, a running balance is tabulated for all liabilities. The farm planner must first enter the beginning balances from the balance sheet into the totals column. These balances then are modified monthly according to principal payments made on old loans and new borrowing and payments. The amount of principal payments made on new borrowings, and associated interest, is made at the discretion of the planner.

TABLE 5.4A

Balance Sheet for the Total Farm Budget Using the Enterprise Approach

Name: Bar X Ranch **Date: November 1, YRX0**

Assets			**Liabilities**		
Description	**Cost Basis ($)**	**Market Basis ($)**	**Description**	**Cost Basis ($)**	**Market Basis ($)**
I. *Current Assets:*			I. *Current Liabilities:*		
Cash, checking	3,500	3,500	Payable accounts	960	960
Business savings	0	0	Crop production loans	0	0
Marketable securities	28,000	35,000	Feeder livestock loans	0	0
Insurance, cash value	9,000	9,000	Feed purchase loans	2,400	2,400
Receivables	1,750	1,750	Principal payable in 12 mo:		
Prepaid accounts	0	0	Intermediate liabilities	2,600	2,600
			Fixed asset loans	11,232	11,232
Grain and feed inventory	28,500	28,500	Accrued unpaid interest:		
Unharvested crops	1,350	1,350	Current liabilities	288	288
Supplies inventory	1,800	1,800	Intermediate liabilities	550	550
Market livestock inventory	35,817	35,817	Fixed asset loans	7,494	7,494
			Unpaid taxes:		
			Property	932	932
			Income	0	8,052
Total current	$109,717	$ 116,717	Total current	$ 26,456	$ 34,508
II. *Intermediate Assets:*			II. *Intermediate Liabilities:*		
Breeding livestock	84,985	84,985	Breeding livestock loans	2,000	2,000
Machinery and equipment	24,350	60,200	Machinery and equipment loans	1,600	1,600
Other	0	0	Other intermediate loans	0	0
			Deferred tax liability	XXX	23,847
Total intermediate	$109,335	$ 145,185	Total intermediate	$ 3,600	$ 27,447
III. *Fixed Assets:*			III. *Long-term Liabilities:*		
Buildings	56,150	93,500	Mortgages	168,485	168,485
Improvements	47,950	74,750	Contracts	0	0
Land	456,665	679,998	Deferred tax liability	XXX	57,497
Use rights	10,860	13,126			
Total fixed	$571,625	$ 861,374	Total fixed	$168,485	$225,982
IV. *Total Assets*	$790,677	$1,123,276	IV. Total Liabilities	$198,541	$287,937
			V. Equity or Net Worth	$592,136	$835,339
Analysis:					
Liquidity = CA/CL	4.14	3.38	Debt structure = CL/TL	0.13	0.12
Solvency = TA/TL	3.98	3.90	Leverage = TL/E	0.34	0.34

Gross Margins and Linear Programming

Gross margins and linear programming (LP) are two other means to develop farm plans. The objectives of both these planning methods are to develop a plan that makes the most efficient use of available resources as well as satisfying other constraints placed on the farm business. The starting point for both these methods is for the manager to identify alternative enterprise choices and prepare unit budgets for each enterprise alternative.

TABLE 5.4B

Summary Income Statement for the Total Farm Budget Using the Enterprise Approach

Name: Bar X Ranch | **Date: October 30, YRX1**

Description	Total	General	1. Alfalfa	2. Barley	3. Pasture	4. Other	5. CowClf	6. Heifers	7. Calves	8. Steers
	I. RECEIPTS									
A. Sales:										
1. Crops	7,380	0	7,380	0	0	0				
2. Livestock	75,458	0	0	0	0	0	11,699	6,187	0	57,572
3. Products	137	0	0	0	0	0	137	0	0	0
Total sales	82,975	0	7,380	0	0	0	11,836	6,187	0	57,572
B. Other receipts	3,535	3,535								
C. Total receipts	86,510	3,535	7,380	0	0	0	11,836	6,187	0	57,572
	II. INVENTORY ADJUSTMENTS									
A Perpetual accounting:										
1. Product to inventory	139,267	0	16,004	11,016	6,797	0	39,732	11,616	54,102	0
2. Value of inventory sold	–65,123	0	–7,380	0	0	0	–14,025	–4,783	0	–38,935
3. Value of inventory moved	–48,235	0	0	0	0	0	0	–8,503	–39,732	0
4. Death and loss adjustment	–1,440	0	0	0	0	0	–753	0	–298	–389
Total perpetual	24,469	0	8,624	11,016	6,797	0	24,954	–1,670	14,072	–39,324
B. Purchases to inventory:										
1. Crops and livestock	2,925	0	0	0	0	0	2,925	0	0	0
2. Purchase offset	–2,925	0	0	0	0	0	–2,925	0	0	0
C. Total adjustment	24,469	0	8,624	11,016	6,797	0	24,954	–1,670	14,072	–39,324
	III. EXPENSES									
A. Operating:										
1. Hired labor	2,024	165	653	726	0	0	480	0	0	0
2. Equipment operation	6,773	999	1,841	1,455	0	0	900	198	792	588
3. Crop costs	1,878	0	664	1,214	0	0	0	0	0	0
4. Irrigation	1,904	0	816	1,088	0	0				
5. Livestock costs	2,660	100					1,615	191	264	490
Feed from inventory	26,254	0					16,608	2,129	5,782	1,735
Feed direct purchase	5,159	0					2,454	316	1,166	1,223
6. Utilities	328	328	0	0	0	0	0	0	0	0
7. Miscellaneous	0	0	0	0	0	0	0	0	0	0
Total operating	46,980	1,592	3,974	4,483	0	0	22,057	2,834	8,004	4,036

TABLE 5.4B

Summary Income Statement for the Total Farm Budget Using the Enterprise Approach—*Continued*

Name: Bar X Ranch										Date: October 30, YRX1
Description	**Total**	**General**	**1. Alfalfa**	**2. Barley**	**3. Pasture**	**4. Other**	**5. CowClf**	**6. Heifers**	**7. Calves**	**8. Steers**
	III. EXPENSES *(cont.)*									
B. Fixed costs:										
1. Taxes, insurance, repair	4,233	2,962	0	0	1,271	0	0	0	0	0
2. Business and overhead	2,443	1,000	0	136	276	0	309	66	264	392
3. Interest on debt	19,952	19,952	0	0	0	0	0	0	0	0
4. Depreciation	11,750	11,750	0	0	0	0	0	0	0	0
Total fixed	38,378	35,664	0	136	1,547	0	309	66	264	392
C. Total expenses	85,358	37,256	3,974	4,619	1,547	0	22,366	2,900	8,268	4,428
	IV. NET INCOME									
A. Net income (IC + IIC – IIIC)	25,621	–33,721	12,030	6,397	5,250	0	14,424	1,617	5,804	13,820
B. Allocation of expenses	0	–15,662	4,398	4.026	1,927	0	2,131	412	1,441	1,327
C. Net after allocations	25,621	–18,059	7,632	2,371	3,323	0	12,293	1,205	4,363	12,493

TABLE 5.4C

Summary Cash-Flow Statement for the Total Farm Budget Using the Enterprise Approach

Name: Bar X Ranch **Date: October 30, YRX1**

Description	Total	November-December	January-February	March-April	May-June	July-August	September-October
	I. CASH RECEIPTS						
A. Sales:							
1. Crops	7,380	0	0	0	0	7,380	0
2. Livestock	75,458	0	0	75	0	17,811	57,572
3. Products	137	0	137	0	0	0	0
Total sales	82,975	0	137	75	0	25,191	57,572
B. Other business income	3,535	566	566	566	566	566	705
Total operations	86,510	566	703	641	566	25,757	58,277
C. Capital sales	0	0	0	0	0	0	0
D. Transfer to business	3,000	480	480	480	480	480	600
Total cash receipts	89,510	1,046	1,183	1,121	26,237	26,237	58,877
	II. EXPENDITURES						
A. Operations:							
1. Hired labor	2,024	104	152	293	617	706	152
2. Equipment operations	6,773	506	506	759	1,895	1,897	1,210
3. Crop costs	1,877	0	0	1,533	172	172	0
4. Irrigation	1,904	0	0	0	952	952	0
5. Livestock costs	2,660	314	838	314	210	323	661
6. Feed purchases	5,159	1,174	1,566	2,294	0	0	125
7. Utilities	328	76	78	51	29	38	56
Total operating costs	20,725	2,174	3,140	5,244	3,875	4,088	2,204
B. Fixed costs:							
1. Real property	4,233	1,412	248	1,767	248	279	279
2. Business overhead	2,443	402	427	390	390	444	390
3. Interest	19,106	0	750	384	0	0	17,972
4. Principal payments	16,232	0	3,400	1,600	0	0	11,232
Total fixed expenditures	42,014	1,814	4,825	4,141	638	723	29,873
C. Livestock purchases	2,925	0	2,925	0	0	0	0
D. Capital purchases	0	0	0	0	0	0	0
Total business expenditures	65,664	3,988	10,890	9,385	4,513	4,811	32,077
E. Transactions out of business	21,052	2,080	8,522	3,690	2,080	2,080	2,600
Total expenditures	86,716	6,068	19,412	13,075	6,593	6,891	34,677
	III. NET CASH FLOW						
A. Net operations (IAB – IIABC)	20,846	–3,422	–10,187	–8,744	–3,947	20,946	26,220
B. Net capital (IC – IID)	0	0	0	0	0	0	0
C. Net transfers (ID – IIE)	–18,052	–1,600	–8,042	–3,210	–1,600	–1,600	–2,000
D. Net all cash	2,794	–5,022	–18,229	–11,954	–5,547	19,346	24,200
	IV. BORROWINGS & PAYBACKS						
A. Beginning balance	3,500	3,500	478	249	295	748	4,242
B. Borrowings	38,000	2,000	18,000	12,000	6,000	0	0
C. Payback/principal	35,000	0	0	0	0	15,000	20,000
D. Payback/interest	1,533	0	0	0	0	852	681
E. End balance	7,761	478	249	295	748	5,242	7,761
	V. LIABILITY SUMMARY						
A. New loan principal	XXX	2,000	20,000	32,000	38,000	23,000	3,000
B. Interest rate/amount	8.00	13.333	133	213	253	153	20
C. Total liabilities	198,541	200,541	215,141	225,541	231,541	216,541	185,309

Once the set of alternative enterprise choices is selected and unit budgets prepared, the manager needs to consider the types of farm planning constraints. These constraints can be classified as follows:

1. *Physical and financial resource availability.* Most farms have a given amount of land, machinery, and buildings, and a limit on the amount of available operating funds. Labor in combination with suitable field days may be in short supply. Fixity in the supply of physical resources is usually the most important constraint when planning in the short run.
2. *Institutional factors.* Often, particular farm activities are restricted by institutional rights, duties, contract obligations, etc., such as water rights, production quotas, or government regulations.
3. *Psychological and personal restraints.* The manager's preference for certain activities influences his or her selection of enterprises or their efficiency. For example, a cattle rancher with a strong preference for a particular cattle breed may refuse to change breeds despite economic considerations favoring this shift.

After the manager specifies enterprise choices, develops unit budgets, and identifies constraints, he or she can proceed to develop a plan by use of gross margins or LP. An illustration of the gross margin method is made followed by a LP example.

Gross Margins

Gross margins are defined as enterprise receipts less operating expenses. Gross margins measure the return to the fixed factors of production. Gross margins are usually expressed per unit of some common resource. Because land is a common limiting factor to most crop enterprises, gross margins are expressed per acre of land. If, however, available irrigation water is an important limitation to crop production, the gross margin could be expressed per acre-foot of water. For livestock, gross margin per head, gross margin per hundredweight, or gross margin per litter are common measures to determine the return per unit of activity to the fixed factors used in livestock production.

EXAMPLE 5.5 Gross Margin Example

A farmer has 640 acres of cropland that is suitable for corn or soybeans. The farmer also has a feedlot for finishing steer calves or yearling steers. The feedlot has a capacity of 150 head at any one time and is used continuously. Other limiting factors besides land and the feedlot are labor and operating capital. The farmer has 2,100 hours of available labor and $136,000 of operating capital. Gross margin per acre for crop enterprises and gross margin per head are calculated on the unit budgets (Table 5.5a). Resource requirements per unit of enterprise are also from unit budgets.

Information from the budgets is summarized in Table 5.5b. The corn column indicates a gross margin of $130 per acre and resource requirements of an acre of land, 2.1 hours of labor, and $50 of capital per acre of corn production. The steer calf column indicates a gross margin of $45 per head and that the finishing of one steer calf requires 3 hours of labor, $350 dollars of capital, and one feedlot space for 0.64 year (230 days/360 days).

Once the gross margin data are summarized in a tableau or a spreadsheet, the selection of the farm plan can begin. The procedure is to select the enterprise with the largest gross margin then proceed to produce as much as allowed by the constraints. Then select the enterprise with the second highest gross margin and add it until a constraint becomes limiting. In plans with mixed enterprises (i.e., crops and livestock), the procedure could be modified to select first the set of crop enterprises with the highest gross margins, then select the livestock enterprises with the highest gross margin. In this example, the manager would select corn to enter the plan. The

TABLE 5.5A

Unit Budgets for Gross Margin and Linear Programming Examples

Corn and Soybean Budgets

	Corn on Class 1 Land, Mulch Tillage				Soybeans on Class 1 Land, Mulch Tillage			
	Unit[a]	Quantity	Price	Values/ Acre	Unit	Quantity	Price	Values/ Acre
Revenue	bu	105	$ 2.40	$252.00	bu	33	$ 6.30	$207.90
Operating Costs								
Seed	1,000 ker	23	$ 0.85	$ 19.55	bu	1.25	$ 7.60	$ 9.50
Fertilizer	lbs	181	$ 0.14	$ 25.34	lbs	65	$ 0.17	$ 11.05
Chemicals	lbs	8.75	$ 1.75	$ 15.31	pts	2	$11.55	$ 23.10
Crop Insurance			$ 6.00	$ 6.00			$ 5.50	$ 5.50
Machinery repair, etc.			$20.22	$ 20.22			$11.91	$ 11.91
Harvest costs			$10.30	$ 10.30			$ 6.41	$ 6.41
Labor	hr	2.1	$10.00	$ 21.00	hr	1.4	$10.00	$ 14.00
Interest	$	50	9%	$ 4.54	$	36	9%	$ 3.21
Total Operating Costs				$122.26				$ 84.68
Gross Margin				$129.74				$123.22
Gross Margin, Excluding Labor and Interest				$155.28				$140.43
Fixed Costs								
Depreciation				$ 40.00				$ 40.00
Property taxes				$ 10.00				$ 10.00
Interest on long-term debt				$ 15.63				$ 15.63
Total Fixed Costs				$ 65.63				$ 65.63
Total Costs				$187.88				$150.30
Net Farm Income per Acre				$ 64.12				$ 57.60

Beef Feedlot Budgets

	Steer Calf Fed on Corn–Hay Ration					Yearling Steer Fed on Corn–Hay Ration				
	Unit	Quantity	Loss	Price	Value	Unit	Quantity	Loss	Price	Value
Revenue	cwt	11	0.01	$72.00	$784.08	cwt	12.5	0.01	$72.00	$891.00
Operating Costs										
Purchase costs	cwt	5		$85.00	$425.00	cwt	7.5		$78.00	$585.00
Feed										
Hay	tons	0.65		$70.00	$ 45.50	tons	0.35		$70.00	$ 24.50
Corn	bu	61		$ 2.50	$152.50	bu	63		$ 2.50	$157.50
Supplement	lb	190		$ 0.17	$ 32.30	lb	120		$ 0.17	$ 20.40
Mineral	lb	40		$ 0.09	$ 3.60	lb	23		$ 0.09	$ 2.07
Veterinary and medicine	hd	1		$10.00	$ 10.00	hd	1		$ 8.00	$ 8.00
Fuel and equipment	hd	1		$16.00	$ 16.00	hd	1		$ 9.00	$ 9.00
Interest	$	350		9%	$ 31.47	$	343		9%	$ 30.88
Labor	hr	3		$ 7.50	$ 22.50	hr	2		$ 7.50	$ 15.00
Total Operating Cost					$738.87					$852.35
Gross Margin					$ 45.21					$ 38.65
Gross Margin, Excluding Labor, Interest, and Corn					$251.68					$242.03
Fixed Costs										
Buildings and equipment					$ 21.00					$ 15.00
Total Costs					$759.87					$867.35
Net Farm Income per Head					$ 24.21					$ 23.65
Days on Feed	days	230				days	180			

[a]Abbreviations: bu, bushels; cwt, hundredweight; hd, head; hd, head; hr, hours; ker, kernels; lb, pounds; pt, pints.

TABLE 5.5B

Gross Margin Tableau

Rows	Corn (acre)	Soybean (acre)	Steer Calf (head)	Yearling Steer (head)	Resource Use and Farm TGM	Resource Availability	Remaining Resource Balance
Gross margin	$130	$123	$45	$39			
Cropland (acres)	1	1				640	
Labor (hours)	2.1	1.4	3	2		2,100	
Capital (annual $)	50	36	350	343		136,000	
Feedlot (year/head)			0.64	0.50		150	
Selection of enterprise							
Total Gross Margin							

TABLE 5.5C

Corn Selected to Enter Total Farm Plan

Rows	Corn (acre)	Soybean (acre)	Steer Calf (head)	Yearling Steer (head)	Resource Use and Farm TGM	Resource Availability	Remaining Resource Balance
Gross margin	$130	$123	$45	$39			
Cropland (acres)	1	1			640	640	0
Labor (hours)	2.1	1.4	3	2	1,344	2,100	756
Capital (annual $)	50	36	350	343	32,264	136,000	103,736
Feedlot (year/head)			0.64	0.50	0	150	150
Selection of enterprise	640						
Total Gross Margin	$83,034				$83,034		

TABLE 5.5D

Steer Calf Selected to Enter Total Farm Plan

Rows	Corn (acre)	Soybean (acre)	Steer Calf (head)	Yearling Steer (head)	Resource Use and Farm TGM	Resource Availability	Remaining Resource Balance
Gross margin	$130	$123	$45	$39			
Cropland (acres)	1	1			640	640	0
Labor (hours)	2.1	1.4	3	2	2,049	2,100	51
Capital (annual $)	50	36	350	343	114,443	136,000	21,557
Feedlot (year/head)			0.64	0.50	150	150	0
Selection of enterprise	640		235				
Total Gross Margin	$83,034	$0	$10,624	$0	$93,658		

resource requirements for corn and resource availability suggest that land is the most limiting factor. The manager would enter 640 in the selection of enterprise row in the corn column, and then calculate the amount of resources used. The amount of resource use and the remaining resource balance could be calculated simultaneously in a spreadsheet. The total gross margin of the plan appears in the last row. The results of selecting corn for the farm plan appear in Table 5.5c.

The manager now selects a livestock enterprise because cropland has been fully used. The manager selects the steer-calf enterprise because it has the largest gross margin. Examining the remaining resource balance and the resource requirements for steer calves indicates that 235 (150/0.64) head could enter. This number actually implies that 150 head were finished and another group of steer calves is 130 days in feedlot by the end of the accounting year. Table 5.5c shows the results after the addition of the steer-calf enterprise. The solution in Table 5.5d would be the final farm plan because cropland and feedlot facilities are fully used.

The main advantage of gross margins as a planning tool is its simplicity in determining income and feasibility. Determining feasibility is especially important when designing farm plans that consider new enterprises. A new enterprise may be profitable but not feasible given the resources available to the farm manager.

A weakness of the gross margin technique is that it does not guarantee the most efficient plan. This failure is due to a selection based on a single criterion such as gross margin per acre or gross margin per head. This single criterion results in formulating a plan that maximizes returns to a single factor. In the current example, returns are maximized with respect to land and per head or animal unit. Closer examination of the livestock enterprise and its use of feedlot space indicates that the yearlings would provide a higher return to the feedlot space. Yearlings only occupy the feedlot for half the year, whereas steer calves occupy a space for 0.64 year. The gross margin per yearly head space for yearlings is $78 ($39 gross margin/yearling ÷ 0.5 year/head space/yearling) compared with $70 for a steer calf ($45 gross margin/calf ÷ 0.64 year/head space/calf).

This simple gross margin planning model indicates the complexities of farm planning. At times when a manager is faced with a large number of enterprise choices and a large number of constraining factors, a more sophisticated planning procedure is required. LP is one such procedure available to farm managers.

Linear Programming

Linear programming is a mathematical technique for solving profit-maximizing or cost-minimizing problems. Applied to farm planning it permits determination of the optimal mix of enterprises that maximize total gross margin or some measure of net income.

The profit-maximizing objective of LP is similar to that of budgeting or gross margin analysis, but the solution of LP problems depends on a set of mathematical rules that ensure an optimal solution rather than semiformal rules and subjective judgements used in budgeting and gross margin analysis. This does not rule out the importance of subjective judgement in developing an LP model or in critically evaluating the solution to an LP farm plan.

An added feature of LP is that it provides additional information to the manager to evaluate the selection of enterprises and the use of resources. LP not only finds the best allocation of resources but also assigns values to scarce farm resources. These values and the opportunity costs of suboptimal enterprises (enterprises not included in the optimal farm plan) are referred to as shadow prices. The importance of shadow prices is illustrated later.

The word *linear* arises from the fact that the program does not automatically adjust for changing input–output ratios as the size of the enterprise or activity increases. The particular input–output ratio is held constant until the manager builds into the model a stop beyond which alternative activities (budgets) are considered.

Thus, the problem of diminishing returns is handled by building into the model a different activity for each input–output combination to be tested. For example, corn fertilized at 100 pounds per acre is a different activity from corn fertilized at 200 pounds per acre, and cattle marketed at 1,150 pounds is a different activity from cattle marketed at 1,050 pounds. Likewise, it is possible to test the use of different combinations of inputs to produce a product. Minimum tillage corn is a different activity from a conventional method of production. Sows farrowed in the winter is a different activity from sows farrowed in the spring. And cattle fed silage is a different activity from cattle fed hay. Linearity, then, is not a difficult problem if the modeler understands the mechanics of the programming model and the physical and economic relationships of agricultural production.

The information needed for the LP method of farm planning is exactly the same as required for budgeting and gross margin analysis. However, more attention is given to the specification of restrictions caused by resource limitations and other forces. Unit budgets (the requirements for producing one unit of product) are developed for each activity to be tested, whereas in regular budgeting, activities and input requirements may be combined.

As with other farm planning, the importance of farm records is apparent. They are particularly useful in furnishing information for specifying resource restrictions and developing activity budgets. A good test of the LP model is to run last year's farming activities to see if the net income calculated is near what was actually realized. This approach helps to discover errors and keeps the programmer realistic in making projections.

It is not the purpose of this discussion to give detailed instructions on the technique of LP, but to provide enough information to influence readers to seek out a good reference containing tabulation techniques. Many business operations management textbooks detail LP procedures.[3] A computer program is required to solve the linear model. Most spreadsheet software programs have LP capabilities. It is therefore not essential for those wishing to use LP to have a detailed understanding of the method of solution. They do need to recognize a problem that is suited to an LP solution, prepare budgets for the basic enterprises, determine constraints, and interpret results.

Two problem situations are used for illustration. The first problem illustrates the mathematics of LP. Note the linear algebraic equations and the graphical solution in Example 5.6. The second problem in Example 5.7 revisits the gross margin analysis.

EXAMPLE 5.6 Selecting Crop Enterprises Using LP

Farmer Mac has cropland suitable for corn and soybean production. The available land and capital limit the amount of crop production. Fifty acres of land and $4,500 of capital are available. A one-acre unit of corn production requires an acre of land and $120 of capital. A one-acre unit of soybean production requires an acre of land and $60 of capital. Mac estimates the gross margin per acre to be $100 for corn and $75 for soybeans based on yield and price expectations used in Mac's unit budgets. Mac's objective is to maximize total gross margin for the farm operation.

A general formulation of an LP problem is as follows:

Objective: maximize or minimize: $Z = C_1 X_1 + C_2 X_2$

subject to: $a_{11} X_1 + a_{12} X_2 \leq b_1$
$a_{21} X_1 + a_{22} X_2 \leq b_2$

Constraints: Z = value of objective function
X_j = enterprises or activities
C_j = values (net prices or gross margin)
b_i = fixed constraint value
a_{ij} = constraint requirement

An applied algebraic model to Mac's farm indicates that the objective is to find a combination of corn (X_C) and soybeans (X_S) that maximizes total gross margin. The constraint equations indicate that the quantity of land and capital used by corn and soybeans cannot exceed the available land and capital resources.

Objective:	maximize	$TGM = 100\ X_C + 75\ X_S$	
Constraints:	Land	$1\ X_C + 1\ X_S$	≤ 50 acres
	Capital	$120\ X_C + 60\ X_S$	$\leq \$4{,}500$
	TGM = total gross margin		
	X_C = acres of corn		
	X_S = acres of soybeans		

FIGURE 5.1

Graphical Representation of Linear Programming Model

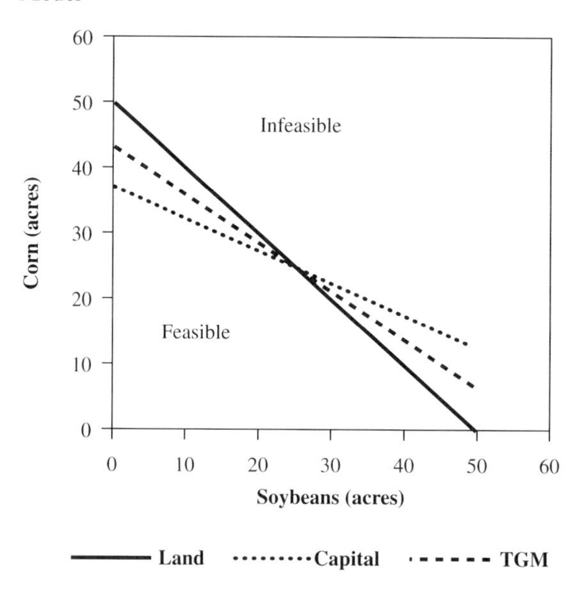

In the graphical representation of Mac's farm (Figure 5.1), the axes represent acres of corn and soybeans in production. The land and capital constraints define the feasible region of production. If corn were selected to enter the farm plan, capital would limit corn to 37.5 acres. This would be one possible solution—37.5 acres of corn and 0 acres of soybeans. Another possible solution would be to substitute soybeans for corn by moving along the capital constraint line until land becomes a limiting factor. This solution would be a profitable substitution because total gross margin would be increasing. Only 0.5 acre of corn is replaced for each acre of soybeans that is substituted. This substitution rate is due to the fact that soybeans require half the capital as does corn. The gross margin for soybeans is $75, which is greater than one-half the forgone gross margin of corn, which is $50. Another possible solution would be to substitute soybeans for corn and move along the land constraint, but this approach would not be profitable. Farmer Mac would gain $75 but lose $100. Therefore, Mac would maximize total gross margin at 25 acres each of corn and soybeans. Recognize that the reasoning used in this analysis is similar to the product–product economic principle discussed in Chapter 6.

EXAMPLE 5.7 An LP Analysis of the Gross Margin Problem

The gross margin example 5.4 of selecting crop and livestock alternatives is reformulated in Table 5.6a as an LP model. The LP model is in a tableau form instead of an algebraic format as in Example 5.5. The activities (enterprise choices) and resource limits are the columns in the tableau. The objective function and constraints are the rows in the tableau. Note that the columns are different from the gross margin example in Table 5.5a. The crop and livestock columns in the gross margin example included the production and marketing function. The marketing function is separated from the production function in Table 5.5. Three additional rows are added to link the production activities to the marketing activities. These rows are referred to as transfer rows. The purpose for this separation is twofold. For corn it provides for alternative use for the corn crop. Corn can be transferred to the livestock enterprises or corn can be marketed. The solution will determine the optimal allocation of corn. The second reason for separating the marketing function is that it facilitates "what if" questions and price sensitivity analysis. The coefficients in the tableau are from the unit budgets in Table 5.5. The cost of resources treated as rows in the tableau are not included in the objective function coefficients because LP imputes a value to the resources defined as constraints.

The development of the coefficients for the corn and steer-calf enterprises is discussed for clarification. The objective function coefficient for the corn column indicates the cost incurred for each acre of corn that is included in the plan. This cost is the total operating cost from the unit budget less the cost of labor and interest on operating capital ($122.26 – 21.00 – 4.54 = $96.72). The resources required for corn production are indicated in each constraint row. The requirements for cropland, labor, and capital are from the unit budget and are the same as for the gross margin analysis. A negative 105 is entered in the corn transfer row. This number represents the expected corn yield as indicated on the budget. The coefficient is negative because if it were transferred to the other side of the equation it would increase the amount of corn in the available resource column.

The objective function coefficient for the steer-calf column indicates the cost incurred for each steer calf that is purchased and finished. This value is total operating costs less the cost of labor, interest on operating capital, and cost of corn ($738.87 – 22.50 – 31.47 – 152.50 = $532.40). The resource requirements for each steer calf are 3 hours of labor, $350 of capital, 61 bushels of corn, and 0.64 year of feedlot space. The final product of the steer-calf finishing activity is an 11-hundredweight steer. The final product for steer calves is entered as a negative 11 in the fat steer transfer row. The 11 hundredweight would be transferred to the fat steer marketing activity and sold at $72 per hundredweight if the steer-calf activity enters the solution. Finally, the sum of the resources used by the crop and livestock activities cannot exceed the available resources shown in the far right column.

The solution to the LP model provides an estimate of total gross margin, identifies the best mix of enterprises, shows which constraints are limiting, and provides shadow prices and the applicable range for the shadow price.

Shadow prices are of two types. Each limiting resource has a value imputed to it indicating the contribution made to the objective function by the last unit. This value is referred to as marginal value product (MVP). This value indicates the resources strategic value (at the margin) in the optimal farm plan. Given the linearity assumption, the MVP of all nonconstraining resources is equal to zero in the optimal solution. Enterprises not included in the farm plan are assigned shadow prices referred to as income penalties. Income penalties indicate the amount by which total gross margin would decrease if one unit of the excluded enterprise is forced into the plan. An alternative

TABLE 5.6A

Linear Programming Farm Planning Model

Rows	Corn (acre)	Soybean (acre)	Steer Calf (head)	Yearling Steer (head)	Corn Marketing (bu)	Soybean Marketing (bu)	Fat Steer Marketing (cwt)	Row Type	Available Resources
Objective function	–$96.72	–$67.47	–$532.40	–$648.97	$2.40	$6.30	$72.00	Max	
Cropland (acres)	1	1						≤	640
Labor (hours)	2.1	1.4	3	2				≤	2,100
Capital (annual $)	50	36	350	343				≤	136,000
Feedlot (year/head)			0.64	0.500				≤	150
Corn transfer (bu)	–105		61	63	1			≤	0
Soybean transfer (bu)		–33				1		≤	0
Fat steer transfer (cwt)			–11	–12.5			1	≤	0

TABLE 5.6B

Linear Programming Solution to Farm Plan

Name	Final Plan	Shadow Prices Income Penalty	Objective Coefficient	Allowable Increase	Allowable Decrease
Objective function: TGM	$129,327				
Enterprises:					
Corn (acres)	640	0	–96.72	∞[a]	14.85
Soybean (acres)	0	0	–67.47	14.84	∞
Steer Calf (head)	0	–14.36	–532.40	14.36	∞
Yearling Steer (head)	300	0	–648.97	∞	11.24
Corn Marketing (bu)	48,300	0	2.40	0.74	0.14
Soybean Marketing (bu)	0	–0.45	6.30	0.45	
Fat Steer Marketing (cwt)	3,750	0	72.00	∞	2.89
Resource constraints:	**Resource Use**	**Marginal Value Product**	**Available Resources**		
Cropland (acres)	640	155.28	640	16.0	460
Labor (hours)	1,944	0.00	2,100	∞	156
Capital ($)	135,195	0.00	136,000	∞	805
Feedlot (year/head)	150	199.66	150	1.17	150

[a]The infinity symbol implies that the allowable increase or decrease is unlimited.

interpretation is that it shows the necessary increase in the gross margin of the excluded enterprise required to bring that enterprise into the optimal farm plan. Ranges provide additional information about shadow prices. Without ranges, the shadow price would only be applicable to the last unit. Ranges on enterprises indicate the allowable increase and allowable decrease in gross margins that would not alter the optimal plan. Ranges on resource limits indicate the allowable increase and allowable decrease in the available resource limit for which the shadow price remains constant.

The LP solution is summarized in Table 5.6b. The total gross margin for the plan is $129,327. The plan consists of 640 acres of corn and 300 head of yearling steers. The plan markets 48,300 bushels of corn and 3,750 hundredweight of slaughter yearlings. The soybean and steer-calf enterprises did not enter the plan. Soybean costs would have to decline $14.85 (allowable increase for soybeans) per acre, or the soybean marketing price would have to increase by $.45 (income penalty for soybean marketing activity) per bushel before soybeans would be a competitive enterprise.

Land and feedlot capacity are the limiting factors. Land has a shadow price of $155 per acre. The allowable increase column indicates an additional 16 acres would contribute this value. Feedlot has a shadow price of $199.66, but this value is only applicable for an additional 1.17 units. These shadow prices and ranges provide information to the manager on the potential profitability of expansion. The manager has an incentive to acquire additional resources if the acquisition cost is less than the shadow price.

A projected net farm income statement is developed from the information provided by the LP model and the unit budgets, Table 5.6c. From the income statement, we can see that the corn enterprise accounts for 72 percent of net farm income but only 30 percent of the total receipts.

TABLE 5.6C

Projected Net Farm Income Statement for YRXX from Linear Programming Solution

	Whole Farm	Corn	Soybeans	Steer Calves	Yearling Steers
Farm Operating Revenues					
Crop sales	$115,920	$115,920	$0		
Livestock sales	$270,000			$0	$270,000
Total Receipts	$385,920	$115,920	$0	$0	$270,000
Inventory Adjustment					
Production	$431,280	$161,280	$0	$0	$270,000
Sales	–$385,920	–$115,920	$0	$0	–$270,000
Total Adjustment	$ 45,360	$ 45,360	$0	$0	$ 0
Gross Revenue	$431,280	$161,280	$0	$0	$270,000
Livestock Purchases and Feeds Fed					
Livestock purchases	$175,500			$0	$175,500
Feed purchases	$ 14,091			$0	$ 14,091
Feeds raised	$ 45,360			$0	$ 45,360
Total Purchases and Feeds Fed	$234,951			$0	$234,951
Value of Farm Production	$196,329	$161,280	$0	$0	$ 35,049
Farm Operating Costs					
Variable Costs					
Crop expense	$ 42,370	$ 42,370	$0		
Machinery repair, etc.	$ 19,533	$ 19,533	$0		
Livestock expense	$ 5,100			$0	$ 5,100
Labor	$ 0				
Total Variable Costs	$ 67,002	$ 61,902	$0	$0	$ 5,100
Gross Margin	$129,327	$ 99,378	$0	$0	$ 29,949
Fixed Costs					
Depreciation machinery and equipment	$ 25,600	$ 25,600	$0		
Depreciation buildings	$ 4,500			$0	$ 4,500
Property taxes	$ 6,400	$ 6,400	$0		
Interest on long-term debt	$ 10,000	$ 8,000	$0		$ 2,000
Total Fixed Costs	$ 46,500	$ 40,000	$0	$0	$ 6,500
Total Costs	$113,502	$101,902	$0	$0	$ 11,600
Net Farm Income	**$ 82,827**	**$ 59,378**	**$0**	**$0**	**$ 23,449**

Notes

1. Ownership and operating costs are often referred to as fixed and variable costs. Generally, the distinction is user based. Any technical difference is not usually recognized. In this book the terms may be used interchangeably.
2. *K-farm* is available from the Department of Agricultural Economics, Cooperative Extension Service, Manhattan, Kansas.
3. The book by Beneke, Raymond, and Ronald Winterboer, *Linear Programming Applications in Agriculture,* Iowa State University Press, 1973, is still useful in outlining LP procedures for farmers.

CHAPTER 6

Economic Principles of Production

Concepts in Measuring Production
How Much Resource to Use (Factor–Product)
How Much Product to Produce (Product–Cost)
How to Produce (Factor–Factor)
What to Produce (Product–Product)

Concepts in Measuring Production

The questions of what to produce, how to produce, how much resource to use, and how much to produce were introduced in Chapter 1. These same questions are the topics of this chapter. Farm production records, agricultural experiment station scientists, and private businesses, such as seed companies and feed manufacturers, provide the production relationships needed for making these analyses.

Agricultural products are the result of combining factor inputs. For example, corn seed planted in tilled soil that is kept free of excess weeds, fertilized, and watered produces additional corn. Milk is the product from dairy cows, feed, housing, and labor. The inputs are the factors of production. The product is the output. Inputs combine to produce outputs. Different inputs produce different products, and the amount of output is related to the level of inputs used.

These input–output relationships can be placed in mathematical terms for clarity and general applications as follows:

$$Y = f(X_1, X_2, X_3, \ldots, X_n)$$

Where Y = output of any product
X = any input required by Y for its production
(Subscripts differentiate the inputs.)
f = functional relationship of X to Y

Subscripts are used to identify specific inputs and products (e.g., X_1 is different from X_2 , and Y_c is different from Y_h). As the amount of any input is changed its proportion in the input mix also changes, and the amount of product may be affected. For example, if fertilizer is increased when producing corn the yield is likely to increase. Likewise, if the amount of grain in a feeder cattle ration is reduced the rate of weight gain is likely to decline. The letter "f" specifies that Y is a function of X. The output variable Y is said to be the dependent variable and X the independent variable. The output of Y is dependent upon the input of X.

It is often desirable to learn of the productivity of just one or two particular factor inputs. This is done by holding all factors constant (at the same level) except for the one (or more) factors being measured. Factors held constant are said to be fixed, whereas those being changed are said to be variable. The separation of variable from fixed inputs is shown in the mathematical model by a vertical bar as follows:

$$Y = f(X_1, X_2 \mid X_3, \ldots, X_n)$$

Because the output of Y is now a function of only the variable inputs, the mathematical formula can be reduced to $Y = f(X_1, X_2)$. The fixed factors, $X_3 \ldots X_n$, are understood to be present and required. Two determinations are used to define variable and fixed inputs. Some inputs are fixed or constant by nature as described in previous chapters. Depreciation, property taxes, and farm liability insurance are examples of overhead factors that are fixed by nature.

Other inputs may be variable in nature, but held constant by experimental design or managerial choice. Often only one or two factors are changed, while all other inputs are held constant so as to measure the specific contribution of the variable factors to production. For example, to measure the relationship of corn output to N fertilizer input requires all other inputs (e.g., other fertilizers, seed, cultural practices, and water) be held at a constant level, while the level of nitrogen (N) is increased in the experimental plots or fields. In this way a specific functional relationship can be established for a defined level of fixed factors. A similar example could be stated for livestock where they are divided into groups that are treated all alike, except for one or two feed elements, or other factors. Allowing too many inputs to change in the same experiment may make it difficult, or even impossible, to measure the contribution of any one item.

The experiment is continued until the experimenter (farmer or scientist) is satisfied that the results are sufficiently accurate for decision making. This usually involves replications and repetitions of the experiment over several years. When the experimenter is sufficiently confident with the results the experiment is concluded. The level of confidence needed is specified by the experimenter and user (e.g., if the experiment were repeated the experimenter is confident that 90 [95 or 99] percent of the time the results would support the previous conclusion). Individual farmers may be willing to make changes (i.e., adopt a new variety, apply added fertilizer, feed higher levels of protein, or add new vitamins or antibiotic) at a lower level of confidence than required for mass recommendation to all farmers. It is important to look at the consequences of being wrong when adopting a new method of production. Risk is be covered in Chapter 8.

Most experiments can be stated in precise mathematical terms—typically by an algebraic equation. Example equations are the polynomial and Cobb–Douglas, or power function. The polynomial equation at the third degree follows for a single variable input ($Y = f[X_1]$):

$$Y = a + bX_1 + cX_1^2 + dX_1^3$$

Where
- Y = output or product
- X_1 = level of variable factor applied
- a = output of product when $X_1 = 0$
- b,c,d = constants determined statistically

This polynomial equation allows both increasing and decreasing returns. Whether the output increases or decreases with increases in X_1 depends upon whether the constants (b, c, or d) are positive (+) or negative (–). If + there is an increasing influence and if – a decreasing influence. The power of the factor (X) indicates the degree of influence of the variable factor. Consider, for example, the equation $Y_1 = 40 + 2X_1 + 0.5X_1^2 - 0.05X_1^3$. There are 40 units of Y_1 produced when $X_1 = 0$. Because X_1 did not contribute to production of these 40 units, they can be dropped from the equation when measuring the contribution of X_1. (The output of Y_1 when $X_1 = 0$ is important to the overall productivity but not in deciding how much X_1 to apply.) Table 6.1 is tabulated by solving for Y_1 at different levels of X_1. Table 6.1 is illustrated in Figure 6.1.

The influence of the constants in the equation in relation to the power of the associated variable factor is as follows:

- The X (first power) has a linear influence, and the sign of the constant number preceding it indicates whether the influence is positive or negative. The linear constant in this equation is +2 so the upward influence is 2 to 1.
- The X^2 (squared power) is stronger. The +0.5 constant contributes an upward influence at an increasing rate.
- The X^3 (cubic power) is even more powerful. The –0.05 constant contributes a strong downward influence at an increasing rate.

The change in the variable factor (X_1) in this example is one unit, beginning at zero and increasing to 10. This change is designated by the Greek symbol Δ, or delta. The marginal product (MP) measures the change in output (Y_1) as the input (X_1) increases by one unit (i.e., $MP = \Delta TP/\Delta X_1$).[1] The average product (AP) measures the contribution that each unit of input makes to output at each input level and thus measures input productivity and is tabulated by dividing the total output (TP), less the output when the variable input level is zero, by the units of input (i.e., $AP = TP/X$ or Y_1/X_1).

TABLE 6.1

Input–Output Table for the Function: $Y_1 = 40 + 2X_1 + 0.5X_1^2 - 0.05X_1^3$

Total			Marginal				Average
X_1	Y_1	Y_{X_1}[a]	ΔX_1	ΔY_1	$\frac{\Delta Y_1}{\Delta X_1}$[b]	$\frac{dY_1}{dX_1}$[b]	AP_{X_1}[c]
0	40.00	0					
			1	2.45	2.45		
1	42.45	2.45				2.85	2.45
			1	3.15	3.15		
2	45.60	5.60				3.40	2.80
			1	3.55	3.55		
3	49.15	9.15				3.65	3.05
			1	3.65	3.65		
4	52.80	12.80				3.60	3.20
			1	3.45	3.45		
5	56.25	16.25				3.25	3.25
			1	2.95	2.95		
6	59.20	19.20				2.60	3.20
			1	2.15	2.15		
7	61.35	21.35				1.65	3.05
			1	1.05	1.05		
8	62.40	22.40				0.40	2.80
			1	–0.35	–0.35		
9	62.05	22.05				–1.15	2.45
			1	–0.25	–0.25		
10	60.00	20.00				–3.00	2.00

[a]Output produced by the addition of X_1, or $Y_1 - 40$.
[b]Each of these columns are measures of the marginal product (MP). The first is tabulated by dividing difference in Y and X_1, whereas the second is tabulated by evaluating the derivative for specified values of X_1.
[c]Tabulated by dividing the total output by the total input, or Y_{X_1}/X_1.

FIGURE 6.1

Graphical Representation of the Function $Y = 40 + 2X_1 + 0.5X_1^2 - 0.05X_1^3$

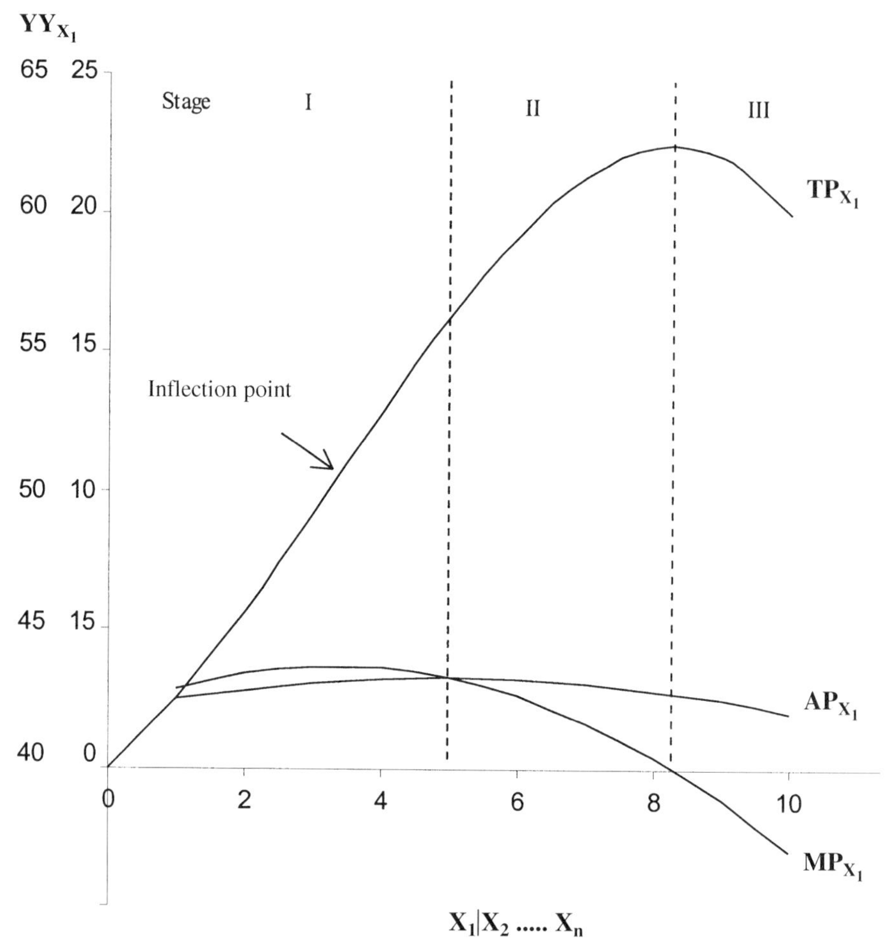

Note that the productivity of the input factor (X_1) changes as the proportion of the variable factor (X_1) changes (increases) in relation to the fixed factors ($X_2, \ldots, X_n$). AP is a measure of the physical production efficiency of the variable input. The production efficiency is greatest when AP is at its highest level. Furthermore, the variable factor can often be added in very small units. At the limit, these changes become so small that it allows connecting the points on the graph into a solid line. This is known as a continuous function. In agriculture, many production functions are continuous.

The polynomial production function illustrated above is for a single variable, but more variables are possible. For example, two variables [$Y = f(X_1,X_2)$] may take this form:

$$Y_1 = a + bX_1 + cX_1^2 + dX_2 + eX_2^2 + fX_1X_2$$

More than two variables are not usually specified by the polynomial equation because of the many possible interactions as defined by fX_1X_2.

The Cobb–Douglas or power function is of the form $Y = kX^n$. This equation does not include the contributions of the fixed factors (i.e., if $X = 0$ then $Y = 0$). The factor "k" is a constant and "n" defines the transformation ratio of X at different magnitudes.[2] Note that this equation in logarithmic form is (Log Y = Log k + n Log X). This logarithmic transformation is important for estimating the constants for the power function by linear regression. Also, note that this equation has only diminishing returns ($0 < n < 1$), and the total product is always increasing. Hence, it is applied only in relevant ranges of inputs. Where there is more than one variable, the formula contains more X_i's, each with its own transformation ratio as follows: $Y = kX_1^n X_2^m$. The mathematical formula selected to represent the experimental data is according to theory of the particular science and the statistical design being applied. Several mathematical formulas may be considered and tested to determine how well each describes the results of the experiment. Using statistical analyses known as regression, the constants are tabulated. Each mathematical formula describes the data differently, some better than others. A "goodness of fit" is tabulated to determine how well the formula fits the data. Generally, the formula that fits the data best is used as the predicting equation. This tells the user what to expect if the experiment were repeated, or practically applied. The degree of nonpredictability is called the error term.

Farmers entering the information age with computers, yield monitors, and variable-rate technology may need to design on-farm experiments and analyze the results with the help of scientists. However, farmers may be content to make their decisions upon less well-founded results. Farmers, for example, may plant their fields to different varieties, use different fertilizer levels, measure growth rates of livestock in different facilities, and feed different rations. These comparisons may not even take place in the same year. Also, farmers may compare their results with those of neighbors. Using these crude means, they draw conclusions about products, varieties, input levels, and methods of production. This is a form of marginal analysis. The units of measure may be crude and imprecise but the application of principles are the same. Marginal analysis is the basis for making important decisions regardless of the information's precision. The economic principles of production occupy the balance of this chapter.

How Much Resource to Use (Factor–Product)

The law of diminishing returns implied in the preceding discussion states the following: "If successive units of one input are added to given quantities of other inputs required in the production of some product, output of product per additional unit of input will reach a point where the addition to product will decline." This definition is present in the mathematical model $Y_1 = 40 + 2X_1 + 0.5X_1^2 - 0.05X_1^3$, used earlier in this chapter, and illustrated in Table 6.1 and Figure 6.1. The MP column and the MP curve illustrate the definition of the law of diminishing returns. Output per unit of additional input begins declining after three units of input (i.e., MP is at its maximum). Output is still increasing but at a diminishing rate. After eight units of input, the total output actually becomes smaller (MP turns negative).

It is important to recognize several things in this illustration when considering agricultural production relationships involving increasing levels of factor input:

- AP is increasing so long as MP is increasing or greater than AP.
- Where MP is at its highest point (largest), the total product (TP) changes from increasing at an increasing rate to increasing at a diminishing rate. This is called the inflection point.

- MP intersects AP at AP's highest point.
- When AP is at its highest level, the productive efficiency of the input factor is at its best.
- The sum of the MPs is equal to TP_x at each input level (i.e., $\Sigma MP = TP_x$). TP_x includes only the product contributed by the factor being increased and does not include the output where $X_1 = 0$.

Three stages of production can be identified. Stage I includes the area between zero input level and where AP is at its maximum, or until the physical efficiency level has been reached. Up to this point the cost per unit of product is getting smaller as additional units of input are being applied. Rather than producing at a variable to fixed-input ratio less than this, it would be better to reduce the fixed factors used, even if it meant idling them. (This explains why some farmers let some parts of their farm go untended, or poorly used. They are then able to spend more time, or use their resources, where the efficiency of production is higher.)

To illustrate that it is irrational to produce in stage I, consider again the equation $Y_1 = 40 + 2X_1 + 0.5X_1^2 - 0.05X_1^3$ (see Table 6.1). Peak physical efficiency is reached at input level 5. At this point $AP_{X_1} = 3.25$, $Y_1 = 56.25$, $Y_{X_1} = 16.25$, and the ratio of variable to fixed factors is 5:1. Now assume there are only three units of variable factor available. If these three units were applied to all of the fixed factors, total output would be 49.15($Y_{X_1} = 9.15$). The ratio of variable to fixed factors is now 3:1, but this ratio is less efficient than the 5:1 ratio. Suppose it were possible to reduce the fixed factors to less than one unit by idling some part of them such that the variable to fixed factor ratio could be 5:1. By idling 0.4 units there would remain 0.6 units. Now if the three units of variable factor X_1 were applied to the 0.6 units of fixed factors, the ratio becomes 3:0.6 or 5:1. The output would then become $0.6 \times 16.25 = 9.75$ ($Y_{X_1} = 49.75$) for the input factor. This is 0.60 units of product larger (49.75 – 49.15) than if the three units were applied to the total of the fixed factors.

Stage III is defined to include the area beyond where MP = 0. (MP would be negative but TP could still be positive.) Producing where MP = 0 is obviously irrational because more output can be had with less input, unless of course someone is paying to have inputs added. This leaves stage II as the only rational area of production, and the area over which prices play a deciding role.

Before leaving this discussion, it is important to understand that for many factor-product relationships, stage I of the production function is not observed. In fact, functions with increasing marginal returns are the exception, and those beginning with diminishing marginal returns the rule. A third possible factor-product relationship may exist over a limited range of input increases that is constant marginal products. A constant marginal-product relationship would imply that each additional unit of input results in a constant increase in output until one of the fixed factors becomes limiting (i.e., an additional unit of the variable factor has a marginal product equal to zero). Nevertheless, it is well to understand these relationships to get a complete picture of production functions. It is the role of the researcher to determine the true nature of the factor-product relationship.

The economic principle governing how much resource to use is stated as follows: Additional units of the variable factor should be added as long as the value of the product added (VMP) is greater than the cost of an additional unit of factor input, or marginal factor cost (MFC) ($VMP = P_Y \times MP$, where P_Y is the price of the product). Thus, output should be increased as long as the value of an additional unit of product is greater than the cost of the additional units of input required to produce it. In symbolic terms, output is increased until VMP = MFC. (Note: If the production function is non-continuous, then the optimization will occur where VMP is slightly greater than MFC.)

Using Table 6.1 as background, Table 6.2 was constructed by adding prices. The cost of adding onc unit of variable factor is called marginal factor cost (MFC). If the price of a unit of X_1 is $1, then MFC = $1. Because this is a marginal concept, MFC is constant at $1 for all input levels. The value of marginal product (VMP) and value of total product (VTP) are tabulated by direct multiplication of MP and TP by the price of a unit of product (P_Y). Assuming P_Y = $2, then VMP and VTP are tabulated by multiplying MP and TP, at all levels of input, by $2. Profit is maximized at eight units of input where TP = 44.80 and profit (income less variable costs) is equal to $36.80.[3] These relationships are illustrated graphically in Figure 6.2.

Suppose the cost of a unit of factor were to increase to $2.50 (MFC = 2.50). At this price seven units would become the most profitable input level. Likewise, a drop in the price of the product would cause the optimum profit level of input to decrease. Changes in the prices of the factor or product causes the optimum profit level of production to change.

TABLE 6.2

Input–Output Table for the Equation $Y_{X_1} = 2X_1 + 0.5X_1^2 - 0.05X_1^3$, with $P_Y = \$2$; $P_{X_1} = \$1$[a]

X_1	TP_{X_1}	MP_{X_1}	MFC_{X_1}	VMP_{X_1}	VTP_{X_1}	TVC_{X_1}	$VTP_{X_1} - TVC_{X_1}$
0	0.00		$1.00		$0.00	$0.00	$0.00
		2.45		$4.90			
1	2.45		1.00		4.90	1.00	3.90
		3.15		6.30			
2	5.60		1.00		11.20	2.00	9.20
		3.55		7.10			
3	9.15		1.00		18.30	3.00	15.30
		3.65		7.30			
4	12.80		1.00		25.60	4.00	21.60
		3.45		6.90			
5	16.25		1.00		32.50	5.00	27.50
		2.95		5.90			
6	19.20		1.00		38.40	6.00	32.40
		2.15		4.30			
7	21.35		1.00		42.70	7.00	35.70
		1.05		2.10			
*8	22.40		1.00		44.80	8.00	36.80
		–0.35		–0.70			
9	22.05		1.00		44.10	9.00	35.10
		–2.05		–4.10			
10	22.00		1.00		44.00	10.00	34.10

[a] TP_{X_1} = total product or output of Y_1 from adding the factor X_1; MP_{X_1} = added or marginal product of Y_1 from X_1; MFC_{X_1} = marginal factor cost of an additional unit of factor X_1 ($MFC = P_{X_1}$); VTP_{X_1} = value of total product ($VTP = P_y \cdot TP$); TVC_{X_1} = total cost of the variable factor ($TVC = P_{X_1} \cdot X$)

*Indicates the profit maximizing quantity of x_1.

FIGURE 6.2

Graphical Representation of the Function $Y_1 = 40 + 2X_1 + 0.5X_1^2 - 0.05X_1^3$ with $P_{Y_1} = \$2, P_{X_1} = \$1$

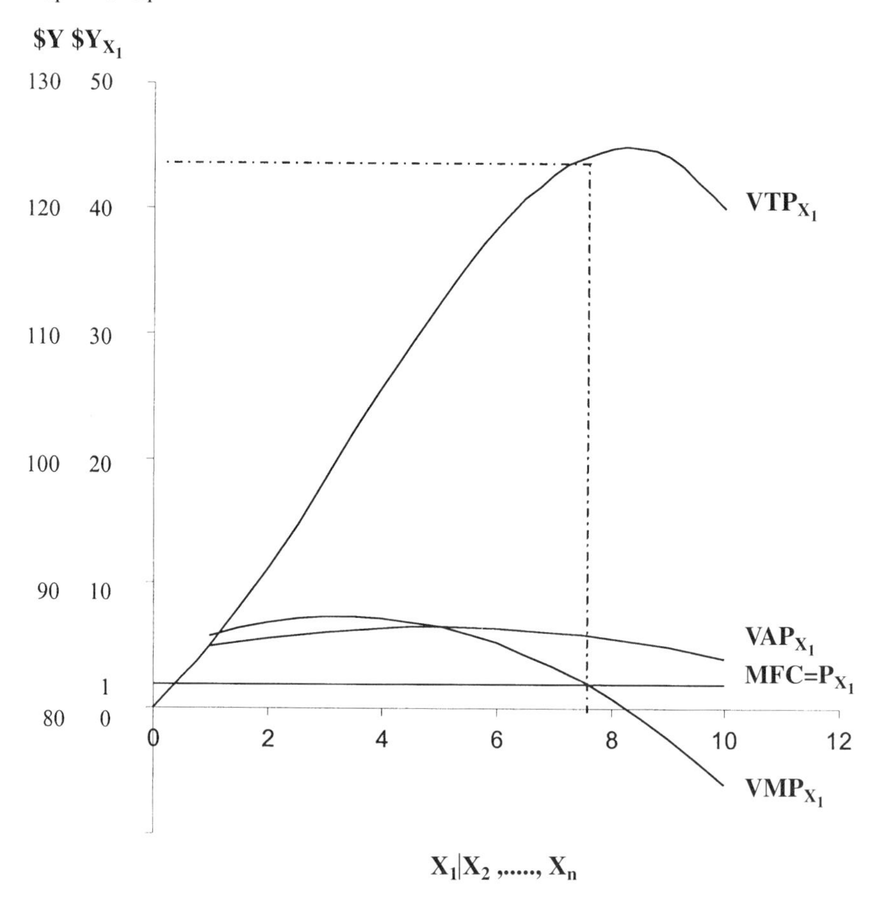

An alternative expression of the decision rule, VMP = MFC, is the factor-product price ratio. Because VMP = MP × P_Y and MFC = P_X and profit is maximized where

$$P_Y \times MP = P_X \text{ or } MP = \frac{P_X}{P_Y}$$

It can now be said that profit is maximized at the level of production where the marginal product equals the factor-product price ratio. The factor-product price ratio is the economic choice indicator. The concern with this indicator moves problems out of the field of pure production into the field of economics.

The use of factor-product price ratios are rather widespread where the physical factor-product relationships are generally known. For example, in the Midwest the corn–hog price ratio is regularly quoted with other market information. The corn–hog ratio measures the number of bushels of corn that are equal in value to 100 pounds of pork. Farmers use this ratio to determine whether to sell corn or feed it to hogs, and if it is fed to hogs, they use this ratio to determine the most profitable weight to market hogs. In the northeastern states where dairying is a major enterprise, the feed–milk ratio is used for much the same purpose. If farmers know their factor-product conversion ratios then these factor-product price ratios become important to monitor. Following are a few generalizations about factor and product prices:

- An increase in the price of a factor lowers the amount of the factor it pays to use and hence, the amount of the product it pays to produce and vice versa.
- An increase in the price of product would increase the amount of factor it pays to use and hence, the amount of product it pays to produce and vice versa.
- All changes in the use of a factor as a result of price changes are limited to stage II of the production function in which AP > MP and MP > 0 and VMP > 0.

EXAMPLE 6.1 Corn Yield Response to Nitrogen by Soil Types and by Growing Conditions

Is there an economical optimum nitrogen (N) rate and do soil type and growing conditions affect the optimum level of N fertilizer for corn? Figure 6.3a and b show the relationship between corn yield and applied N for different soil types and growing conditions. Figure 6.3a shows estimated yield responses to N on different Wisconsin soil types. Note that for some soils the MP is changing rapidly, whereas others exhibit a near linear change. Table 6.3a illustrates how the optimum yield varies by soil type and price for Fayette and Withee soils, two of the soil types illustrated in Figure 6.3a. In Table 6.3a, N rates are shown at 25-pound increments. If the price of corn is $2 per bushel and the price of N is $0.20 per pound then the profit maximizing N rate is 150 pounds of N for Fayette soils and 100 pounds of N for Withee soils. These conclusions can be made by observing the N rate in which income over costs is the greatest or observing the last increment for which VMP is greater than MFC. If the price of N was $0.13 per pound then it would be profitable to increase the N rate by 25 pounds for both Fayette and Withee soils. The amount of N and the change in N in response to N price changes varies by soil type.

Figure 6.3b shows corn yield response to N classified as high- and low-yielding years in Lancaster, Wisconsin. It is apparent that growing conditions affect overall yield outcomes, but note that the slopes or MPs for each response are very similar. Should farmers anticipating good growing conditions apply more N? The answer is illustrated by the information shown in Table 6.3b. If the corn price is $2 per bushel and the price of N is $0.20 per pound, the profit-maximizing level is 150 pounds N per acre for either growing condition. Although resulting yields for the growing conditions are different, 147 bushels per acre in high-yielding years and 109 bushels per acre in the low-yielding years. However, if the price of N were to decrease to $0.15 per pound, the profit-maximizing level would increase to 175 pounds N. The implications of the Vanotti and Bundy studies are that farmers do not need to adjust N rates in anticipation of an expected growing condition, but they do need to adjust rates for soil type and prices.

FIGURE 6.3A AND B

Corn Yield Response to Nitrogen Fertilizer for (a) Different Wisconsin Soil Types and (b) High-Yielding and Low-Yielding Years, Lancaster, Wisconsin

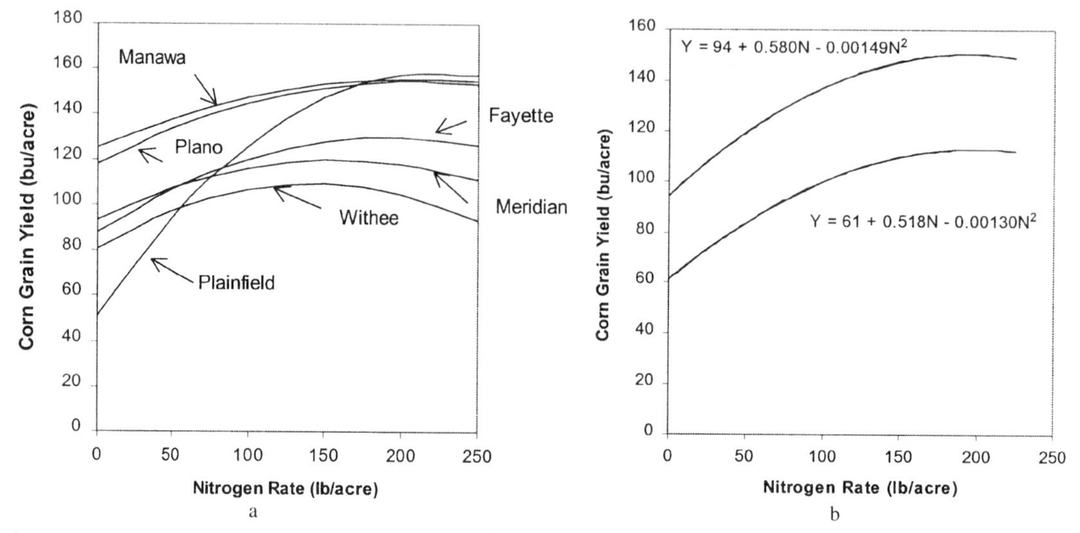

Source: (a) Vanotti, M. B. and L. G. Bundy. 1994. Corn nitrogen fertilizer recommendations based on yield response. *J. Production Agriculture* 7:249–256. (b) Vanotti, M. B. and L. G. Bundy. 1994. An alternative rationale for corn nitrogen fertilizer recommendations. *J. Production Agriculture* 7:243–249.

TABLE 6.3A

Estimates of Corn Yield Response to Nitrogen Fertilizer for Fayette and Withee Soils in Wisconsin

	Fayette					Withee				
N Rate (lb/acre)	TP (bu/acre)	MP (bu/lb)	VMP (per lb)	MFC (per lb)	Income Over Cost	TP (bu/acre)	MP (bu/lb)	VMP (per lb)	MFC (per lb)	Income Over Cost
0	88				$176.00	81				$162.00
		0.405	$0.81	$0.20			0.358	$0.72	$0.20	
25	98				$191.25	90				$174.89
		0.349	$0.70	$0.20			0.289	$0.58	$0.20	
50	107				$203.70	97				$184.35
		0.293	$0.59	$0.20			0.221	$0.44	$0.20	
75	114				$213.35	103				$190.39
		0.237	$0.47	$0.20			0.152	$0.30	$0.20	
100	120				$220.20	107				$193.00
		0.181	$0.36	$0.20			0.084	$0.17	$0.20	
125	125				$224.25	109				$192.19
		0.125	$0.25	$0.20			0.015	$0.03	$0.20	
150	128				$225.50	109				$187.95
		0.069	$0.14	$0.20			–0.053	–$0.11	$0.20	
175	129				$223.95	108				$180.29
		0.013	$0.03	$0.20			–0.122	–$0.24	$0.20	
200	130				$219.60	105				$169.20
		–0.043	–$0.09	$0.20			–0.190	–$0.38	$0.20	
225	129				$212.45	100				$154.69

Source: Vanotti, M. B. and L. G. Bundy. 1994. Corn nitrogen recommendations based on yield response data. *J. Production Agriculture*. 7:249–256.

Fayette model $Y = 88 + 0.433N - 0.00112N^2$

Withee model $Y = 81 + 0.392N - 0.00137N^2$.

TABLE 6.3B

Estimates of Corn Yield Response to N in High- and Low-Yielding Years for Rozetta Soil at Lancaster, Wisconsin

	High-Yield Years					Low-Yield Years				
N Rate (lb/acre)	**TP (bu/acre)**	**MP (bu/lb)**	**VMP (per lb)**	**MFC (per lb)**	**Income Over Cost**	**TP (bu/acre)**	**MP (bu/lb)**	**VMP (per lb)**	**MFC (per lb)**	**Income Over Cost**
0	94				$188.00	61				$122.00
		0.543	$1.09	$0.20			0.486	$0.97	$0.20	
25	108				$210.14	73				$141.28
		0.468	$0.94	$0.20			0.421	$0.84	$0.20	
50	119				$228.55	84				$157.30
		0.394	$0.79	$0.20			0.356	$0.71	$0.20	
75	129				$243.24	93				$170.08
		0.319	$0.64	$0.20			0.291	$0.58	$0.20	
100	137				$254.20	100				$179.60
		0.245	$0.49	$0.20			0.226	$0.45	$0.20	
125	143				$261.44	105				$185.88
		0.170	$0.34	$0.20			0.161	$0.32	$0.20	
150	147			$0.20	$264.95	109				$188.90
		0.096	$0.19				0.096	$0.19	$0.20	
175	150			$0.20	$264.74	112				$188.68
		0.021	$0.04				0.031	$0.06	$0.20	
200	150			$0.20	$260.80	113				$185.20
		−0.053	−$0.11				−0.035	−$0.07	$0.20	
225	149			$0.20	$253.14	112				$178.48

Source: Vanotti, M. B. and L. G. Bundy. 1994. An alternative rationale for corn nitrogen fertilizer recommendations. *J. Production Agriculture*. 7:243–249.
High-yield model $Y = 94 + 0.580N - 0.00149N^2$; Low-yield model $Y = 61 + 0.518N - 0.00130N^2$

The yield responses for the different soil types in Figure 6.3a indicate that the nature of the yield response to N varies by soil type. Economists use the term *elasticity of production* to indicate how responsive changes in output are to changes in input. Elasticity of production (E_p) is defined as the percentage change in output compared with a corresponding percentage change in input.

$$E_p = \frac{\text{percentage change in output}}{\text{percentage change in input}} = \frac{\Delta Y / Y}{\Delta X / X} \text{ or } \frac{\Delta Y / \Delta X}{X / Y} \text{ or } \frac{MP}{AP}$$

Ep is related to the three stages of production as defined for the single factor-product relationship in Table 6.1. In stage I, E_p is greater than one because MP is greater than AP. In stage II E_p is less than or equal to one and greater than or equal to zero. In stage III E_p is negative.

TABLE 6.4

Elasticities of Production for Corn Yield Response to Nitrogen

	Fayette Soils			Withee Soils		
N Rate (lb/acre)	MP	AP	E_p	MP	AP	E_p
0	0.4	0.4	1.00	0.4	0.4	1.00
50	0.3	0.4	0.85	0.3	0.3	0.79
100	0.2	0.3	0.65	0.1	0.3	0.46
150	0.1	0.3	0.37	0.0	0.2	–0.10
200	0.0	0.2	–0.07	–0.2	0.1	–1.32
250	–0.1	0.2	–0.83	–0.3	0.0	–5.92

EXAMPLE 6.2 Elasticity of Production

- E_p may be used to discuss production efficiency. Only production in stage II, where $1 > E_p > 0$, would be considered economically efficient.
- E_p can be used to describe the response of farmers to changes in factor or product prices. Consider the following:
  - Production functions with a constant and high elasticity coefficient (say, 0.9) have a steep slope. Thus, the MP curve is rather flat. Hence, factor-price changes affect greatly the optimal amount of factor to use. The same is true for E_p coefficients that are high and decline slowly.
  - Production functions with a constant and low E_p (say, 0.4) have a rather gradual slope. Thus, the M_p curve is very steep in its beginning stages and drops to a rather low level before leveling out. In this case, changes in factor price generally do not affect greatly factor use. The same is true for production functions whose E_p values start high and then decline rather rapidly.

The elasticity of production was calculated (Table 6.4) for the Fayette and Withee soils illustrated in Figure 6.3a and in Table 6.3a. The elasticity of production changes rapidly for both soil types, but more so for the Withee soil. Note that the elasticity of productions corresponding to the optimum N rates of 150 pounds N for Fayette soils and 100 pounds for Withee soils are low, 0.37 and 0.46, respectively. This low E_p implies that N rates are not that sensitive to changes in the price of N.

EXAMPLE 6.3 Optimum Marketing Weight for Hogs

Alice has the problem of deciding at what weight to market her hogs. From past records the amount of feed required to produce hogs to various weight levels is known. Prices of feed and hogs are anticipated based upon various outlook materials. Assuming all other input requirements for producing hogs are held constant, Table 6.5 is developed.

TABLE 6.5

Feed Costs for Producing Hogs to Heavier Weight Levels

(1) Total Hog Weight (lb)	(2) Marginal Feed Required (lb)	(3) Total Feed Required (lb)	(4) Marginal Feed Price ($/cwt)	(5) Marginal Feed Cost ($)	(6) Total Feed Cost ($)	(7) Hog Price (cents/lb)	(8) Total Hog Value ($)	(9) Marginal Hog Value ($)	(10) Net Income Over Feed Costs ($)
20	—	15	14.25	2.18	2.18		—	—	—
40	36	51	11.35	4.09	6.27		25.00	—	—
60	50	101	8.00	4.00	10.27	a	—	—	—
80	60	161	8.00	4.80	15.07	a	—	—	—
100	68	229	8.00	5.44	20.51	a	—	—	—
120	74	303	7.45	5.51	26.03	a	—	—	—
140	80	383	7.45	5.96	31.98	a	—	—	—
160	84	467	7.45	6.26	38.24	a	—	—	—
180	87	554	7.45	6.48	44.72	42.0	75.60	—	30.88
200	90	644	7.45	6.70	51.42	41.8	83.60	8.00	32.18
220	93	737	7.45	6.93	58.35	41.6	91.52	7.92	33.17
240	97	834	7.45	7.23	65.58	41.3	99.12	7.60	33.54
260	102	936	7.45	7.45	73.18	41.0	106.60	7.48	33.42

a, No prices given for these weight levels.

- Column 2 shows the amount of feed required to increase the weight of the hog from the previous weight level shown in column 1 to the weight level opposite the feed quantity; e.g., increasing the weight from 20 to 40 pounds required 36 pounds of feed.
- Column 3 summarizes the total feed required to produce a hog to each of the weight levels shown in column 1. When the pigs are small, the higher protein and more refined feeds cost more money per unit than at heavier weights when the feed is mostly made up of concentrates (grains).
- Column 4 shows the per unit feed prices for the type of feed fed at each weight level.
- Columns 5 and 6 summarize the values of the feed fed (columns 2 and 3). Column 5 is tabulated by multiplying column 2 by column 4. Column 6 is tabulated by accumulating the values in column 5.
- Column 8 shows the value of a hog at different levels and is tabulated by multiplying the total weight (column 1) by the price (column 7).
- Column 9 shows the change in hog value as heavier weights are reached. As long as the marginal hog value (column 9) is greater than the marginal feed cost (column 5), it pays the farmer to feed to higher weight levels. For this situation the optimum marketing weight is reached at 240 pounds. This is verified by the income figures shown in column 10.

Examples 6.1 and 6.3 illustrate some important points about record keeping. To be useful for specifying optimum production levels, data must be kept in such a manner as to associate production levels with particular input levels. Any one yield level experienced on a farm represents only one level of input. Other input levels give different yield levels. If farmers are to generate their own yield estimates, they must be able to identify the quantities of each factor used to produce each level of output. If too many input factors are allowed to change at one time, it may not be possible to identify which influenced the production change. Once the data are kept, they must be organized to associate the amount of output increase caused by each one unit of input increase (or the amount of input or cost increase required to increase output by one unit). Thus, the step increases should not be very large. Several land-grant universities provide publications and advice to farmers on how to conduct their own on-farm research (e.g., Miller et al. *On-Farm Testing: a Grower's Guide,* EB1706 Cooperative Extension Washington State University). Farmers should watch for published research helpful in setting

FIGURE 6.4

(a) Input–Output Curve Showing the Relationship of Physical Output to Costs of the Factor Input. (b) The Inverse of (a) Showing Total Variable Cost and Total Revenue Plotted Against Output

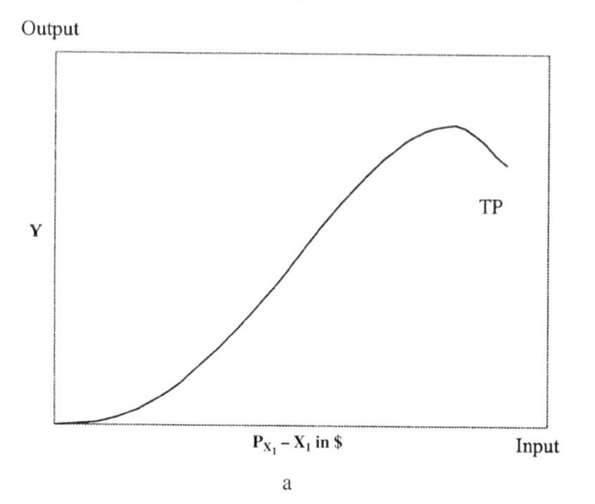

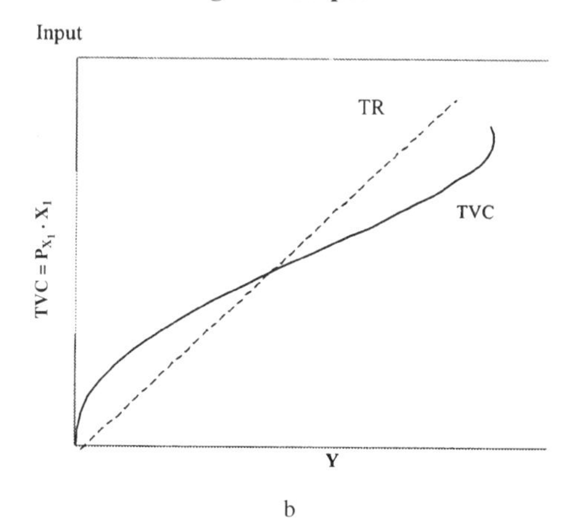

input–output relationships. For example, Pioneer Hi-Bred International issued *Pioneer's Agronomy Research Update* in the fall of 1998. It contained corn response functions to N fertilizer for several of their seed varieties and corn response functions to plant populations. Drs. Leon Orme and Sydney James analyzed growth rates for several classes of feedlot steers fed a high grain ration and a high roughage ration. Animals were slaughtered at 28-day intervals and their carcasses evaluated. It was found that the high roughage ration was not economical and the high-grain optimal marketing weight was 1,250 pounds. This kind of research is adaptable to farm situations.

How Much Product to Produce (Product–Cost)

The level of production can be approached from the standpoint of costs. The production relationship is still $Y = f(X_1)$. The discussion of factor–products has been on the basis of units of output per unit of input. But factor inputs cost money, and thus it may be more meaningful for some production relationships to consider the added cost to produce an additional unit of product. Rather than looking at the problem from the factor side, it is looked at from the product side. The principles are the same.

Consider the typical factor–product curve (inputs are on the horizontal axis and outputs are on the vertical axis) shown in Figure 6.4a. Now multiply the factor input by its price. The horizontal axis now becomes a dollar axis on which the cost of the factor can be measured. Total factor costs are now associated with different levels of product output on the vertical axis. If the axes are exchanged so the costs are on the vertical axis and the output on the horizontal axis, the inverse of the production function (TP) becomes a total variable cost curve (TVC), showing the quantities of product that can be produced from different expenditures, Figure 6.4b.

Farmers often consider their production from the output side. For example, grain yields are measured in bushels per acre, and farmers want to know their production costs per bushel. Similarly, livestock feeders want to know how much it costs to add 100 pounds of grain. Cost of production relationships are of this nature.

If a total revenue (TR) line is drawn on Figure 6.4b, it is possible to visualize the amount of net income (income over variable costs) at each level of output. Total revenue is the gross income received from selling different amounts of product. The horizontal axis is now a dollar axis so it is possible to show the revenue received from sell-

FIGURE 6.5

General Form of the Total Cost Curve

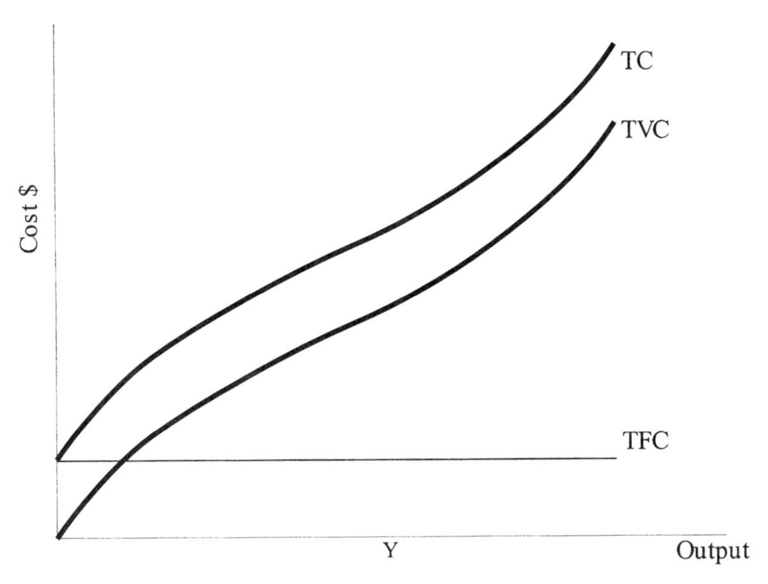

$Y = f(X_1|X_2 \ldots X_n)$; $TVC_{x_1} = X_1P_{x_1}$; $TFC = X_2P_{x_2} + X_3P_{x_3} + \ldots + X_nP_{x_n}$; $TC = TVC + TFC$

ing different amounts of product. The revenue curve is a straight line if the price received is constant at all output levels.

The place on the curves where TR minus TVC is the greatest is where profit is maximized. Profit is maximized where the marginal cost is equal to the marginal revenue (MC = MR), assuming continuous input levels. This is equivalent to saying that profit is maximized where VMP = MFC. This relationship is explained in the cost concepts named below.

Total costs (Figure 6.5):

- Total variable cost ($TVC = X_1 \cdot P_{x_1}$) measures the cost of adding the variable inputs factor and is tabulated by multiplying the units of factor used by its respective price. This is similar to the operating cost on the income statement.
- Total fixed cost ($X_2 \cdot Px_2 + \ldots X_n \cdot Px_n$), respectively for each fixed factor, measures the cost of all inputs held constant or fixed at all variable input levels. Hence, it is a straight horizonal line.
- Total cost (TC = TVC + TFC) measures the total cost of production. This is similar to the total cost on the income statement (TBD) although from an economic accounting perspective all unpaid factors are accounted for elsewhere.

Average costs (Figure 6.6):

- Average variable cost (AVC = TVC/TP) measures the average cost of the variable factor for producing one unit of product. It is a similar concept to tabulating the average cost of seed per bushel of grain produced.
- Average fixed cost (AFC = TFC/TP) measures the allocation of the fixed factor costs to the output of product. It has the form of a rectangular hyperbola.
- Average total cost (ATC = TC/TP) measures the average total cost for producing a unit of product. It is the figure often quoted by policy makers when trying to impress someone of the high cost of production. Any product price above this would be profit.

FIGURE 6.6

General Form of the Average Cost Curves

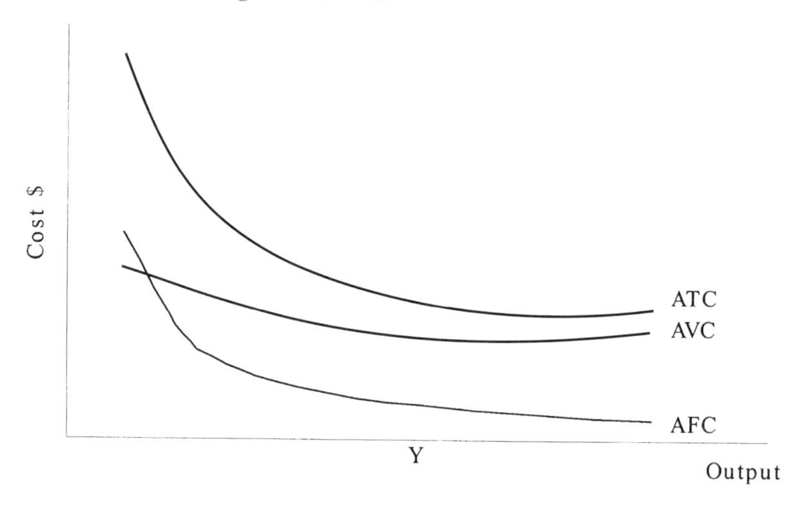

AFC = TFC/TP; AVC = TVC/TP; ATC = AFC + AVC

FIGURE 6.7

General Form of the Marginal Cost Curve and Its Relation to the Average Cost Curves

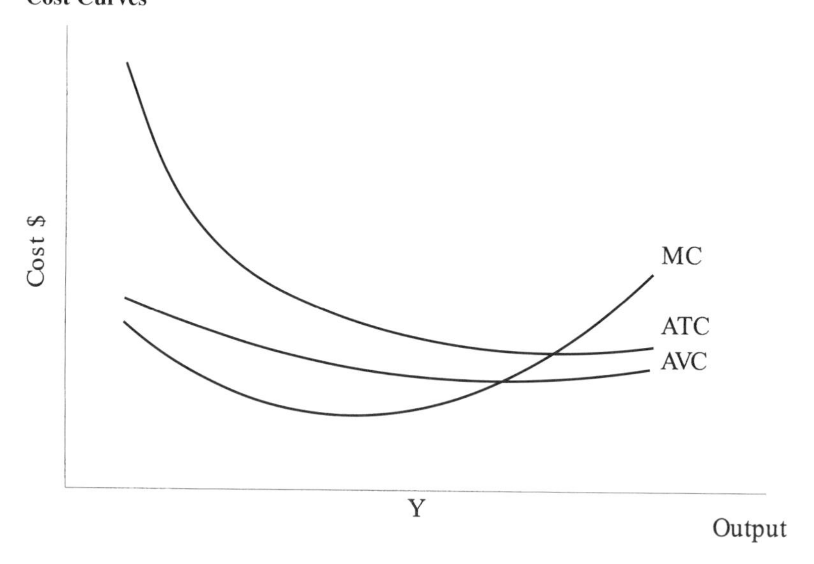

$$Y = f(X_1 \mid X_2 \ldots X_n);\ MC = \Delta TC / \Delta Y$$

Marginal costs (Figure 6.7):

- Marginal cost ($MC = \Delta TC/\Delta Y$) measures the additional cost required to add an additional unit of product output.

Note that the AVC curve and the MC curve are the inverse of the AP and MP curves.[4]

There are three revenue concepts (Figure 6.8).

FIGURE 6.8

General Form of the Revenue Curves

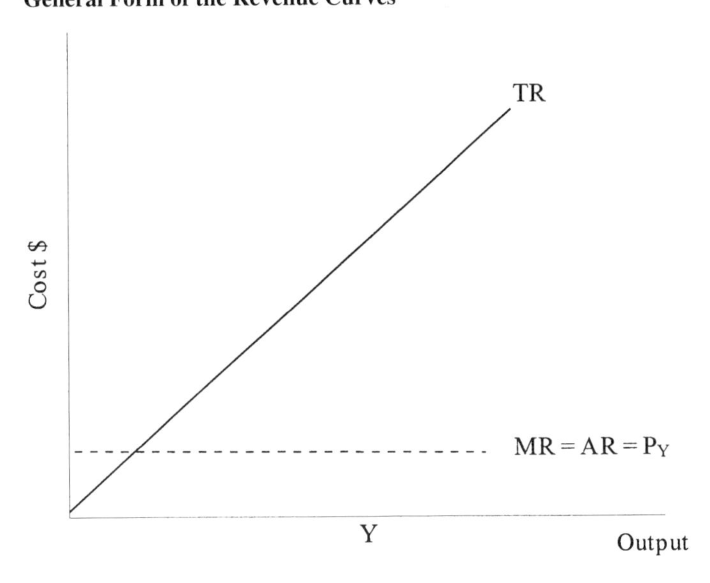

- Marginal revenue ($MR = P_Y$) is the income received from selling one unit of product. So long as the price received for a unit of product is the same for all levels of output, the MR and product price are the same.
- Total revenue ($TR - TP \times P_Y$) is the gross revenue received from selling all of the product produced at each output or cost level. So long as P_Y is constant, it is a linear increasing line. It is similar to gross income (TBC) on the income statement.
- Average revenue ($AR = TR/TP$) is the average amount received per unit of product sold. If P_Y is constant, $AR = P_Y = MR$.

Profit is maximized where MR = MC, which is another way of saying that more units of variable factor input are added to the fixed factors so long as the value of a unit of product exceeds the added cost of producing it. This is illustrated in Figure 6.9.

Several conclusions can be reached by studying these cost relationships:

- Production takes place where MC = MR. In this illustration, profit is maximized at output level Q_3 when $P_Y = P_3$.

  $TR = Q_3(P_3)$

  $TC = Q_3(\text{ATC at } C_2)$

  $TVC = Q_3(\text{AVC at } C_1)$

  $TFC = Q_3[(\text{ATC at } C_2) - (\text{AVC at } C_1)]$

  $\text{Profit} = [Q_3(P_3)] - [Q_3(\text{ATC at } C_2)]$
- If P_Y drops to P_2, production will still take place in the short run but output will drop to Q_2. All variable costs can be paid, but not all of the fixed costs can be paid. At this price level production can continue only until the fixed factors need to be replaced.
- If P_Y drops to P_1, it will not be profitable to produce at any level because total revenue will not cover total variable costs.

This reasoning can be applied to a cattle feeder who has new feed lots but cannot pencil enough returns to cover feed and other variable costs.

FIGURE 6.9

Cost Curves Showing Profit Maximization and Income Distribution

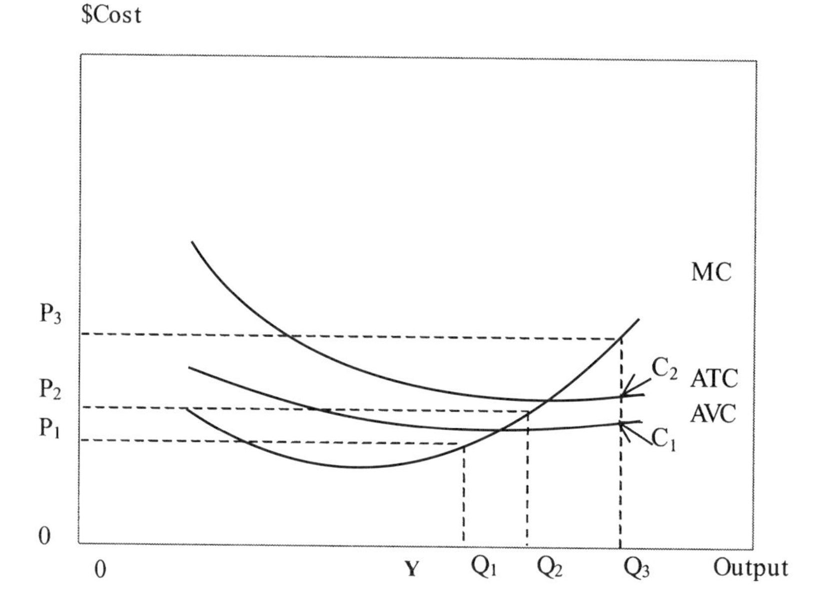

Cost-of-production concepts can be expanded to include more than one variable factor. This application is considered briefly following the factor–factor discussion.

EXAMPLE 6.4 Optimal Cattle Stocking Rate for a Pasture

A farmer who is required to keep a large part of his farm in permanent pasture is deciding how many beef cows he or she should have. The farmer determines that the average stocking rate is about 2 acres per cow. Data from other sources provide background for estimating the total pounds of beef marketed from different stocking rates as shown in Table 6.6. Observe the following:

- Variable costs increase on a per head basis as more cows are added.
- Fixed costs are constant over all stocking-rate levels.
- There are no decreasing marginal costs (increasing marginal products), but in this example there are some constant costs when stocking rates are low.
- Average variable costs continue to increase; whereas, average fixed costs continue to decline at a reducing rate.
- Average total costs are lowest at 110 cows.
- Marginal costs continue to increase and are equal to ATC at near 110 cows.
- Marginal cost is equal to marginal revenue at 120 cows. Profit is maximized between 110 and 120 cows.
- Beef prices could drop to near 50 cents/pound before this farmer would stop producing beef but production would be cut back at each price decrease. Production will continue as long as AVC can be covered.

TABLE 6.6

Determining the Most Profitable Stocking Rate for a Midwest Cow-Calf Operation

Given the following:

a. Total acres in farm = 500 acres, with 200 acres in permanent pasture. Total value of the farm including buildings = $400,000; pasture valued at $400 per acre.

b. Variable costs per cow:

Cow depreciation	$ 60
Death loss	5
Bull cost	10
Winter feed	60
Supplements	10
Health care	5
Interest on cow	10
Power and fuel	5
Miscellaneous	5
Total variable	$170

c. Fixed cost for pasture:

Interest on land	$6,000
Fence repair and depreciation	500
Taxes	1,000
Pasture maintenance	500
Total fixed costs	$8,000

d. Sales price = $0.85/lb

Total Cows (head)	Total Production (lb)	Change Total Production (lb)	Total Variable Cost ($)	Total Fixed Cost ($)	Total Cost ($)	Average Variable Costs ($/lb)	Average Fixed Costs ($/lb)	Average Total Costs ($/lb)	Marginal Cost ($/lb)	Marginal Revenue ($/lb)	Total Revenue ($)	Net Income ($)
0	0	0	0	8,000	8,000							–8,000
20	8,400	8,400	3,400	8,000	11,400	0.405	1.354	1.357	0.405	0.85	7,140	–4,260
40	16,800	8,400	6,800	8,000	14,800	0.405	0.881	0.881	0.405	0.85	14,280	–520
60	25,200	8,400	10,200	8,000	18,200	0.405	0.317	0.722	0.405	0.85	21,420	3,220
70	29,350	4,150	11,900	8,000	19,900	0.405	0.273	0.678	0.410	0.85	24,947	5,047
80	33,400	4,050	13,600	8,000	21,600	0.407	0.240	0.647	0.420	0.85	28,390	6,790
90	37,200	3,800	15,300	8,000	23,300	0.411	0.215	0.626	0.447	0.85	31,620	8,320
100	40,000	3,400	17,000	8,000	25,000	0.419	0.197	0.616	0.500	0.85	34,510	9,510
110	43,600	2,800	18,700	8,000	26,700	0.431	0.184	0.615	0.607	0.85	36,890	10,190
120	45,400	2,000	20,400	8,000	28,400	0.449	0.176	0.625	0.850	0.85	38,590	10,190
130	46,400	1,000	22,100	8,000	31,100	0.476	0.172	0.648	1.700	0.85	39,440	9,340
140	46,200	–200	23,800	8,000	31,800	0.515	0.173	0.688	–0.118	0.85	39,270	7,470

Note: one unit of factor = 1 cow; one unit of product = 1 pound.

TABLE 6.7

Cost of Production for Alternative Pork Production Systems

	150 Sow Low Technology	150 Sow High Technology	300 Sow High Technology	600 Sow High Technology	1,200 Sow High Technology
Total feed cost ($/cwt)	21.66	19.80	19.80	18.56	18.56
Total operating cost ($/cwt)	25.33	23.29	23.37	22.07	22.07
Total ownership cost ($/cwt)	22.55	17.25	15.26	13.64	12.17
Total cost ($/cwt)	47.88	40.54	38.63	35.72	34.25
Cost difference vs. 1,200 sow	+13.63	+6.29	+4.38	+1.47	—

Source: Hurt, Chris, et al. *Positioning Your Pork Operation for the 21st Century*. Chapter 14, Purdue Cooperative Extension Service. 1995.

EXAMPLE 6.5 Long-Run Planning for Swine Enterprise

In the long run a farm manager must cover all costs. Consider a manager of a swine enterprise who expects the long-run price of hogs to be P_2 on Figure 6.9. At this price the manager would not be able to cover all of his or her costs. What options does our swine manager have for the long run? One option is to exit the business. Another option is to improve the efficiency of the fixed factors by expansion or adopting new technologies. A study from Purdue University compared pork production costs for five alternative systems. Table 6.7 summarizes the costs of five alternative pork production systems. Table 6.5 compares total feed costs, total operating costs (variable costs), and total ownership costs (fixed costs). All costs are reported per hundredweight. The Purdue study concluded that the larger system could produce pork at lower costs per hundredweight because of better feed efficiency and lower ownership costs. The improved feed efficiency for the large systems is due partly to newer technologies that can be adopted plus the use of superior breeding stock that large producers obtain at lower prices than do small producers. The major difference is in ownership costs in which the larger systems are able to lower costs for buildings, equipment, and labor by having greater throughput of animals, resulting in more output per dollar of investment. Smaller producers either have to become larger to be competitive, or they may achieve some of these economies of size by forming networks with other producers.

How to Produce (Factor–Factor)

Considered in this section is the principle of determining how best to produce any given level of output. The principle can be used to determine the combination of fertilizers to apply in producing crops, how to balance plant population (seeding rate) with fertilizers, how livestock feeds should be blended, and the combination of tillage practices to follow. There is a least-cost way of doing things, and when combined with the previously discussed maximization principles, it provides additional insight into maximizing income.

A production function with only two variable input factors is presented to simplify the discussion. The relation is $Y = f(X_1, X_2 | X_3 \ldots X_n)$, or $Y = f(X_1, X_2)$. Consider for illustrative purposes the data in Figure 6.10. For the present, ignore the curved lines. Look first at the rows one at a time and consider each independently of the others. For $X_2 = 0, 10, 20, 30, 40$, the numbers show the product output as increasing levels of X_1 are applied. There are five factor–product relationships shown, one for each level of X_2 input ($Y = f(X_1 | X_2)$). For $X_2 = 0, 10, 20, 30$, the product responses of increasing levels of X_1 are shown in Figure 6.11.

Likewise, if the columns in Figure 6.10 are looked at in a similar manner, the production amounts shown represent output as X_2 is increased and X_1 is held constant

FIGURE 6.10

Output of Y from a Contribution of X_1 and X_2 Factors Illustrating the Development of the Isoquant or Isoproduct Curve

X_2 \ X_1	0	10	20	30	40
0	0	0	0	0	0
10	0	40	60	70	75
20	0	70	100	115	120
30	0	90	125	140	140
40	0	95	135	145	145

Y = 50

Y = 100

Y = 140

Horizontally $Y = f(X_1 | X_2)$, Vertically $Y = f(X_2 | X_1)$, Totally $Y = f(X_1, X_2 | X_3, \ldots, X_n)$

FIGURE 6.11

Factor Product Curves Illustrating Different Levels of Two Variable Factors

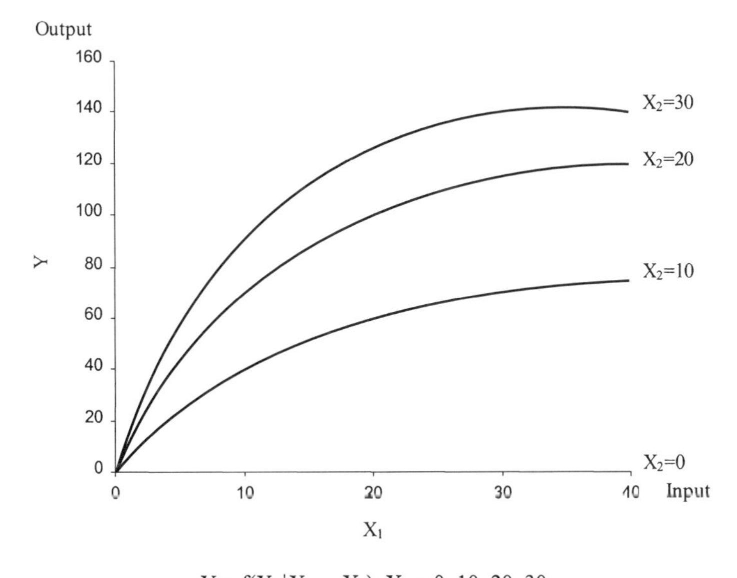

$Y = f(X_1 | X_2, \ldots, X_n)$; $X_2 = 0, 10, 20, 30$.

($Y = f(X_2 | X_1)$). Diminishing returns are shown as vertical distances between curves become smaller at higher levels of X_2 input.

Return now to Figure 6.10 and observe the curved lines superimposed on top of the production data. It should not be difficult to see that each of these lines represents a constant level of production forthcoming from different combinations of the two factors X_1 and X_2. Production levels of 50, 100, and 140 are illustrated, but more levels could be drawn. Each level of production is called an isoquant or iso, or equal, product curve. It is like a contour line on a topographic map, which shows equal elevation.

FIGURE 6.12

Three-Dimensional Diagram Showing Yield as a Product of Two Factor Inputs Y = 100

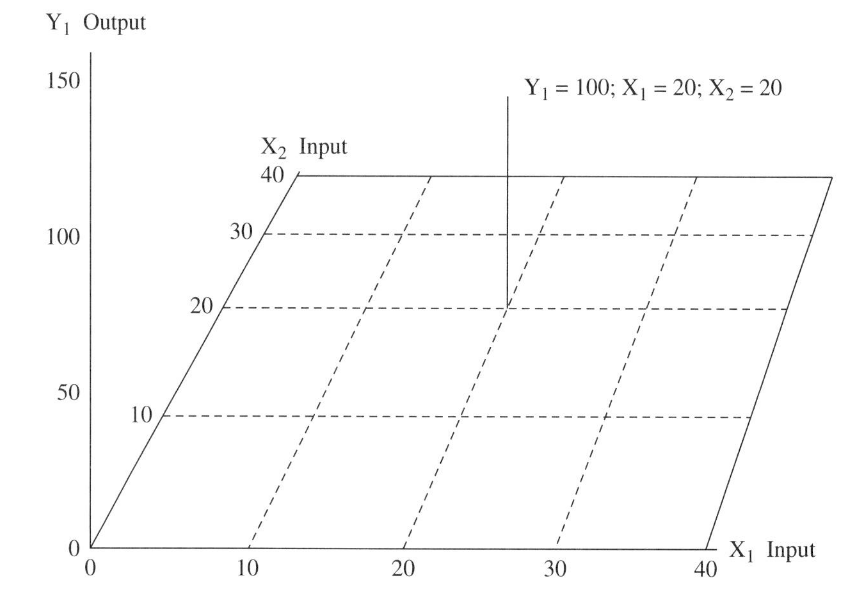

FIGURE 6.13

Three Dimensional Diagram Showing Output Levels for Different Combinations of Two-Factor Inputs X_1 and X_2

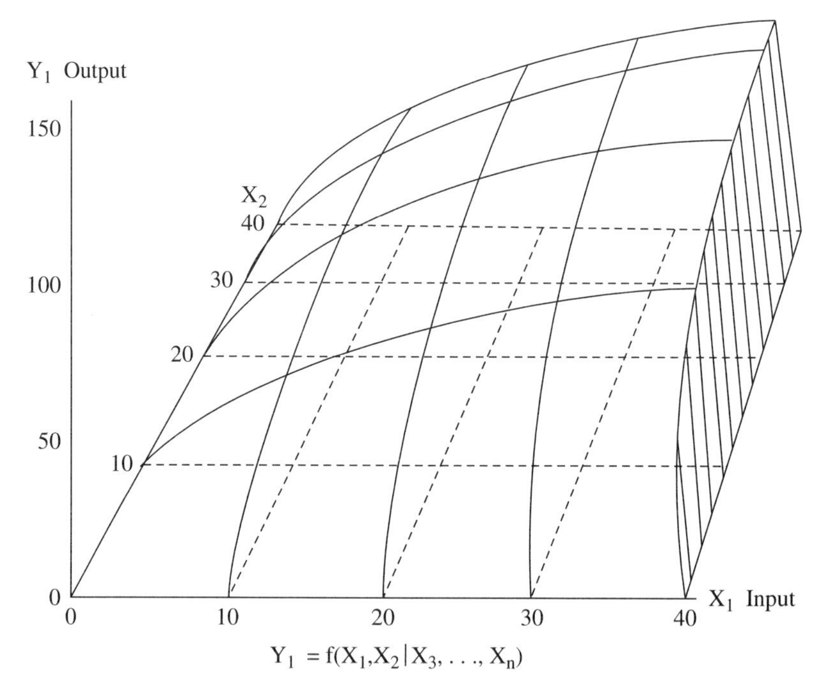

$Y_1 = f(X_1, X_2 | X_3, \ldots, X_n)$

The idea of a topographic map is further illustrated in Figures 6.12 and 6.13, shown in three dimensions (i.e., width, depth, and height). The width and depth represent the inputs X_1 and X_2, respectively, and the height measures the output of Y. The upper figure shows only one combination of factors (i.e., $X_1 = 20$ and $X_2 = 20$) producing 100 units of product. The lower curve shows all possible combinations of X_1 and X_2 with the resultant outputs, assuming X_1 and X_2 are continuous (can be divided into infinitesimally small units). The output is represented by a solid surface much like a hill in topography. The 100-unit level of production is marked and other input continuations can be seen. Equal levels of production are marked by imaginary lines drawn around the elevated production surface.

Diminishing marginal productivity causes the factors to substitute for each other as follows: As more of one factor is substituted for another factor in producing a given or constant amount of product, the number of units of the second factor that is replaced by one unit of the first factor declines.

Assume ΔX_1 is one unit in each of the two locations shown in Figure 6.14. Note the difference in the size of ΔX_2. As more X_1 is added, X_2 becomes smaller, while TP remains the same.

Next, study Figure 6.15 to understand what is taking place with the marginal products of X_1 and X_2 as X_1 is increasing and X_2 is decreasing. Note that their respec-

FIGURE 6.14

Factor–Factor Curve Showing the Principle of Diminishing Rate of Factor Substitution

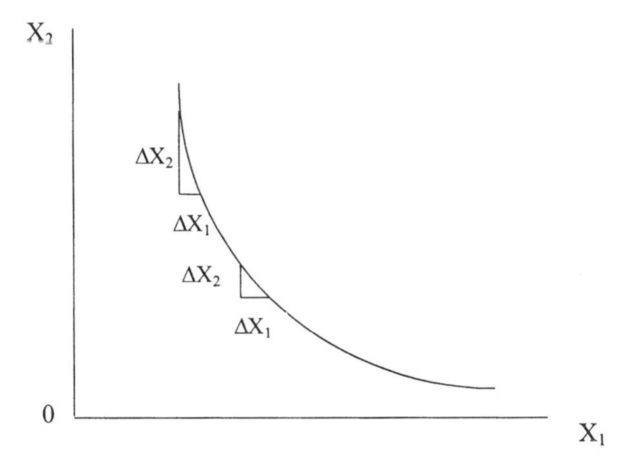

FIGURE 6.15

Factor–Product Curves Showing the Relationship of the Change in Factor Inputs to Their Respective Marginal Products

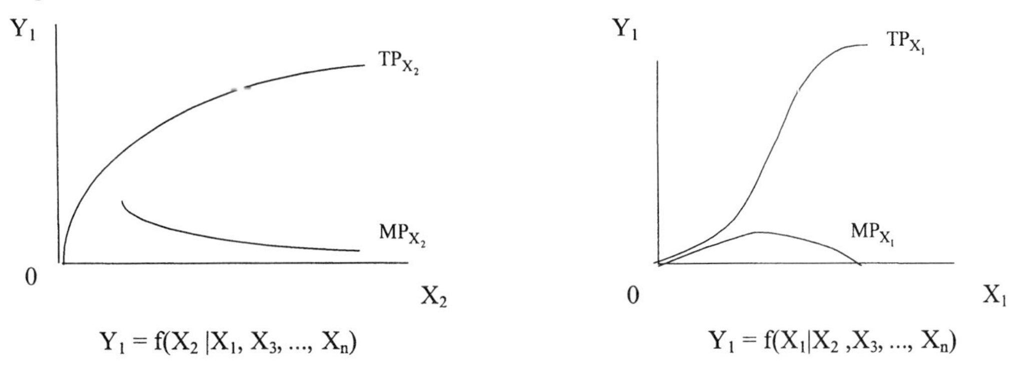

tive MPs are moving in opposite directions. As X_1 is increasing, its MP is getting smaller and this effect is explained by the law of diminishing returns. The opposite is true for X_2. This observation provides the following identities:

- As X_1 is increased, MP_{X_1} becomes smaller.
- As X_2 is decreased, MP_{X_2} becomes larger.
- The amount of product remains the same.
- The decrease in output of product resulting from smaller amounts of X_2 input is exactly offset by an increase in output of product caused by larger amounts of X_1 input.
- Symbolically the substitution ratio is as follows:

  $\Delta X_2 \cdot MP_{X_2} = -\Delta Y$

  $\Delta X_1 \cdot MP_{X_1} = +\Delta Y$

  Because $-\Delta Y = \Delta Y$ in quantity, then

  $-\Delta X_2 \cdot MP_{X_2} = \Delta X_1 \cdot MP_{X_1}$ in quantity and

  $$\frac{\Delta X_2}{\Delta X_1} = \frac{-MP_{X_1}}{MP_{X_1}}$$

  The substitution ratio is inversely equal to the marginal product ratio.

Factor–factor relationships can be shown to have the same three stages of production as factor–product relationships (Figure 6.16). Stage III is where the same output can be produced with more of both inputs. Stage III begins where the isoproduct curve changes from a negative slope to a positive slope. Stage I is harder to detect. If it exists, it is where X_1 and X_2 are small and the isoproduct curve would curve out from the origin rather than curve in toward the origin. Stage II is the rational area of production or the rational zone of input substitution.

The cost-minimizing principle is as follows: One factor will continue to be substituted for another factor in the production of some level of output so long as the cost decrease from reducing the second factor is greater than the cost increase of

FIGURE 6.16

Factor–Factor Curves Showing the Location of the Three Stages of Production Efficiency

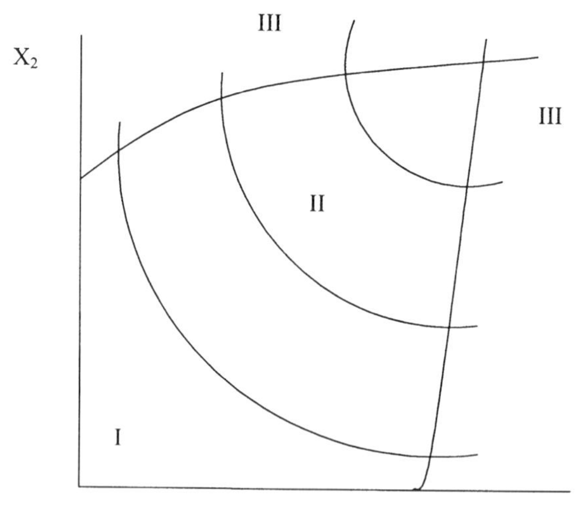

adding the first factor. Symbolically, this principle is represented by the following equation:

$$-\Delta X_2 \cdot P_{X_2} = \Delta X_1 \cdot P_{X_1} \text{ or } \frac{\Delta X_2}{\Delta X_1} = \frac{P_{X_1}}{P_{X_2}} \text{ or } \frac{MP_{X_1}}{MP_{X_2}} = \frac{P_{X_1}}{P_{X_2}} \; or \; \frac{MP_{X_1}}{P_{X_1}} = \frac{MP_{X_2}}{P_{X_2}}$$

If the above equality does not exist it is either possible to increase output for the same cost, or obtain the same output at a reduced cost. This is shown graphically in Figure 6.17.

Note that the factor-price ratio dictates to the producer the desired factor substitution ratio, or inverse marginal product ratio. Also note that where cost is minimized marginal product–price ratios are equal among factors. The price ratio is illustrated in Figure 6.17a as an equal cost line (isocost line). Its slope is the needed marginal product ratio. Assume P_{X_1} = \$3 and P_{X_2} = \$2, then 20 units of X_1 are equal in value to 30 units of X_2. At the point of tangency of the price ratio line and the isoproduct curve their slopes are equal,

$$\frac{P_{X_1}}{P_{X_2}} = \frac{MP_{X_1}}{MP_{X_2}}$$

and the desired equality has been obtained for a least cost of production (Figure 6.17b).

Now consider that there are many isoproduct curves and each is tangent to an equal price ratio line. This gives rise to a line of tangency points and is called the scale line or expansion path (Figure 6.17b). At each point on the expansion path:

$$\frac{MP_{X_1}}{MP_{X_2}} = \frac{P_{X_1}}{P_{X_2}}$$

At what level should production stop? Recall the factor–product maximization criteria:

$$VMP = MFC \text{ or } MP \cdot P_Y = P_X \text{ or } \frac{MP \cdot P_Y}{P_X} = 1$$

FIGURE 6.17

The Equal–Cost Line or Price–Ratio Line Is Developed in (a) for the Input Prices. Factor–Factor Relationships Showing the Optimum Combination of Factor Inputs Are Shown in (b), Including Price–Ratio Lines, Isoquants, and Expansion Path for Different Quantities of Output

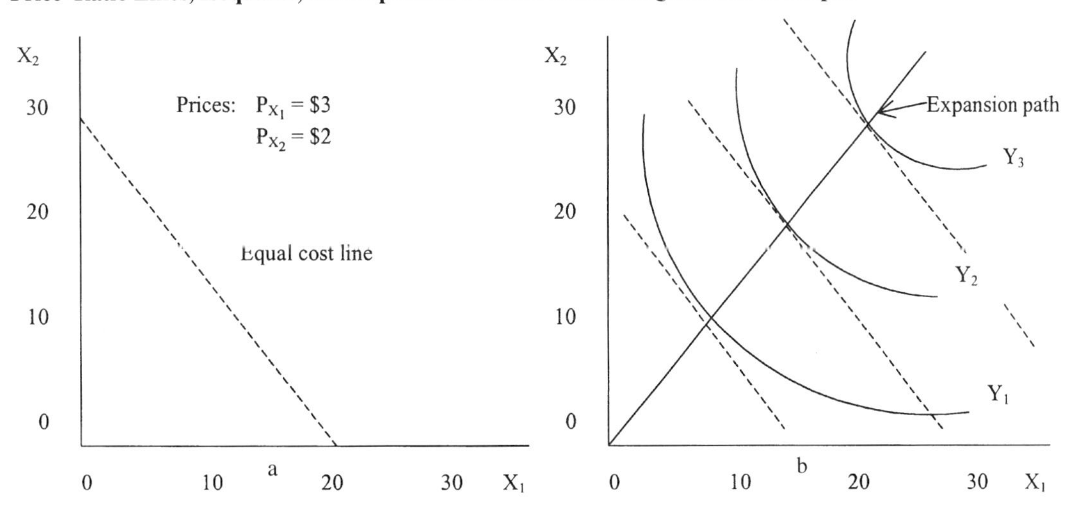

Expanding this concept to include two factors gives the following formula for profit maximization:

$$\frac{MP_{X_1} \cdot P_Y}{P_{X_1}} = \frac{MP_{X_2} \cdot P_Y}{P_{X_2}} = 1$$

This maximizing condition can be further expanded to more than two factors.

The least cost of production principle can now be applied to the cost of production principles discussed earlier in this chapter. As expenditures are increased (up the vertical axis in Figure 6.9) for the factors of production, the combination of inputs purchased (applied) is a least cost mix. Expenditures continue until the added cost for this least cost mix is just equal to the value of production added.

EXAMPLE 6.6 Least Cost Combination of Labor and Herbicide for Weed Control in Sugar Beets

Besides selecting input combinations to improve profit, farm managers also must consider the effects of using inputs that are safe for humans and for the environment. A study by Held et al. examined the trade-offs between hoe labor and herbicides. They evaluated 11 alternative weed management options that used different proportions of hoe labor and herbicide treatments. Herbicide treatments consisted of no-, one-half-, and full-preplant herbicide applications combined with one to four postemergence applications. Hoe labor was used to control the remaining weed populations. There was no significant difference in yield between treatments; therefore, they focused on the cost trade-offs between treatments. Table 6.8 summarizes the findings.

To simplify the results herbicide treatments were combined and costs averaged by the number of postemergence applications. Column 1 indicates the number of postemergence treatments. Column 2 indicates the combined herbicide treatment costs. Column 3 indicates the amount of hoe labor used to control existing weed populations. Columns 4 and 5 show the changes in herbicide cost and labor hours. Column 6 reports the labor-to-herbicide substitution ratio. If hoe labor costs $6 per hour, the herbicide-to-labor price ratio is 0.167 ($1/$6); herbicide should be substituted for labor as long as the substitution ratio is greater than the price ratio. At $6 per hour for labor, the three-times-over herbicide treatment would be the least cost. The wage rate would have to be $5.50 per hour or less to get any further decrease in herbicide use. Wage rates would have to increase to $45 per hour to justify a fourth postemergence treatment.

TABLE 6.8

Herbicide Costs and Hoe Labor Requirements by Number of Postemergence Applications

(1)	(2)	(3)	(4)	(5)	(6)
Post-emergence Treatments (#)	**Herbicide Cost ($/acre)**	**Hoe Labor Use (hr/acre)**	**Herbicide Cost Increase ($/acre)**	**Labor Decrease (hr/acre)**	**Labor Decrease per Herbicide Cost Increase (hr/$)**
1× over	10	8.0			
			10	3	0.30
2× over	20	5.0			
			11	2	0.19
3× over	31	3.0			
			9	0.2	0.02
4× over	40	2.8			

Source: Held, Larry J., K. James Fornstrom and Stephen D. Miller. 1996. Reducing herbicide in sugarbeets: how much is economical? *J. of the American Society of Farm Managers and Rural Appraisers.* 19:122–128.

EXAMPLE 6.7 Least-Cost Ration for Market Hogs

A farmer has the problem of which ration to feed to feeder pigs that weigh about 100 pounds. From an agricultural experiment station publication, information is found describing the quantities of corn and soybean oil meal, representing different protein levels that would be required to increase the weight of pigs from 105 to 125 pounds. These amounts are next compared with farm feed records for verification of protein and other feed amounts. After making adjustments, Table 6.9 is developed to represent the farm situation. Columns 2 and 3 are the basic data. Any combination of corn and soybean oil meal (SBOM) shown (represented by the different protein levels) will increase hog weights from 105 to 125 pounds.

- Each combination of corn and SBOM increased the pig weight by 20 pounds.
- The amount of corn decreased (column 5) as SBOM was increased (column 7) by one-unit drops.
- The decrease in corn becomes smaller as more SBOM is substituted in the ration for corn (column 5).
- Feed cost is minimized (column 9) when the cost decrease in corn (column 6) is no longer greater than the cost increase of SBOM (column 8).
- Feed cost is minimized (column 9 at 13 percent protein) at a level where total feed fed (column 4) is not at its smallest amount (16 percent protein). In this example the feed cost is minimized at 13 percent protein but the results are such that the farmer could use either the 12, 13, or 14 percent level of protein in the ration without much penalty.
- The least-cost ration would shift with changes in either the price of corn or the price of SBOM.

TABLE 6.9

Corn and Soybean Oil Meal (SBOM) Requirements for Producing Hogs from 105 to 125 Pounds

(1) Protein (%)	(2) Corn (lb)	(3) SBOM (lb)	(4) Total Feed (lb)	(5) Decrease – Corn[a] (lb)	(6) Decrease – Corn[a] ($)	(7) Increase – SBOM[b] (lb)	(8) Increase – SBOM[b] ($)	(9) Total Feed Cost ($)
10	88.50	2.64	91.14					3.86
				12.63	0.506	1.52	0.182	
11	75.87	4.16	80.03					3.53
				7.20	0.288	1.42	0.160	
12	68.67	5.58	74.25					3.42
				4.80	0.192	1.40	0.168	
13	63.87	6.98	70.85					3.39
				3.43	0.138	1.43	0.172	
14	60.44	8.41	68.85					3.43
				2.50	0.100	1.44	0.172	
15	57.94	9.85	67.79					3.50
				1.89	0.076	1.51	0.182	
16	56.05	11.36	67.41					3.61
				1.48	0.060	1.67	0.200	
17	54.57	13.03	67.61					3.75

Source: Robert Johnson, Ph.D. dissertation, Department of Animal Science, Iowa State University, 1965.
[a]Corn price = 4 cents/lb.
[b]SBOM price = 12 cents/lb.

EXAMPLE 6.8 Least-Cost Fertilizer and Water Combination for Growing Corn

Alan Kleinman, Iowa State University, developed the following formula for producing corn (C in pounds) from nitrogen fertilizer (N in pounds) and water (W in acre inches):[5] $C = 4.579.50 + 549.20W + 10.90N - 29.96W^2 - 0.03N^2 + 0.06\ WN$. Using this formula and solving for W at several levels of N for two levels of corn production gives the figures in Table 6.10.

Because it is profitable to keep increasing N relative to W so long as the substitution ratio is greater than the price ratio, the least-cost level for producing 100 bushels is 50 pounds N and 1.06 acre in. W. At 130 bushel yield the least-cost level increases to 100 pounds N and 4.6 acre in. W. Note that the least-cost combination of these factors is determined by the prices of the factors.

If W were fixed at 4 acre inches the above formula reduces to the following for N: $C = 6576.5 - 11.2N - 0.03N^2$. Table 6.11 was derived from this formula.

TABLE 6.10

Water and Nitrogen Substitution Rates to Produce 100– and 130–Bushel Corn

C = 5,600 lb = 100 bu			C = 7,280 lb = 130 bu		
N	W	ΔW/ΔN	N	W	ΔW/ΔN
0	2.10		0	7.74	
		0.021			0.034
50	1.06		50	5.98	
		0.012			0.027
100	0.44		100	4.60	
		0.006			0.011
150	0.16		150	4.04	
		0.002			0.001
200	0.06		200	4.01	

TABLE 6.11

Corn Yield Response to Nitrogen (N) with Water Fixed at 4 Acre Inches

N (lb)	C (bu)	ΔN	ΔC	MFC (P_N = \$0.20)	VMP (P_C = \$2.40/bu)
0	117.4				
		50	8.7	$10	19.49
50	126.1				
		50	6.0	10	13.44
100	132.1				
		50	3.3	10	7.39
150	135.4				
		50	0.6	10	1.34
200	136.0				

100 lb of N is selected as the optimal level of application to produce 132 bushel of corn.

What to Produce (Product–Product)

What to produce raises the problem of how many enterprises or products. Some farm advisors recommend that farmers should not "put all their eggs in one basket" but should diversify their production by producing several products. Others recommend that farmers should specialize in only one or two enterprises, thus gaining advantages from greater management skill, increased efficiency in the use of machinery and labor, and volume discounting. These are problems examined in this section. The discussion does not consider the additions of products for reducing production or price risk and uncertainty. These matters are treated in a separate chapter. Price and production are assumed to be known and stable.

The mathematical foundation follows that of a single product, except there are now two products or more. Considering only two products, the following equations apply:

$$Y_1 = f(X_1, X_2, \ldots, X_n, Y_2)$$
$$Y_2 = f(X_1, X_2, \ldots, X_n, Y_1)$$

Note now that the output of Y_1 depends upon the output of Y_2 and vice versa. Where Y_1 and Y_2 both use the same factors of production they become competitive when increasing Y_1 reduces the production of Y_2, and vice versa. In this competitive area of production, assume $Y_1 = f(Y_2)$ and $Y_2 = f(Y_1)$.

The economic principle governing the optimum combination of crops and livestock to produce considers the income added versus the income lost as one product gradually replaces another product in competition for farm resources. There are three major relationships to observe in this regard:

- *Complementarity.* The income from a product already being produced is increased as a result of adding a new product that likewise has a positive net income. This usually happens where the new product furnishes some needed input to the old product. Examples include nitrogen-fixing legumes, crops that break up disease cycles in existing crops, crops that help control erosion, and livestock that furnish manure to crops. Complementarity exists only in the long run in some instances. Erosion control in crops is an example. In the short run (1-year consideration) the same crops may be highly competitive.
- *Supplementarity.* The income from a product already being produced is not affected by the addition of another enterprise or activity. This generally is made possible where the added enterprise uses only resources that otherwise would remain idle (go unused). Examples include adding livestock to use winter labor, adding beef cows or sheep to graze a pasture that cannot be used for commercial crops, working off the farm in slack seasons, and selling feed or seed as a side business.
- *Competitive.* The net income from an existing enterprise or activity is reduced as a new enterprise or activity is added or expanded. In this situation, there is competition for the same farm resources. An increase in either product decreases the other.

All complementary and supplementary products are produced so long as each is independently profitable. Production will take place where enterprises compete for limited resources (i.e., increasing the production of one enterprise reduces the production of another enterprise). Graphically, these three conditions are depicted in Figure 6.18.

FIGURE 6.18

Product–Product Relationships

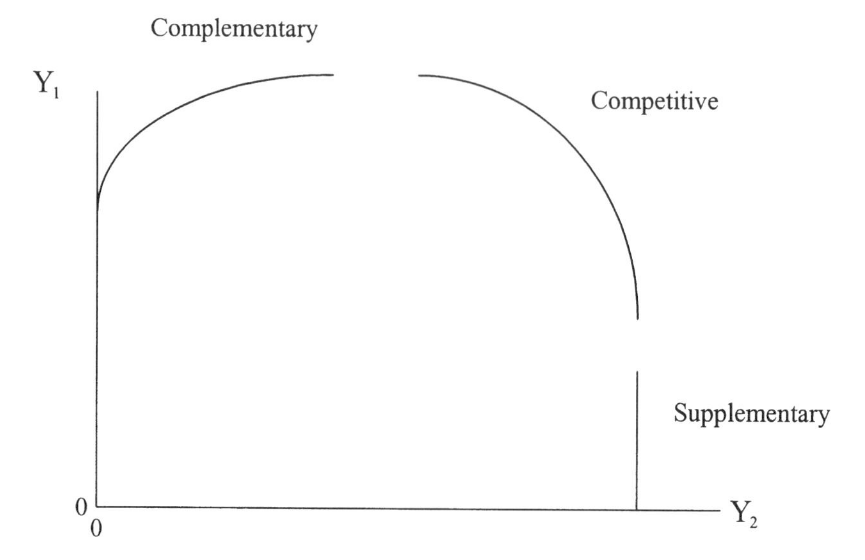

Within the competitive area, the reduction in income from one product as output of another product is increased takes place for two important reasons. First, there is the possibility of diminishing returns. As variable inputs are taken away from the production of one product and added to another, the marginal productivity expands for the first and diminishes for the second, using the reasoning of the factor–production relationships (see Figure 6.1). Second, and perhaps more important, there are resource limitations as illustrated for linear programming discussed later. Each new limitation adds an additional restraint, which produces a similar condition to purely marginal causes.

A new product is expanded so long as its value added is greater than the income reduction of the product replaced. Ignoring signs and assuming Y_2 is replacing Y_1, this replacement takes place until

$$-\Delta Y_1 \cdot P_{Y_1} = \Delta Y_2 \cdot P_{Y_2}$$
$$-MP_{Y_1} \cdot P_{Y_1} = MP_{Y_2} \cdot P_{Y_2}$$
$$-VMP_{Y_1} = VMP_{Y_2}$$

Placing the prices on the same side of the equal sign, the equation becomes

$$\frac{-\Delta Y_1}{\Delta Y_2} = \frac{P_{Y_2}}{P_{Y_1}} \text{ and } \frac{-MP_{Y_1}}{MP_{Y_2}} = \frac{P_{Y_2}}{P_{Y_1}}$$

It is better to use the net price (gross income or product price per unit less variable costs of production per unit) when applying this relationship as it considers the change in net income.

Where there are more than two products, the following equation applies:

$$VMP_{Y_{1net}} = VMP_{Y_{2net}} = VMP_{Y_{3net}} \ldots VMP_{Y_{nnet}}$$

and the output of each is optimized where each VMP = 1.

The principle of equating the marginal returns on investments is often referred to as the equal marginal returns principle or opportunity cost principle. The equal margin returns principle uses the ideas developed for finding the least cost of producing a

TABLE 6.12

Two-Way Table Showing Supplementary, Complementary, and Competitive Relationships

	Corn	Soybeans	Alfalfa	Oats	Hogs	Cattle
Corn	—	Co–C	Co	C	S	S
Soybeans		—	C	Co–C	S	S
Alfalfa			—	Co–C	S	S
Oats				—	S	S
Hogs				—	C	
Cattle						—

C, competitive; Co, complementary; S, supplementary.

given quantity of product and for selecting among alternative products to produce. If a farmer or other business person always invests additional money where the returns on the last dollar invested are the highest, then the returns on all investments tend to be equal at the margin and profits are maximized. This means that all expenditures on old and new business activities should be scrutinized against other alternatives (opportunities) to determine if each will return the highest profit available. Diminishing returns spreads investments out among the many opportunities, on and off the farm.

A practical approach to enterprise selection is as follows:

- List all enterprises you think are worth considering and gather relevant data pertaining to each (i.e. production, markets, prices, net income, etc).
- Determine the relationship of each enterprise with respect to the other enterprise selections. Identify supplementary, complementary, and competitive relationships. Table 6.12 illustrates some possibilities.
- Select the one enterprise thought to be the most profitable and make it as large as the most restricting resource allows. This gives economies of scale and efficiency.
- Add supplementary and complementary enterprises until they become competitive with the first enterprise selected. (This means that further additions of the supplementary or complementary enterprises will reduce the production of the first enterprise selected.)
- Select the enterprise thought to be the next most profitable (this could be one of the supplementary or complementary enterprises), and see if its addition increases income. Keep increasing this second enterprise as long as the value added is greater than the value given up from the first. Stop increasing the second enterprise when the value added equals the value given up as indicated by the following expressions:

$$(VMP_{Y_1} = VMP_{Y_2}) \text{ or } -\Delta Y_1 \cdot P_{Y_{1net}} = \Delta Y_2 \cdot P_{Y_{2net}} \text{ or } \frac{\Delta Y_1}{\Delta Y_2} = \frac{P_{Y_{2net}}}{P_{Y_{1net}}}$$

- Consider adding additional enterprises to further use resources and reduce income variability.

Budgeting, gross margins, and linear programming are the tools most often used in applying production selection principles. Example 6.9 illustrates the product–product concept using linear programming. Example 6.10 provides an applied example of the product–product concept.

EXAMPLE 6.9 Selecting Crops to Produce Using Linear Programming

This example considers two activities, corn and soybean productions, and four effective resource limitations, land, capital, and April and November labor. Unit budgets and specifications are shown in Table 6.13. The maximum production columns are tabulated by dividing the limited resource by the activity requirement and multiplying by the yield. Because there are several restrictions, the alternatives are graphed to give a clearer picture of the alternatives (Figure 6.19). Each line is straight because of the linearity assumption. The alternative points for consideration are A, B, C, D, and E. The alternative production levels are shown in Table 6.14. The substitution ratio measures the bushels of corn that must be given up to produce one more bushel of soybeans. Because the net profit of soybeans is \$2.66 per bushel and corn is \$1.10 per bushel, it is profitable to keep substituting soybeans for corn until the substitution ratio reaches 2.42 (\$2.66/\$1.10). Because the substitution ratio is less than 2.42 when going from alternative B to C, it is greater when going from C to D. Alternative C is selected to give the most profitable combination (this converts to 27 acres of corn and 73 acres of soybeans). However, in the long-run situation where all prices become variable (new machinery must be purchased, storage facilities built, etc.), the net prices of corn and soybeans drop to \$0.10 per bushel and the price ratio is 1.00. Considering the long-run, situation B is selected. Going from alternative A to B is a supplementary condition.

FIGURE 6.19

Production–Feasibility Curve (A-E) for Corn and Soybean Production. May, June, and October Constraints Are Not Shown Because These Constraints Are Not Limiting

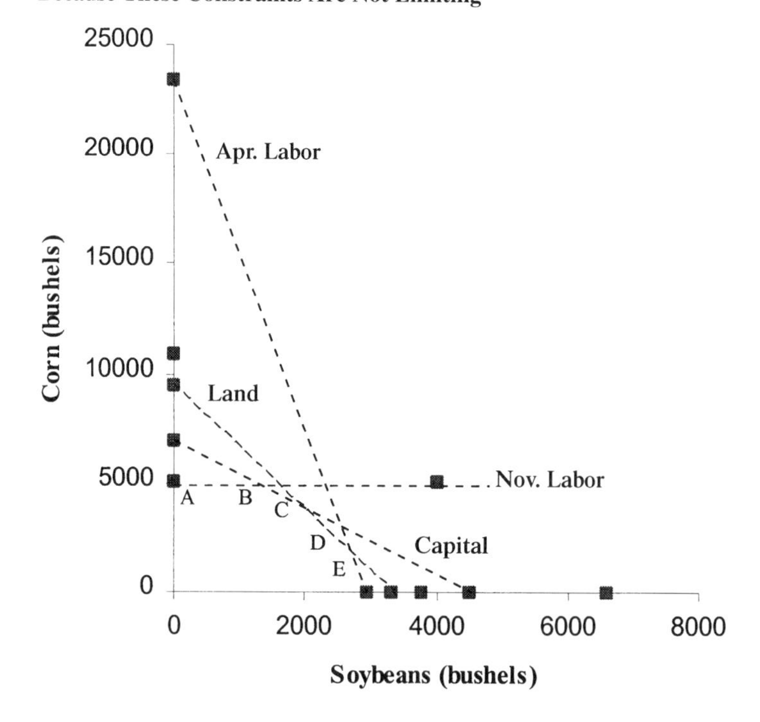

TABLE 6.13

Unit Budgets and Resource Limitations for Producing Two Products

Item	Corn: Resource Requirements (1 acre)	Corn: Maximum Production (bu)	Soybeans: Resource Requirements (1 acre)	Soybeans: Maximum Production (bu)
Yield (bu)	95	—	33	—
Resource requirements (limitation)				
Land (100 acre)	1	9,500	1	3,300
Labor				
Apr. (80 hr)	0.3	25,333	0.9	2,933
May (80 hr)	0.8	9,500	0.5	5,280
June (80 hr)	0.7	10,857	0.4	6,600
Oct (80 hr)	1.1	6,909	0.4	6,600
Nov (80 hr)	1.5	5,066	—	—
Capital requirement (8,000 dollar limit on funds for variable cost)	109.62[a]	6,933	70.58[a]	3,740

	Corn Unit Budget (1 acre): Fixed	Corn Unit Budget (1 acre): Variable	Soybeans Unit Budget (1 acre): Fixed	Soybeans Unit Budget (1 acre): Variable
Seed, fertilizer, chemicals	$ —	$ 57.00	$ —	$38.48
Machinery and fuel	10.50	13.42	10.18	8.80
Harvest and storage	8.00	23.40	3.30	9.66
Labor	9.30	15.80	6.60	13.64
Overhead	67.20	—	64.30	—
Total/acre	$95.00	$109.62	$84.38	$70.58
Total/acre (fixed + variable)	$204.62		$154.96	
Total/bu	$1.00	$1.15	$2.56	$2.14
Total/bu (fixed + variable)	$2.15		4.70	
Unit profitability (fixed – variable cost)				
Corn at $2.25/bu	$104.50/acre	$1.10/bu		
Soybeans at $4.80/bu			$87.78 acre	$ 2.66 bu

TABLE 6.14

Alternative Production Levels of Corn and Soybeans

Alternative	Corn	Soybeans	Substitution Ratio[a]: ΔCorn	÷	ΔSoybeans	=	Ratio
A	5,066	0					
			0		1,000		0
B	5,066	1,000					
			2.766		1,500		1.84
C	2,300	2,500					
			700		250		2.80
D	1,600	2,750					
			1,600		183		8.74
E	0	2,933					

[a]Although the change in corn is negative, the ratio (ΔCorn/ΔSoybeans) is expressed in positive terms.

The computer reaches the above conclusions by solving a series of mathematical equations. The example could be formulated as follows:

$1.0C + 1.0S + 1.0X_1 = 100$ acres of land
$0.3C + 0.9S + 1.0X_2 = 80$ hours of April labor
$0.8C + 0.5S + 1.0X_3 = 80$ hours of May labor
$0.7C + 0.4S + 1.0X_4 = 80$ hours of June labor
$1.1C + 0.4S + 1.0X_5 = 80$ hours of October labor
$1.5C + 0.0S + 1.0X_6 = 80$ hours of November labor
$109.62C + 70.58S + 1.0X_7 = 8{,}000$ dollars of operating capital
$104.50C + 87.78S = Z$

The symbol C refers to acres of corn and S to acres of soybeans. X_1 through X_7 are nonuse activities that are added to allow resources to remain idle if that is more profitable. On the right are shown the resource limitations. The corresponding C and S coefficients show the quantity of that resource required to produce one unit (acre) of corn or soybeans. Representing net profit over variable costs, Z is to be maximized.

EXAMPLE 6.10 Selecting Profitable Enterprises to Control Soil Erosion

Table 6.15 reports research by Beaulieu et al. (1998) on developing representative farm planning model for the Big Creek watershed in the Cache River basin of southern Illinois. Besides the economic impacts of conservation practices for individual farms, conservation practices can have beneficial impacts on local water sheds. The Cache ecosystem has been identified as a national and international natural landmark. The researchers evaluated alternative crop enterprises, tillage systems, and participation in the conservation reserve program (CRP). Farm plans for three levels of erosion control are reported: none, 2T, and T. The T value is considered the maximum amount of soil loss to maintain productivity.

Note first that as erosion control is intensified, the conservation tillage row crop acres decreased and no-till row crop acres and CRP acres increased. Gross margins are decreased by $2,000 to achieve 2T, and reduced another $3,000 to achieve T. It would appear to achieve further soil reductions farmers would have to receive incentives to achieve lower soil loss.

TABLE 6.15

Income and Land-Use Effects of Alternative Soil Loss Constraints on Southern Illinois Farms

		Land Allocation			
		Row Crops			
Erosion Control	Annual Gross Margin ($)	Conservation (acres)	No-Till (acres)	Alfalfa (acres)	CRP (acres)
None	90,954	540	0	60	0
2T	88,930	338	166	60	36
T	85,900	286	144	60	110

Source: Beaulieu, Jeffrey, Steven Kraft, and Philip Letting. "Combining farm management and ecosystem management: Insights from representative farm models in a conservation priority watershed." Department of Agribusiness Economics, Southern Illinois University, paper presented at Southern Agricultural Economics Association Annual Meeting, Little Rock, Arkansas, Feb. 1998.

Notes

1. Marginal product (MP) can be tabulated arithmetically by dividing the change in output by the change in input ($\Delta Y \div \Delta X$). MP can be derived algebraically by taking the first derivative of the equation as follows: (Given: $Y = kX^n$, and k and n are constants.)

   $$\frac{dY}{dX} = nkX^{n-1}$$

   Applying this formula to equation $Y_1 = 40 + 2X_1 + 0.5X_1^2 - 0.05X_1^3$ gives:

   $$\frac{dY}{dX} = 2 + X_1 - 0.15X_1^2$$

   This formula prescribes a precise, or point, tabulation for any level of X_1. The difference procedure ($\Delta Y \div \Delta X$) measures the average increase over the units of factor increase. Thus, these two methods of tabulating MP may not give precisely the same value as illustrated in Table 6.1. The precision of the difference procedure is sufficient for most farm decisions and thus will be used in further illustrations.
2. The value of n is less than one ($0 < n < 1$) and measures the elasticity of production. The parameter n is constant over the entire range of production. Elasticity of production will be treated later in this chapter.
3. The optimum level of production can be tabulated mathematically from the marginal product equation shown in note 1. The equation developed was

   $$MP = \frac{dY}{dX} = 2 + X_1 - 0.15X_1^2$$

   If this equation is multiplied by the product price (P_Y), the VMP is tabulated. Setting VMP equal to (P_{X_1}) and solving for X_1 the optimum level of X_1 can be solved for. For $P_Y = \$2$ and $P_X = \$1$ this tabulation follows:

   $$2\,(2 + X_1 - 0.15X_1^2) = 1$$
   $$4 + 2X_1 - 0.30X_1^2 = 1$$

   Setting the equation equal to zero and reordering the terms the formula becomes:

   $$0.30X_1^2 - 2X_1 - 3 = 0$$

   The formula for solving quadratic equations of form $aX^2 + bX + c = 0$ can now be used:

   $$X = \frac{-b \pm \sqrt{b^2 - 4ac}}{2a} = \frac{2 + \sqrt{2^2 - 4(0.30[-3])}}{2(0.30)} = 7.92$$
4. $Y = f(X)$

   $AVC = TVC/TP$

   $TVC = P_X \cdot X$

   Then: $AVC = P_X \cdot X / TP = \dfrac{P_X}{TP/X}$

   But: $TP/X = AP$

   Therefore: $AVC = \dfrac{P_X}{AP}$

   A similar proof can be offered to show

   $MC = \dfrac{P_X}{MP}$
5. Source: Kleinman, A. F., *The Production Function and the Imputation of the Economic Value of Irrigation Water,* Ph.D. dissertation, Iowa State University, 1969.

CHAPTER 7

Developing Marketing Plans

Introduction

The pricing of products and inputs has thus far been taken for granted in preceding chapters. Budgeting procedures assumed that prices of both the output being sold and the inputs being purchased were known. This assumption also was made when applying economic principles to solve the problems of what to produce, how much to produce, and which production technology to use. This practice is dangerous and if followed will surely lead to income reductions and probably business failure. Prices are at least as important as production efficiency in determining farm profitability. Although using up-to-date technology and making sound production decisions are still very important, these alone are not sufficient to ensure success. Effective management of capital and marketing are at least equally important and must be integrated into business plans. The strongest commercial businesses are those with superior marketing skills.

Marketing historically has been ignored in farm management courses. Two major reasons accounted for this oversight. First, marketing is a major subject matter specialty and there are specific course offerings in this area. Second, it has been assumed that farmers were largely price takers and could do little to influence the prices received or paid, aside from those built into the market, such as volume adjustments. Pricing of outputs and inputs can no longer be left out of business planning and not integrated into production plans. Farmers often are able to choose among price alternatives for products sold, and in most situations they can reduce the costs of purchased inputs. The price improvement may not be large for any one item, but accumulatively can make a big difference. Consider for example the effect of a $0.10 per bushel increase in the price received per bushel of corn harvested. If a farmer had 600 acres that yielded 125 bushels, the increase in income would be $7,500. For the farmer who was able to purchase hay for his dairy cows at $5 less per ton, the cost savings would be $6,000 if he or she had 200 cows and 6 tons of hay were fed per head. A $1,000 increase in income from selling at a higher market

FIGURE 7.1

Annual Price Differentials for Hogs, Steers, Corn, and Wheat, 1988–1997

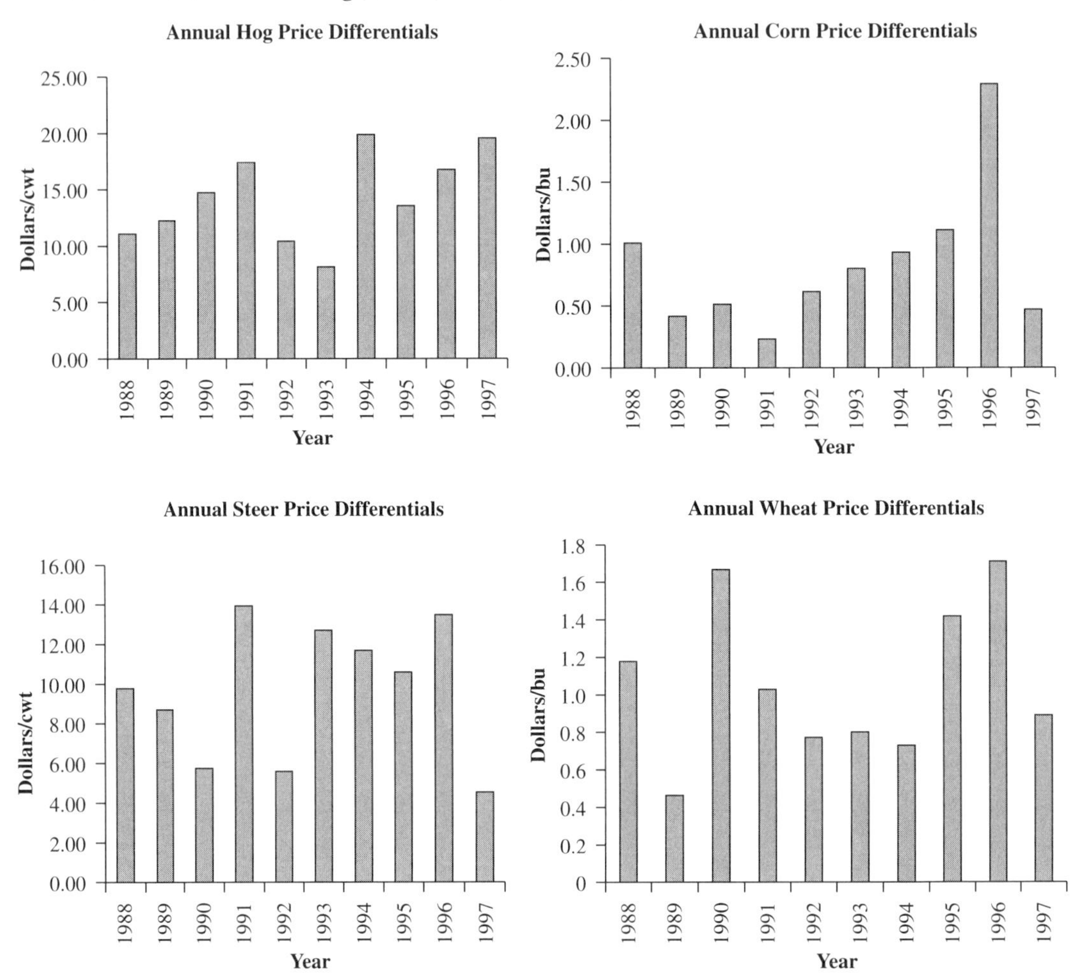

Sources: USDA, *Livestock, Meat and Wool Market News,* for livestock prices. USDA, *Wheat or Feed Grains Situation and Outlook,* for wheat and corn.

price or from buying at a reduced cost has a present value of $12,500, assuming an infinite discount period at an 8 percent return. In this one chapter there is not room, nor is it our intent, to provide a complete treatment of prices and marketing matters. This topic should be selected as a separate course of study. The purpose of this chapter is to call attention to important market considerations when doing farm business planning.

Farmers have long been faced with decisions of which marketing method to use, where to sell, when to sell, and in some cases, what form of product to sell. But timing of product pricing, tax consequences, and the most appropriate marketing strategy for particular situations have taken on new significance. Changing lifestyles, dietary health, aging of the population, and expansion of international trade are examples of important market influences. The payoff for opportune timing of crop and livestock sales can offer a much greater benefit now than in the past. Consider for example the amount of variation in market price for each of four commodities in each of 10 years (Figure 7.1). The average annual range of price differentials (annual high minus annual low) for corn from 1988 to 1997 (10 years) was $0.83 per bushel; for wheat $1.06 per bushel; for hogs $14.36 per hundredweight; and for market steers $9.61 per hundredweight. For a farmer with 600 acres of corn at 125 bushels per acre yield, this annual price differential amounts to more than $62,250; for a dryland farmer with 1,000 acres with a 35-bushel wheat yield, this differential amounts to $37,100; for a swine producer marketing 2,400 market hogs per year (a 160-sow unit), the income differential amounts to $82,113; and for a cattle feeder finishing 1,000 head per year to 1,100 pounds, this differential amounts to $105,710. It cannot be interpreted that a farmer could achieve these kinds of gains by improving his or her marketing, but the magnitude by which income can be improved is significant. It has been estimated that for grains, about two-thirds of production is sold in the bottom one-third of the price range. Brock and Hanson tabulated the difference in corn price received by farmers selling in the upper one-third of all producers in Illinois and those of farmers selling in the lower one-third, between 1979 and 1989. These price differences were then applied to 500 acres of land producing 130 bushels per acre. The accumulated gross income difference over this period was more than $400,000. The same tabulation for soybeans showed an accumulated difference of $300,000, and for Kansas wheat the income difference was $150,000.[1] Consider what a 10 percent improvement in the prices received would do for farm profits. Marketing is important and should be included in every production plan.

Market plans must be integrated with production and financial plans. The long-run production plan, which involves the selection of products and financing investments, must project both input and output prices. Trends in supply and demand must be assessed. Intermediate-run planning requires an evaluation of price cycles, and year-to-year planning considers seasonal variations of prices. Price or market uncertainties also are important variables to consider in selecting a product to produce. Price assessment is at least half of the planning process. This chapter can only outline key components.

Marketing Decisions

What and How Much to Produce?

These topics were discussed in Chapter 6 when considering the economic principles of production. They are introduced here to emphasize the importance of pricing in these decisions. There is no benefit from producing products that cannot be profitably marketed. Finding these markets before production takes place is vital. For many products, markets are well established and no special arrangements are needed, but for some products there are contracts, agreements, bases, and other special arrangements needed. Most vegetables for canning and freezing are produced under contract. Milk and some fruits are produced under state or federal marketing orders and agreements where it is necessary to have a base to sell in that marketing area. Many eggs and most broiler chickens are produced under contract with a processor, and even government programs may be important in what and how much to produce. Corn, barley, sorghum, cotton, and wheat may be eligible for a USDA commodity loan price program. Participation not only assures a bottom price but also is important in financing the production.

For most products the quantity a producer sells has little or no effect on the market price, but this is not true for all products. There are limited markets in which the quantity offered for sale does affect significantly the price received. Examples include specialty products such as organically grown fruits and vegetables, you-pick fruits and vegetables, roadside stands, and branded beef. For these products it is important to assess the market potential and the effect of volume on price.

Where to Sell?

For some products, where to sell may not be a relevant decision because there may be only one or two outlets. Other products may have multiple buyers. A standard rule is to check with at least two buyers and reaffirm the price periodically.

After checking the most relevant markets, add transportation and other delivery costs to determine the highest price at the farm gate. When these procedures are practiced by most of the farmers in an area, prices are kept in line for all producers with no buyer making a large profit at the detriment of their suppliers. This activity is called "arbitrage." Price differentials tend to be transportation differences. Shrink and spoilage are important considerations when marketing. The distance hauled is a factor in the amount of shrink. For livestock fill shrink is normal and hard to control. Tissue shrink results from the animal being off feed and water for extended periods (more than 12 hours) or stressed from sorting, loading, heat, and shipping. Fruits and vegetables may spoil with the heat and time in transit. It may be useful to draw circles of varying distances from your farm and make a list of the outlets or dealers within each of these areas.

There may be fees or marketing charges assessed so it is useful to make this determination when comparing alternatives and establishing the farm-gate price. These fees might be for grading, sorting, packaging, and storage. There may be quantity as well as quality discounts or add-ons. If products are to cross state lines, there may be inspection fees, disease-free certification, and bills of sale required.

The ability of the buyer to pay for the product has become increasingly important. Many producers have been kept waiting weeks, months, and sometimes years for their payment, and some few have not been paid at all. If a purchaser becomes bankrupt the seller may be treated as an unsecured creditor and thus does not have a very high priority in the settlement process. Purchase money held back is similar to a no-interest loan by the seller to the buyer.

How to Sell?

The decision of how to sell can be involved. Most grains and soybeans are sold to local elevators and at terminals for posted cash prices derived from major futures markets. Livestock may be sold at terminals, through local auctions, directly to packers, and to local consumers. Most milk is marketed through farmer cooperatives at prices established by a system of market orders and government price supports. Fresh produce is sold through several channels, including cooperatives, brokers, terminal markets, direct delivery to wholesalers and retailers, and sometimes directly to consumers through roadside stands and you-pick fields. Some farmers use a combination of these methods.

Basic to the question of how to sell is how much of the marketing function should be performed by the producer. The answer relates to financial position, management skills, and personality traits of the producer in addition to the nature of the product for sale and local conditions. There seems to be a lot of innovative skill used by those who are successful in marketing their own product. Even though this may be an option for particular products many producers do not feel they have the time and resources, or the interest, so they use the services of a commission house to perform the selling function. The process by which one firm performs several of the basic functions of producing, processing, and selling the product to the consumer is called vertical integration.

Using the commodity futures market is a possibility for some products (e.g., wheat, feed grains, soybeans, and livestock [cattle, hogs]). Futures market pricing includes hedging and option trading. Trading takes place at selected locations such as the Chicago Board of Trade and Chicago Mercantile Exchange. Placing of bids to buy or sell is by open outcry designed to foster competition. The trading is over contracts called futures, which are legally binding to deliver or take delivery of a given quantity and quality of a commodity at a specified price, during a specified future month, and at a specific location. In reality few deliveries are made. Instead, buyers and sellers "offset" their purchases and sales by appropriate sales and purchases before the delivery date. Hence, the purchase of sale becomes a paper transaction. Nevertheless, the buying and selling gains and losses are real. What usually happens with the farmer who hedges his or her production is that he or she sells the hedged product at the local market and buys a contract near the delivery date to fulfill the commitment to deliver. Thus, if the price has dropped, the gain on the futures trade is used to support the local price. But if price has risen the loss in the futures trade reduces the local cash price, and the farmer receives a price near the original futures value, plus or minus the pricing differential between the board of trade and the local market. This difference, mostly related to transportation costs, is called the "basis." Due to imperfections in the marketing process and economic forces in the market, the basis fluctuates for each locality and among market locations. Understanding these basis differences is important in using the hedge to lock in a price in advance of delivering the product for actual sale. The out-of-pocket cost for hedging is the margin and brokerage fee. To initiate the hedge the farmer must put up a portion of the commodity price as a margin. If it is a sale and the price rises to the level of the margin money, the broker will call for more margin money and if not paid will sell the commodity.

Options are another opportunity in futures trading available to farmers since 1984. They may be used to insure a farmer against the loss in the hedge when the sale is covered with a purchase at a higher price. Options in commodity trading are similar to options to buy in the real estate market. The prospective buyer gives a deposit of cash to insure his or her serious intentions and the property is pulled off the market. If the buyer completes the purchase the deposit becomes part of the down payment. But if the buyer changes his or her mind the deposit is lost to the seller. In commodity trading an option is the right to buy or sell a hedge. Purchasing the right to buy futures is a call option; purchasing the right to sell futures is a put option. For this right the farmer pays a premium to a brokerage firm. By not exercising the option the most the farmer can lose is the premium.

Contracting is a marketing strategy used by farm producers in a variety of ways. For some products, such as cannery and frozen foods, contracting is the only marketing avenue available. Specialty grains such as bioengineered grains are more likely to be contracted. The contractor maintains title to these products in many of these contracts. Therefore, the producer is paid for services rendered rather than the actual crop. Livestock producers have long contracted the delivery of their feeder cattle at a specified price and quality. Local elevators started contracting to receive a farmer's grain crop since about 1970. Today local elevators offer a number of contract options ranging from forward cash contracts to hedge-to-arrive contracts. These contracts offer several pricing and risk-management alternatives to farm managers. Common terms include specification of quantity, quality, grading procedures, timing of delivery,

delivery location, price or formula to determine net price, terms of dispute settlement, and method and form of payment. It may be important to include a provision to cover the situation if a farm producer is not able to deliver. Contracts can lock in a favorable price to the manager, but they can increase the risk exposure as well. A list of alternative grain pricing alternatives and areas of risk exposure is shown in Table 7.1.

What Grade, Quality, or Form?

This is a production decision but also affects the price received. The price for most products that are not market contracted are set to reflect differences in quality and grade. For example, livestock are discounted if too fat or too lean, too heavy or too light, or overaged. Prices may be $3 to $4 per hundredweight higher for choice-grade steers than good-grade steers. Some cattle producers try to move their young cows to market early enough for them to be graded as heifers rather than cows. Grains are discounted if too dirty, moldy, or wet. Apples, oranges, and other fruits and vegetables are sorted for specialized markets, with some being processed and others sold on the fresh market. The profit level is a function of the interrelationships of prices received, marketing costs, and costs of production. The important point is that the market grade and quality standards are important and need to be part of the production plan and marketing agreements.

Form has to do with how the item is packaged. Nitrogen and other fertilizers can be purchased in bulk or bags; in dry, liquid, or gaseous applications; and combined in various formulations. The applications may have purpose in different stages of plant growth. Animal concentrate feeds are sold in bags or bulk and also in various combinations with different supplements. The efficiency of these various forms needs to be weighed against their costs and benefits. The input costs of these different forms may vary considerably with the same results. For example, on a farm in Idaho that one of us is acquainted with, nitrogen fertilization costs on 488 acres of central-pivot irrigated grains were reduced $3,368 by switching from dry application to water application. Animals may be marketed live-weight or on a grade and yield basis, which may affect the price received. These decisions need to be evaluated as to their profit effectiveness.

When to Sell?

Commodities produced under contract are sold when the contract is validated. Milk and eggs are examples of continuous production in which lasting purchasing agreements are made. Grains and livestock are examples of products for which there are several decision points in the production process when sales, or agreements to sell, can be made. Table 7.1 listed several pricing alternatives for grain farmers. Figure 7.2 illustrates some of these choices facing grain producers and Figure 7.3 illustrates the same for some beef cattle producers. Note that in both illustrations there are preproduction, harvest, and postproduction choices that can be made. Many producers use only the cash sale at harvest, whereas there are other possibilities. Also, note that these figures incorporate some of the other marketing choices listed previously. Where there are alternatives the producer should evaluate each of them as to their feasibility, profitability, desirability, uncertainty, and any other related feature. For example, if the crop or livestock product is contracted or hedged before harvest the actual quantity produced may be less than the quantity sold. Some producers use a variety of these methods. Climatic, political, and economic conditions may cause a producer to use a different set of marketing alternatives each year.

TABLE 7.1

Comparison of Grain-Pricing Alternatives with Areas of Risk Exposure

	Areas of Risk Exposure										
Pricing Alternatives	**Price Level**	**Basis**	**Price Spread**	**Cash Flow**	**Options Volatility**	**Counter-Party**	**Tax**	**Control**	**Production**	**Quality**	**Risk Rating**
Preharvest and Harvest Choices											
1. Cash Sales	x	x							x[1]	x[2]	Moderate Large
2. Forward Contracts						x			x[1]	x[2]	Low to Large[3]
3. Short Futures		x		x				x	x[1]	x[2]	Low to Large[3]
4. Buy Put Options	x[4]	x				x		x	x[1]	x[2]	Low
5. Minimum Price Contracts	x[4]					x		x	x[1]	x[2]	Low
6. Nonroll Single or Multiyear H-T-A		x				x		x	x[1]	x[2]	Low to Large[3]
7. Intrayear Rolling H-T-A		x	x					x	x[1]	x[2]	Low to Large[3]
8. Interyear Rolling H-T-A		x	x			x	x	x	x[1]	x[2]	Extreme
Postharvest Choices											
1. Storage	x	x	x								Moderate
2. Sell Grain, Buy Futures	x		x	x			x	x		x[2]	Moderate Large
3. Basis Contract	x	x	x			x		x			Moderate Large
4. Delayed Pricing Contracts	x		x			x		x			Moderate Large
5. Sell Grain, Buy Call Options	x				x		x	x			Low
6. Minimum Price Contracts	x				x	x		x			Low

Source: Adapted from Good, Darrel and Robert Wisner. "Module 7: Risk Exposure and Risk Management with Various Grain Pricing Tools" in *Managing Risks and Profits.* Internet URL: http://www.agecon.ag.ohio-state.edu/mrp/module7.htm. 6 Dec. 1998.

H-T-A hedge to arrive contracts

1. If priced before harvest
2. If stored on the farm
3. If priced before harvest without crop insurance
4. Downside risk limited to premium paid
5. If priced with forward contract rather than cash sale

FIGURE 7.2
Grain Marketing Alternatives for Farmers

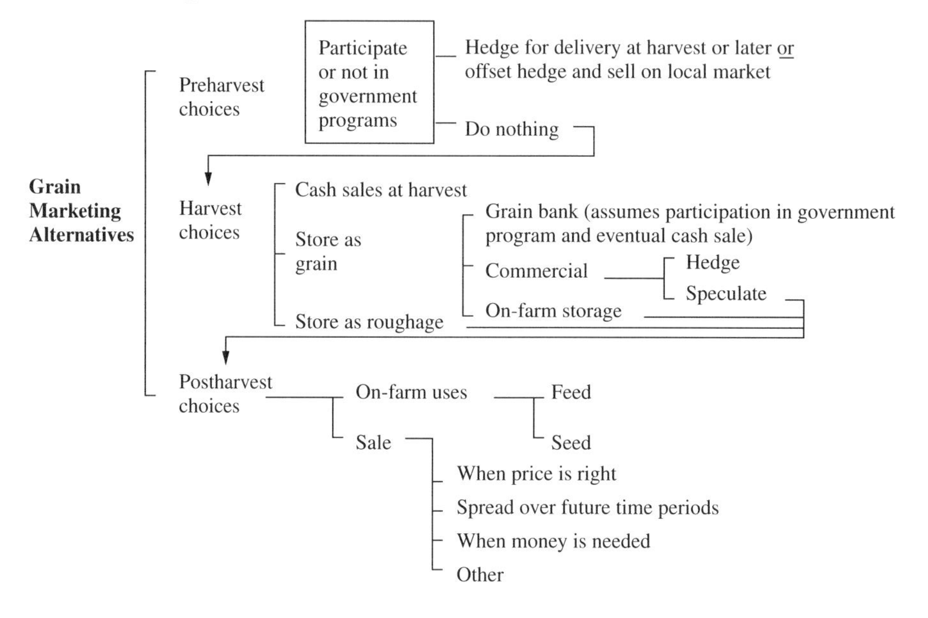

FIGURE 7.3
Livestock Marketing Alternatives for Farmers

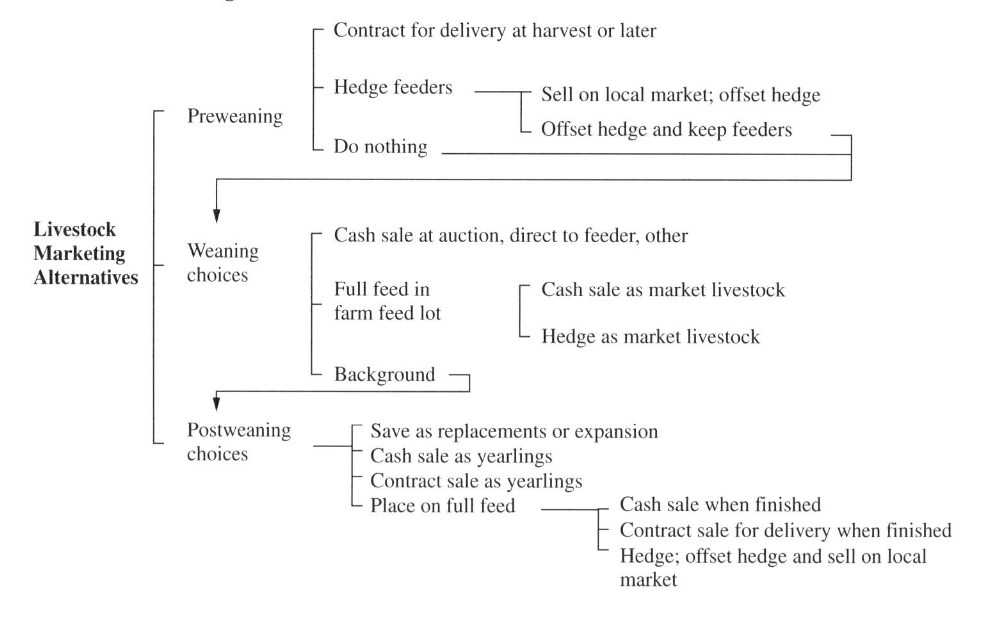

When to Deliver?

When the product is delivered for sale is important in some situations. A livestock marketing example has the owner holding his or her finished cattle or hogs because the price is down and it surely will rise in the next week or so. This situation might happen, but it might not and there are costs of waiting. The animal's cost of gain may exceed the added value even if the price does increase. It is not just the added weight

TABLE 7.2

Cattle Feeder Costs of Marketing at Heavier Weights

Weight of Animal (lb)	Marginal Feed Required (lb)	Marginal Feed Cost[a] ($)	Price of Steers ($/lb)	Marginal Value of Steers ($)
900				0.65
950	320	22.40	0.65	32.50
1,000	345	24.15	0.67	52.50
1,050	375	26.25	0.68	44.00
1,100	410	28.70	0.69	45.00
1,150	450	31.50	0.70	46.00
1,200	495	34.65	0.70	35.00
1,250	545	38.15	0.68	10.00
1,300	600	42.00	0.65	–5.00

[a]Feed cost was tabulated using 7¢ per pound. This is a combined cost of feed plus feeding and yardage.

that must be considered. The market price of the whole animal changes and it is this change in receipt that must be considered against the added cost of holding the animal. Suppose, for example, a cattle feeder has a pen of steers that have reached 1,200 pounds, the anticipated optimal market weight, but the price is now $5 per hundredweight below the anticipated price of $70 (The operator has already passed the optimal marketing weight for this lower price.) (Table 7.2). Should these cattle be held for a higher price or should they be sold because the cost of gain is greater than the market price? The answer is not known but there are at least three questions that should be considered: (1) Will the price really increase?, (2) What is the cost of carrying the steers to a heavier weight?, and (3) What are the market discounts for the heavier weights? Whether the price will increase is related to the seasonal price pattern and will be discussed later. Assuredly, the efficiency of converting feed into meat is decreasing. And probably there will be market-price discounts for the heavier weights. Any price increase must overcome both of these factors. Using the figures in Table 7.2 the market price would need to increase by $2.50 per hundredweight to justify carrying the animals to 1,250 pounds and $6.60 per hundredweight for the 1,300-pound weight.

For grains, beans, and other dried products, the time to sell is related to costs of storage. In addition to the direct costs of providing protection from the elements, including rodents and insects, is the change in weight of the product called shrink. Shrink might amount to from 3 to 5 percent for some grains. Storage costs might be fixed or variable, depending whether the storage is farm-producer owned or contracted for through an elevator. Once the structure is built, depreciation, repair, and taxes become fixed and not related to the content of the structure. These same items may be part of variable cost considerations if hired. There is an interest cost even if funds are not borrowed. If the products were sold at harvest the cash received could be used to pay off contracted debt; used to meet operating expenses, which otherwise would have sustained debt; or profitably invested. As a minimum there is an opportunity cost, which is as real as if an interest expense had actually occurred. These direct and indirect storage costs need to be evaluated when determining whether to sell at harvest or store for a later sale. Income taxes are another important factor in deciding when to sell. These taxes become especially important if taxable income is tabulated by using the cash system. It is possible that 2 year's grain crops could be taxed in the same year with no taxes in another year. This situation could potentially increase or decrease

TABLE 7.3

Farmer Costs of Storing Shelled Corn from Harvest to Varying Delivery Times

	Months Grain Stored			
Cost Item	1	2	3	4
	On-farm storage (cents/bushel sold)			
Storing:				
Annual bin costs	8.0	8.0	8.0	8.0
Extra handling labor and shrink	2.0	2.0	2.0	2.0
Total storing	10.0	10.0	10.0	10.0
Holding:				
Insurance and conditioning	0.1	0.3	0.4	0.5
Interest @ 10% ($2.50/bu)	2.1	6.2	12.5	18.7
Extra shrink (15.5%–13.0%)	?	?	7.2	7.2
Extra drying	?	?	1.7	1.7
Total holding	2.2	6.5	21.8	28.1
Total storing and holding	12.2	16.5	31.8	38.1
Rented Storing and Holding:	Off-farm storage (cents/bushel sold)			
Storage service ($0.10 min. + $0.015/mo)	10.0	10.0	13.7	18.2
Shrink (15.5%–14.0%)	4.4	4.4	4.4	4.4
Interest at 10%	2.1	6.2	12.5	18.7
Total storing and holding	16.5	20.6	30.6	41.3

Source: *Price Forecasting and Sales Management.* Cooperative Extension Service, University of Illinois, Urbana.

average taxes paid. Some grain farmers have found that prices improve during seasons when farmers are too busy to attend to marketing their crops and market supply is down.

Estimated costs for storing grain are illustrated for corn in Table 7.3. The price increase required to break even is significant, from 5 to 7 percent for the first month and it increases by about 1.5 percent for each month thereafter. Storage costs need to be evaluated against the anticipated seasonal price increase and whether the harvest price is unusually depressed. Seasonal price movements are discussed later in this chapter, and risk and uncertainty considerations are covered in the next chapter. Results from storing corn and soybeans in Iowa are shown in Figure 7.4.

Marketing Objectives

Objectives define the desired result. Without these clearly in mind there is no guiding focus. Objectives provide the philosophy in developing a marketing strategy. Developing clear and understandable objectives that are obtainable is not an easy task, but it must be done. It involves the conversion of personal values or quality-of-life concerns into explicit market goals.

Marketing objectives fall primarily into two groups: those that are production oriented and those that are not. Those that are related to production are based upon cost. Alternatively, objectives may be price specific and not related to production.

FIGURE 7.4

Average, Maximum, and Minimum Unhedged On-Farm Storage Returns for Northwestern Iowa Soybeans and Corn, 1979–1996

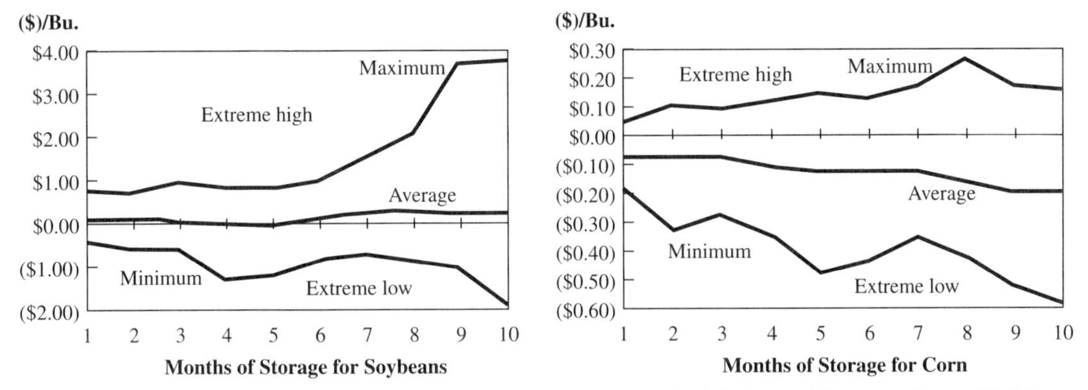

Source: Balwin, Dean and Robert Wisner. "Module 8: Basis, Localizing 1997 Grain Prices, and Returns to Storage." in Managing Risks and Profits. Internet URL: http://www.agecon.ag.ohio-state.edu/mrp/module8.htm. 6 Dec. 1998. Figures 2 and 3.

Price-Specific Objectives

Earlier it was stated that as many as two-thirds of the grain producers sold their crops in the lowest one-third of the annual price range. Price objectives have to do with selling at the highest possible price, selling above the average price, or above the midpoint of the year, selling in the upper one-third of the price range, or some other similarly conceived goal. It cannot be said that these objectives are not concerned with profit levels, only that marketing is considered separately from production decisions. To sell at the highest price assumes a zero cost of storage for most commodities and probably is not attainable. Selling at the highest price may require unseasonably high costs of production. A lesser goal may be easier to achieve. Knowing the highs and lows of the year and their causes and consistency is very important and should be diligently studied. Then farm producers can plan marketing strategies. A major problem with these price-based strategies is that they are highly speculative and fail to signal the proper time to sell.

Profit- or Cost-Based Objectives

Profit-based objectives require that the producer knows what his or her costs are and what the break-even price must be. This knowledge requires that an accounting system is in place to gather variable and fixed costs of production for all enterprises having products for sale. Furthermore, it requires that budgets have been prepared for the products being priced to know the expected costs of production. How can anyone plan to sell at a profit if costs of production are not known? Table 7.4 illustrates projected break-even budgets for hay and grain. To cover operating expenses for hay, our farmer would need $40.37 per ton. To cover all costs, including a 10 percent return on investment, our farmer would need $67.64 per ton. For grain, the farmer would need $2.51 a bushel to cover all costs.

Objectives that read, "price to obtain a fair return" are too vague. What is fair? It is better to be able to measure the results. A profit of 10 percent on the investment, $150 return per $100 feed fed for livestock, or $1.21 per bushel above variable costs

TABLE 7.4

Projected Costs and Returns for a 1,000-Acre Farm Producing Grain and Hay

	Hay-500 Acres 5.5-Ton Yield			Small Grain-500 Acres 120-Bushel Yield		
Item	**Total**	**Per Acre**	**Per Ton**	**Total**	**Per Acre**	**Per Bu**
			Dollars			
Operating Expenses:						
Machine operations	55,000	110	20.00	12,000	24	0.20
Seed, fertilizer, and chemical	18,500	37	6.73	20,500	41	0.34
Irrigation	22,500	45	8.18	15,000	30	0.25
Insurance			0.00	5,000	10	0.08
Harvesting and handling	5,000	10	1.82	18,750	38	0.31
Hired labor	7,500	15	2.73	6,000	12	0.10
Other	2,500	5	0.91	1,000	2	0.02
Break-even O.E. Price			40.37			1.30
Ownership Costs:						
Depreciation	17,500	35	6.36	20,000	40	0.33
Interest (10% on investment)	7,500	15	2.73	5,000	10	0.08
Land	40,000	80	14.55	40,000	80	0.67
Management	10,000	20	3.64	7,500	15	0.13
Break-even O.C. Price			26.27			1.21
Total Costs	186,000	372	67.64	150,750	302	2.51
Projected Sales	192,500	385	70.00	153,000	306	2.55
Return above all Costs	6,500	13	2.36	2,250	4.5	0.045

for grain are examples of measurable objectives. (The $1.21 value is the cost of depreciation, interest, land, and management costs for grain in Table 7.4.) These values give guidance to the person making the decisions of when to sell. When the price reaches a particular level, it is time to act. Knowing the break-even cost of production in all of these objectives is paramount.

Contingency price objectives are necessary in case actual price levels never reach those anticipated, or continually exceed them. Pricing is a process and not an event. It must begin before production is started and be continually evaluated until after the product is sold. One's personal finance might alter these decisions as well. Original plans may have specified that the product should be sold in January but conditions have changed to require an immediate sale.

It may be wise to have different goals for different products and perhaps even for the same product. It may not be the best policy to sell all the product at the break-even price or at the highest price. It may be useful to establish a variety of marketing patterns and practices to reduce income variability that may even enhance the income of the firm.

Establishing Product Prices

The interaction of buyers and sellers establishes a market and sets the price. It is important to distinguish between the market place and the market. A market place is a physical location where the bidding for purchase and sale takes place. The market is the bidding process that establishes the price. A market can occur at any number of locations and even over a computer or telephone network. The market might be fragmented as with livestock auctions and grain elevators and might even change over the season. The bidding for grain may be concentrated in the local elevators at harvest time only to shift to national boards of trade later in the season.

Markets for farm products are considered to be among the freest of all markets. Most products are sold by a large number of producers such that no one of them can individually affect the prices received. This is also true for items purchased. Both buyers and sellers are price takers. In a perfectly competitive system, the bidding is free of any outside pressures and the product is of uniform quality. Furthermore, the buyers and sellers are fully informed about market conditions and there is mobility of goods and services. These conditions do not exist in their entirety, but they exist in sufficient detail to build a framework useful when predicting market prices. It is the competitive market that is described in this section. Its purity could be questioned for most products, but the test is in whether it works.

Demand

Demand is the schedule of quantities that buyers purchase at a specified series of prices, in a given market, at a given time. Demand is not the same as wants. Wants are unrestricted, whereas for demand, quantity is restricted by price or cost, desire for other goods, ability to buy, or income. Demand for the factors of production to farm (e.g., fertilizer, seed, feed, labor, machinery) was introduced in Chapter 6, but it may have not been recognized. Consider the value marginal product (VMP) curve that slopes down from left to right as illustrated in Figure 6.2. As the price of the factor input decreased, the farm producer found it to his or her economic advantage to apply more of that production factor and vice versa. The VMP curve, within stage II of the production function, becomes the farm demand for that production factor. The industry demand function is generated by adding together all of the farm demand functions. The relative prices for the industry would be similar to that of the individual farmer, but the quantities for the industry would be greatly expanded. The demand for farm-produced items is arrived at similarly. If the end use is for the raw product, such as with feed grains and forages where there is not much processing, the farm demand for these products can be measured rather directly. One farmer's demand is another farmer's supply. The same is true of items such as watermelons, squash, and other unprocessed fruits and vegetables produced for human food, except that the demand is not for production but for consumption. The logic of the demand function for human-food consumption is not developed in this text. Suffice it to say that as the price goes up consumers buy less, and as it goes down they buy more. Even though people need certain levels of food items to meet nutritional needs, there is much substituting among products and there is a lot of indifference exercised.

The farm-level demand for most farm-produced items cannot be arrived at directly because of the processing and other marketing functions required. The farm-level demand is a derived function arrived at by subtracting the marketing margin as depicted in Figure 7.5 where D_F is the farm demand schedule and D_R is the retail or nonfarm industry demand. For example, consider wheat for bread making. If bread prices go up relative to other food substitutes consumers buy less bread and switch to other food items. This switch causes bakers to buy less flour and millers to buy less wheat. Only 5 to 10 percent of the cost of bread is in the cost of wheat, even though it is the principal ingredient. A hypothetical demand schedule for wheat is shown in

FIGURE 7.5
Marketing Margins for Farm Products

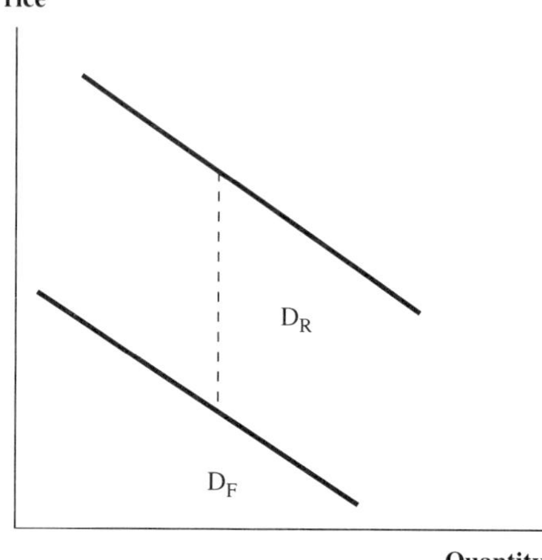

FIGURE 7.6
Domestic Demand for Food Wheat

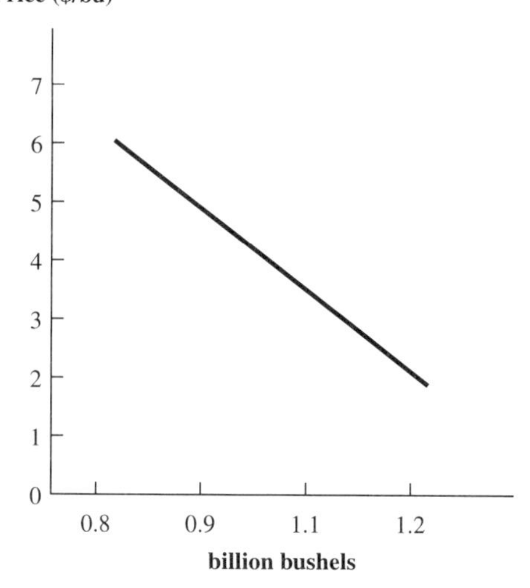

Figure 7.6. The Law of Demand simply states that, "Other things being equal, the higher the price of a good, the lower is the quantity demanded."

The elasticity of demand (E) is an important statistic when forecasting price. It measures the degree with which the quantity changes as price changes and is expressed in percentages. For an inelastic product the percentage change in price is greater than the percentage change in quantity. Interpreted at the farm level, gross farm income will go up in response to a decrease in quantity. Specifically, the formula is as follows:

$$\text{Elasticity (E)} = \frac{\Delta q / q}{\Delta p / p}$$

TABLE 7.5

Retail and Farm Price Elasticities for Selected Foods

	Effect of a 1% Change in Price on the Percentage Change in Quantity Consumed	
Commodity	**Farm Level**	**Retail Level**
Turkey	—	–1.56
Margarine	–0.69	–0.84
Chicken	–0.60	–0.78
Apples	–0.68	–0.72
Butter	–0.46	–0.65
Beef	–0.42	–0.64
Ice cream	—	–0.52
Cheese	—	–0.46
Pork	–0.24	–0.41
Fresh milk	–0.32	–0.34
Eggs	–0.23	–0.31
Potatoes	–0.15	–0.30
Bread, cereals	—	–0.15

Source: George, P. S. and G. S. King, *Consumer Demand for Food Commodities in the United States with Projections for 1980,* University of California, Giannini Foundation Monograph No. 26, 1971.

and can be estimated with

$$\frac{(q_2 - q_1)/(q_1 + q_2)/2}{(p_2 - p_1)/(p_1 + p_2)/2}$$

where p = price and q = quantity. For wheat in Figure 7.6

$$E = \frac{(0.8 - 0.9)/(0.8 + 0.9)/2}{(4 - 3)/(4 + 3)/2} = -0.42$$

Elasticities of 1 (absolute value) are unitary, with those above 1 elastic and those below 1 inelastic. Estimated elasticities for a few farm products are shown in Table 7.5. As can be seen, most agricultural commodities have inelastic demand functions in the relevant price ranges, meaning that prices are very responsive to changes in quantities used. Elasticities differ considerably over the range of farm products and marketing functions. Typically, demand elasticities exhibit the following characteristics:

- Prices are more inelastic at the producer (farm) level than at the processing and retail levels.
- Products with substitutes are more elastic than those with no substitutes.
- Prices for necessities are more price inelastic than for luxuries.
- The demand for high-priced goods is more elastic than for low-priced goods.
- Products with many uses have higher elasticities than products with few uses.
- As the trading territory is broadened the demand elasticity is increased.

Thus, it is important to study all of these market forces when predicting the future price of a product. The cross elasticity of a product is tabulated as follows:

$$\text{Elasticity XY } (E_{xy}) = \frac{\Delta q_x / q_x}{\Delta p_y / p_y}$$

FIGURE 7.7

Supply of Domestic Wheat (Hypothetical Function)

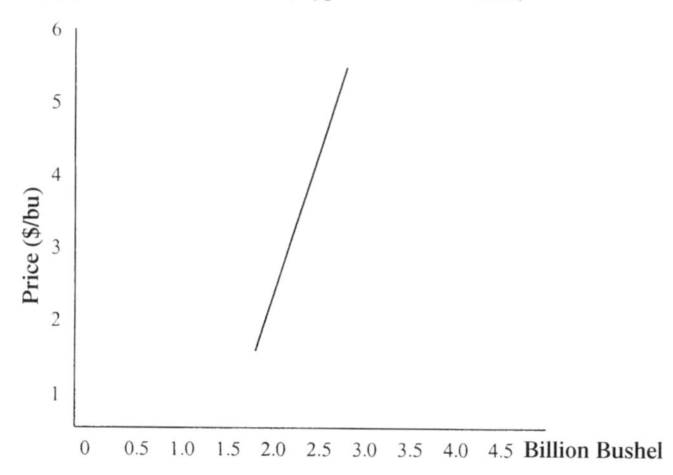

When goods substitute for each other the coefficient is positive. Goods that compliment each other have a negative coefficient.

Supply

Supply is defined to mean the quantity of product suppliers are willing to offer for sale at a specific series of prices. These prices are defined from the costs of production curves as explained in Chapter 6. Consider the cost of production function illustrated in Figure 6.9. Farm producers find their profit-maximizing levels of production where marginal costs (MC) are equal to marginal revenues (MR) (i.e., MR is at or near the price of the product). Thus, the MC curve, above where MC is equal to the average variable cost (AVC), is the farm supply function of the product being produced. This definition is very similar to the one given for demand, except that it is related to sellers rather than buyers, and the slope is in the opposite direction (i.e., increasing from left to right). The farm response to an increase of price is to produce more. A hypothetical domestic supply schedule for wheat is shown in Figure 7.7. Farmers produce more by expanding acres and increasing variable inputs.

Elasticities of Supply

Elasticities of supply are again important in forecasting farm-level prices. The tabulation formula is exactly the same as for demand elasticity. Furthermore, supply elasticities are inelastic for most agricultural products, particularly in the short run (i.e., farmers tend to produce about the same quantity regardless of price). However, in the long run large changes in quantity have been offered for sale in response to price incentives. Consider, for example, the expansion of soybeans, canola, and safflower production during the past 20–30 years. Some supply elasticities are shown in Table 7.6. Note that these elasticities are positive in sign in contrast to the negative elasticities of demand. This change in sign is due to the opposite slopes of their curves.

Price

Price is established in a competitive market at the intersection of the supply-and-demand curves. So long as this market is allowed to operate with no external controls

TABLE 7.6

Estimated Aggregate Supply Elasticity of Farm Output, with Crop and Livestock Components

	Elasticities	
Components	**Short Run (2 years)**	**Long Run (Many years)**
Crops	0.17	1.56
Acreage	0.04	0.10
Yield per acre	0.15	1.50
Livestock	0.38	2.90
Animal units (stock)	0.12	1.80
Yield per animal unit	0.26	1.10
Aggregate supply elasticity	0.25	1.79

Source: Tweeten, Luther G. and C. Leroy Quance, "Positivistic Measures of Aggregate Supply Elasticities," *American Journal of Agricultural Economics*, 51 (May 1969):341–352.

FIGURE 7.8

Demand and Supply for Domestic Wheat (Hypothetical Function)

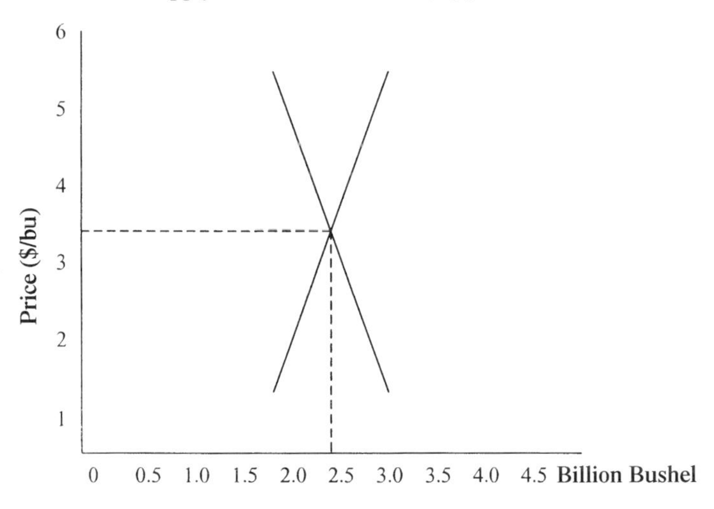

and no regulations on trading, the market price will adjust so as to clear the market of goods by equating the quantity demanded with the quantity supplied. Price discovery is illustrated in Figure 7.8 for domestic wheat where the supply-and-demand curves intersect. Where both of these curves are inelastic, as with most agricultural products, the price is very volatile. Another factor of importance in price forecasting is consumer income. Understanding how consumers react in the market place to increases or decreases in their income is important. Some goods are considered to be superior, meaning that consumers will increase their purchases of them as their incomes increase, whereas the sale of other goods actually decreases as incomes increase, and thus these goods are labeled inferior. Elasticities are again tabulated as a partial means of accessing this condition. For most food items, income elasticities are again inelastic, meaning that the quantity purchased changes less in percentage terms than does income. Some income elasticities are shown in Table 7.7. Meats are much more responsive to increased incomes than potatoes.

TABLE 7.7

Estimated Income and Price Elasticities of Demand, and the United States per Capital Consumption of Selected Food Commodities

	Estimated Elasticities of Demand at Retail	
Food Commodity	**Income**	**Price**
Beef	0.290	–0.644
Pork	0.133	–0.413
Chicken	0.178	–0.777
Fish	0.004	–0.230
Eggs	0.055	–0.318
Cheese	0.249	–0.460
Butter	0.318	–0.652
Margarine	0.000	–0.847
Apples, fresh	0.140	–0.720
Potatoes	0.117	–0.309
Wheat flour	0.083	–0.300
Rice, milled	0.055	–0.320
Sugar, refined	0.032	–0.242
All foods	0.176	–0.237
Nonfoods	1.243	na

Source: Halcrow, Harold G. *Agricultural Policy Analysis,* p. 70, McGraw-Hill, 1984.

Understanding the supply-and-demand relationships of the products you produce are important for predicting future prices, or even interpreting outlook materials developed by others. One of the useful tools used in forecasting prices is to tabulate national and international changes in supply. The procedure for doing this is illustrated in Table 7.8 for soybeans and wheat. The total supply is a sum of the beginning inventory, new production, and imports. The total uses are exports and domestic uses for food, feed, seed, and manufacture. The difference is ending inventory. The percentage changes from one period to the next can be interpreted by the elasticities previously discussed to predict changes in price. Rules-of-thumb have been developed from historical experience, which are useful first estimators. Following are a few of these rules:

Corn: A 1 percent increase in supply will reduce price 2–3 percent.
Soybeans: A 1 percent increase in supply will reduce price by 2–2.5 percent.
Hogs: A 1 percent increase in supply will reduce price 1.7–2.3 percent.
Beef: A 1 percent increase in supply will reduce price 1.6–2 percent.

Forecasting of prices seems to be a combination of experienced judgement, rules-of-thumb, simple guideline formulations, and fitting complicated formulas. Whether the art of forecasting is stronger than science is not known. Science deals with elements that are quantifiable, whereas art deals with hunches and the study of governments and politics. Rules-of-thumb have their foundations in scientific measurements. A not-so-complicated forecasting model of hogs and cattle is illustrated in Table 7.9. Note that several important variables are used in predicting prices. The coefficients in the formula are derived from the price elasticities previously discussed.

TABLE 7.8

Wheat and Soybean Supply and Utilization, 1997–1998 Crop Year

	Wheat			Soybeans	
	Million Bushels	Share of Total Supply (%)		Million Bushels	Share of Total Supply (%)
Beginning stocks	444	14.5	Beginning stocks	200	14.8
Production	2,527	82.6	Production	2,769	93.1
Imports	90	2.9	Imports	5	0.1
Supply, total	3,060	100.0	Supply, total	2,975	100.0
Domestic			Domestic		
Food	915	29.9	Crush	1,600	53.8
Seed	96	3.1	Seed, feed		
Feed and residual	300	9.8	and residual	150	5.0
Domestic, total	1,311	42.8	Domestic, total	1,750	58.8
Exports	1,075	35.1	Exports	830	27.9
Total disappearance	2,386	78.0	Total disappearance	2,580	86.7
Ending stocks	674	22.0	Ending stocks	395	13.3

Source: USDA–ERS, *Wheat Situation and Outlook Yearbook* and *Oil Crops Situation and Outlook Yearbook.*

Predictable Price Patterns

Markct planning involves predicting future prices. There are three predictable price movements to study: trends, cycles, and seasons. For each of these movements there are random elements that sometimes completely overpower the known influences. Random variability elements include the weather; biological diseases and insects; actions of governments; and scientific discoveries that markedly change yields, costs, and marketing. Some variable elements, such as weather, can be insured against. That which is a random element to the individual farmer may be a probability statistic to an insurance company. Insurance and other forms of protection from risk and uncertainty are covered in the next chapter. Trends, cycles, and seasonal patterns of price movement are related to the predictable nature of humans and to the environment. Predictable patterns of price movement are not of equal importance to all products

Trends

Trends are long-run predictable price movements that may be increasing, decreasing, or constant. Trends for a few products and items important in pricing are shown in Figure 7.9. The items illustrated and discussed below are the same ones that influence supply and demand as previously discussed. A few items that cause price trends also are considered.

Population

Population is important nationally and internationally for major world-trade items. The more mouths there are to feed, the more food it takes. In the face of a fixed supply

TABLE 7.9

Price Forecasting Worksheets for Hogs and Cattle

Hog Price					
1. Per capita pork supply					
% change from year ago × price effect:	–8.0	×	–2.00	=	+16.00
2. Per capita beef supply					
% change from year ago × price effect:	+3.0	×	–0.50	=	–1.50
3. Per capita chicken supply					
% change from year ago × price effect:	+2.0	×	–0.50	=	–1.00
4. Per capita disposable income					
% change from year ago × price effect:	+5.0	×	+4.0	=	+2.00
5. Population					
% change from year ago × price effect:	+0.95	×	+1.0	=	+0.95
6. Net percent effect on hog price (sum of 1 through 5)					+16.45
7. Hog price forecast				=	$41.00
price in year-earlier period					
price change = price in year-earlier period × estimated net percent price effect, or	$41.00	×	0.1645	=	+6.74
price forecast = $41.00 + $6.74					$47.74

Cattle Price					
1. Per capita beef supply					
% change from year ago × price effect:	–3.0	×	–1.40	=	+4.20
2. Per capita pork supply					
% change from year ago × price effect:	–6.0	×	–0.40	=	–2.40
3. Per capita chicken supply					
% change from year ago × price effect:	+3.0	×	–0.20	=	–0.60
4. Per capita disposable income					
% change from year ago × price effect:	+5.0	×	+3.0	=	+1.50
5. Population					
% change from year ago × price effect:	+0.95	×	+1.0	=	+0.95
6. Net percent effect on hog price (sum of 1 through 5)					+8.45
7. Cattle price forecast					
price in year-earlier period				=	$59.00
price change = price in year-earlier period × estimated net percent price effect, or	$59.00	×	0.0945	=	+4.99
price forecast = $59.00 + $4.99				=	$63.99

Source: Futrell, Gene A. and Robert N. Wisner, *Marketing for Farmers,* Doanes, 1987.

FIGURE 7.9

Trends for a Few Items Important in Pricing

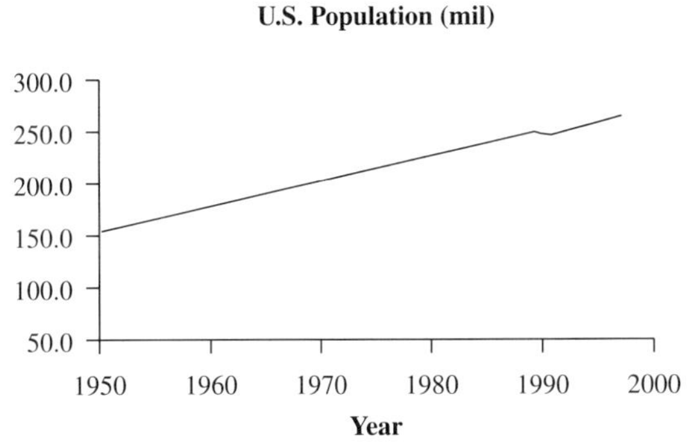

Source: Bureau of the Census, *Current Population Reports.*

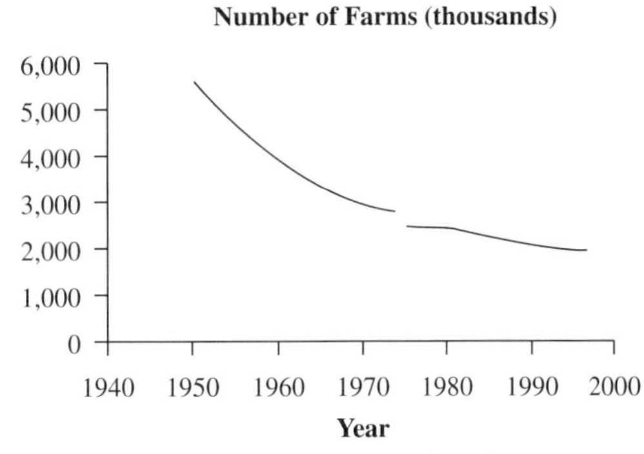

Source: USDA, *Number of Farms and Land in Farms.*

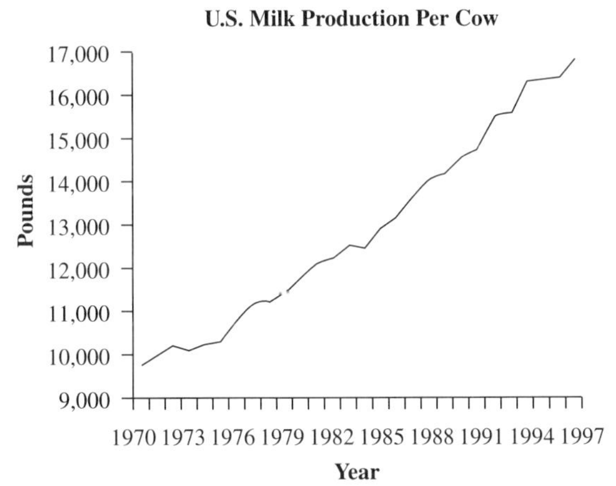

Source: USDA–NASS, *Milk Production.*

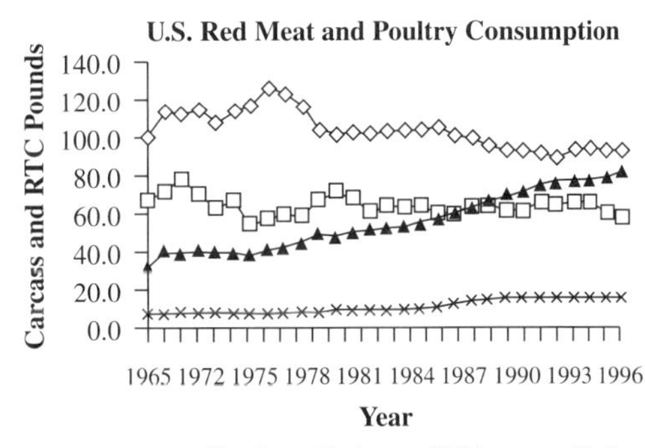

Source: USDA–ERS, Various publications.

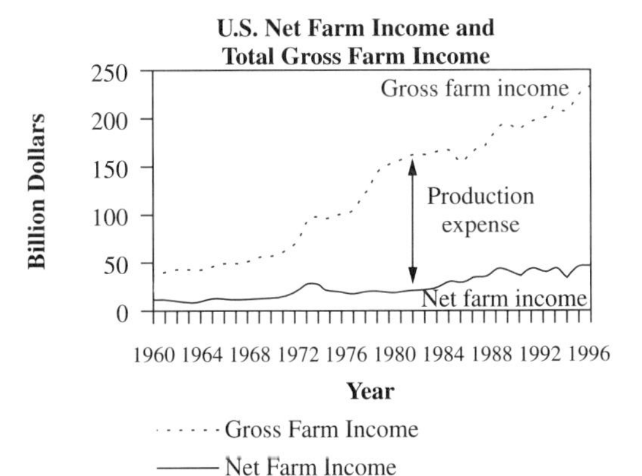

Source: USDA–ERS, *Economic Indicators of the Farm Sector.*

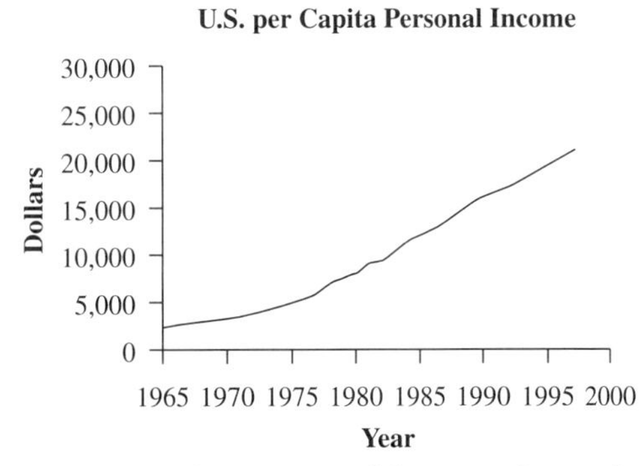

Source: U.S. Department of Commerce, *Survey of Current Business.*

schedule, this increase in demand would cause an increase of price and profit. The demographics of this population are also important. Different age levels do not eat the same products, neither do different ethnic groups.

Disposable Income

Disposable income affects the demand for goods but not in all the same ways as discussed previously. Some goods are inferior and as incomes increase their demands decrease, whereas for superior goods people buy more of them with their increased incomes. It is important to account for inflation when evaluating income effects. An increasing income may be none other than an increase in the general price level. Thus, it is better to consider inflation-adjusted or constant-dollar incomes that reflect the increase in buying power.

Tastes and Preferences

Tastes and preferences relate to the utility that people experience from a good or service. These change over time and are related to the population demographics previously discussed. Advertisements, travel, health discoveries, fads, technology, and a variety of other things affect how much of an item we buy.

Competition and World Trade

Competition and world trade have markedly affected the demand for some farm products, and all products have been affected directly or indirectly. Consider, for example, the import–export changes that have taken place in the past 20 years. Increased competition from South American countries compete with the United States in producing grains, soybeans, and fruits. Russia and the newly formed countries of the former Soviet Union are potential new markets. China, Korea, Japan, and other Asian countries have been important export markets. The signing of the North American Free Trade Agreement (NAFTA) and continuing negotiations on free-trade agreements could increase greatly the exchange of agricultural products. As incomes of underdeveloped countries increase, so also do their imports of agricultural products.

Input Costs

Input costs affect the supply of agricultural products. If costs increase producers may change to a different product or go out of business. If costs decrease producers may increase the production of the affected product. Minimum wage legislation is an example of a direct price increase. Environmental Protection Agency standards and requirements may cause an indirect cost increase. Changes in technology, however, can reduce producers' costs. Note changes in milk production per cow in Figure 7.9 as an example.

Cycles

Price cycles are peculiar to livestock and crops such as tree fruits that require several years before a decision to increase production can actually take place. Cycles are repeating price movements covering a time period longer that 1 year. For some products, price cycles are not well defined. Cattle, hog, and sheep production and price cycles are more apparent. A livestock producer decides to increase his or her

FIGURE 7.10

Total Cattle on U.S. Farms by Cycles

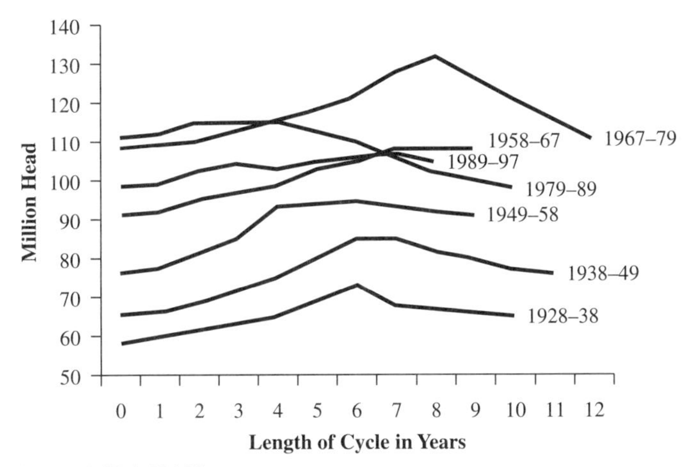

Source: USDA–NASS.

production because of current and anticipated profits. Female animals are held from sale to increase the breeding herd. This scenario reduces the supply of slaughter animals and the price goes even higher. Farmers not producing livestock decide to do so because of the anticipated profit. This ratchet effect continues until the expanded breeding stock furnishes offspring to the market and the increased supply begins to drive the price down. Producers respond to this reduced profitability by saving back fewer females, and they may even cull their breeding herd more severely. They do this not only because of lower profits but also to meet contractual principal and interest payments on loans to finance the expansion. Cycles for beef are shown in Figure 7.10.

It can be argued that producers of beef, hogs, and sheep ought to produce countercyclically. This may be so and should be considered as a means of receiving a higher average price. But from a practical and financial standpoint, it may not be possible. A producer should be very careful not to be misled into making large investments on the upswing with plans for payback that span the down side of the cycle. The number of years in the build-up period is longer than in the selling-off phase as can be seen in Figure 7.10. Droughts and other nonfarm business influences often shorten or lengthen a cycle so they are not always of the same length. Some of these influences are random elements and must be dealt with in a different way.

Seasonal Price Movements

Many prices vary in a predictable manner within the year because of seasonal differences in supply and demand. Most field crops can be harvested in only one season of the year, and there are storage costs for extending the consumption into other months. Many livestock can be produced throughout the year, but the costs of production are much higher in some months than others. There are costs of tempering the elements to allow for an expanded production period. Examples include confinement systems

FIGURE 7.11

Prices Received for Choice Steers in Interior Iowa and Southern Minnesota and Utility Cows in Sioux Falls

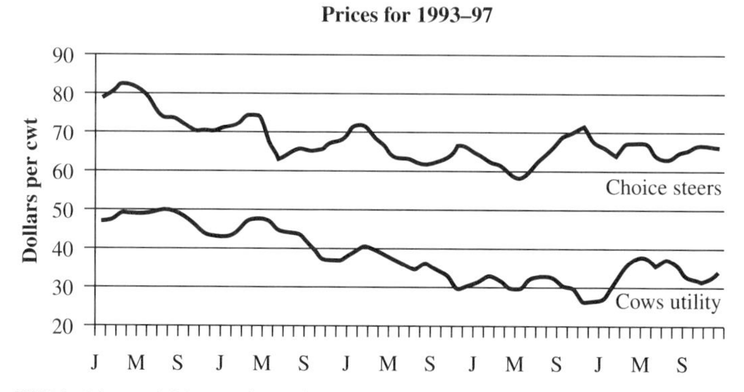

USDA, *Livestock Meat and Wool News.*

of hog farrowing and cattle feeding. Each producer must evaluate the higher prices received with the increased costs of production to determine his or her most profitable alternative. Even if there were not seasonal differences in costs there may be seasonal patterns of demand. Ice cream is not as popular in the winter as in summer, and hot cereal and chili beans are not as popular in the summer. Seasonal patterns of price for cattle are shown in Figure 7.11. Annual irregularities of seasonal price sometimes are removed by tabulating a seasonal index around a seasonal average price of 100. Variability lines of standard deviation above and below the seasonal price line are then tabulated. Historical prices fall within these lines about two-thirds of the time.

These seasonal price influences are generally present even though a seasonal pattern cannot always be detected. Humankind's ability to control nature and expanding imports have tended to reduce the width of the seasonal pattern. The interference of unpredictable elements reduces the significance of seasonal prices. Nevertheless seasonal price movements are real and should be studied in selecting when production should take place and the most profitable time to market.

It should be realized that seasonal price movements are in addition to those patterns already discussed. The seasonal patterns of price overlay any cyclical effect or trend effect. These too may mask some of the seasonal influence and render it less observable.

Seasonal patterns of price can be shown as increases or decreases in price from one yearly time period to the next as illustrated in Tables 7.10 for hogs and Table 7.11 for choice steers.

Developing Marketing Plans

The questions, "Where am I?, Where do I want to go?, and How do I get there from here?" have been alluded to before in this chapter and in previous ones. It is appropriate to raise them again here. There needs to be a market assessment as well as a production assessment and a financial assessment. The financial assessment is contained

TABLE 7.10

Two-Week Change in Hog Price, 1988–1997 Iowa–Southern Minnesota Barrow and Gilts

Week	Average Change (%)	Years Up	Percent Up	Years Down	Percent Down
1	–1.70	3	3.00	7	–3.70
3	3.40	7	5.60	3	–1.60
5	3.90	9	4.50	1	–1.90
7	0.60	6	2.70	4	–2.70
9	–1.60	3	2.90	7	–3.60
11	–0.20	5	3.30	5	–3.60
13	–1.20	3	2.30	7	–2.60
15	0.50	5	2.70	5	–1.70
17	2.00	6	3.90	4	–0.90
19	4.90	10	4.90	0	—
21	6.10	9	6.80	1	–0.30
23	–1.50	3	1.40	7	–2.70
25	1.60	7	4.60	3	–5.70
27	0.40	5	3.50	5	–2.70
29	–0.50	6	1.90	4	–4.00
31	0.00	6	1.40	4	–2.20
33	–0.90	5	2.30	5	–4.00
35	–2.90	4	1.20	6	–5.60
37	–4.70	1	1.50	9	–5.40
39	0.20	4	5.50	6	–3.40
41	0.30	5	2.40	5	–1.80
43	–5.20	1	0.70	9	–5.80
45	–5.20	1	1.10	9	–5.90
47	0.10	5	2.90	5	–2.80
49	2.00	6	4.70	4	–1.90
51	1.80	6	6.40	4	–5.10

Source: John Lawrence. "Iowa Farm Outlook," Iowa State University Extension, Economic Information 1741.

in the balance sheet and the income statement. These financial records need to include a financial history of the performance of the different enterprises and activities. How else can the farm producer know which activities have been profitable and which to expand or contract? They serve as a backdrop for developing future marketing plans. The question of where to go is synonymous with what to produce. The physical evaluation of this question was discussed in Chapter 6 under the title of "What to produce?" Left out of the answer was the level of price to use when making this evaluation. It is clear now that this is a long-run question involving the trend price. The intermediate price, which includes cycles, may have some relevance, but seasonal price patterns probably are too much in the short run. It is generally inappropriate to use current prices when making long-run decisions. The procedure for making these long-range decisions then are as follows:

- Set personal and family goals for the business.
- Follow the economic guidelines for selecting the products to produce. (See Chapter 6.)

TABLE 7.11

Two-Week Change in Fed Cattle Prices, 1988–1997 Iowa–Southern Minnesota Choice Steers 1100-1300 Pounds

Week	Average Change (%)	Years Up	Percent Up	Years Down	Percent Down
1	0.90	7	1.90	3	–1.30
3	1.00	6	2.50	4	–1.40
5	–0.30	6	1.10	4	–2.50
7	1.00	6	2.30	4	–0.90
9	1.10	5	3.10	5	–1.00
11	1.30	7	2.30	3	–0.90
13	–0.20	6	1.40	4	–2.60
15	–0.40	4	1.50	6	–1.60
17	–1.90	2	1.20	8	–2.70
19	0.30	7	2.20	3	–4.20
21	–2.70	1	1.30	9	–3.20
23	–0.80	4	1.10	6	–2.10
25	–1.80	3	1.90	7	–3.40
27	–1.10	2	1.60	8	–1.80
29	0.90	5	3.90	5	–2.00
31	1.00	8	1.80	2	–2.20
33	–0.30	5	2.10	5	–2.60
35	–0.50	3	1.80	7	–1.50
37	0.40	6	1.90	4	–1.80
39	0.50	6	1.80	4	–1.30
41	–0.50	5	0.90	5	–1.90
43	1.90	8	2.50	2	–0.10
45	0.50	6	1.40	4	–0.90
47	1.60	8	2.20	2	–0.50
49	–0.90	2	1.00	8	–1.40
51	0.00	6	1.70	4	–2.50

Source: John Lawrence. "Iowa Farm Outlook," Iowa State University Extension, Economic Information 1741.

- Study trend prices to determine the appropriate level to use in evaluating the products under study.
- Determine if qualified markets exist and their probabilities of continuance.
- Consider how price cycles and seasonal fluctuations will affect the flows of income.
- Study the uncertainty aspects of the products selected and their respective prices. (See Chapter 8.)
- Make the hard decisions and test their feasibilities by using budgets.

The balance of this consideration has to do with shorter-run decisions.

Once the decisions have been made relative to what to produce and the overall business plan, there is a need to develop annual marketing plans. But first a word about cycles. Cycles are of an intermediate nature. With livestock there might be benefits to produce and market countercyclically. This should be studied to determine if it is physically possible and financially feasible. With increasing capital investments in production facilities and equipment, the operating-cost portion is a smaller share of the total cost of production. Thus, there is a greater justification for continuing to produce even when prices are down. This decision can only be made by considering the total length of the cycle and the flow of expected costs and receipts.

The steps in developing an annual marketing plan are as follows:

- Make a list of the alternative marketing strategies for each product.
- Consider the marketing objectives previously discussed.
- Budget the break-even costs of production for each product.
- List the marketing alternatives available for each product.
- Project market prices for each product for each season of the year over which marketing decisions are relevant.
- Determine if a profit can be made by locking in a price through contracting a sale or hedging.
- Consider controlling risk and uncertainty by using more than one marketing strategy for each product and for all products combined.
- Fit the feasible opportunities to the objectives and write them down.
- Do it.

Example of Marketing and Risk Management Planning

Marketing plans to be effective must be integrated with the production and financial plans of the farm business. Table 7.1 illustrated a number of grain-marketing strategies. Table 7.1 also indicated that these alternative strategies can expose the farm to different sources of risk. The Ohio State University has developed a computer software program called *AgRisk* to evaluate alternative crop, crop insurance, and marketing strategies for farms in the North Central region. Farm managers can select from a number of marketing and crop insurance plans by using data for their own farms or by using averages from their county. They also can enter current market data or download current market information. AgRisk then estimates gross revenue and risk levels for each strategy. An illustration of AgRisk is presented in Figure 7.12. The information from AgRisk can then be used to evaluate the effects of alternative strategies on the financial position of the farm business. To this point, the consequences of risk have not been discussed. Chapter 8 examines decision making in an environment of risk and uncertainty.

FIGURE 7.12

AgRisk, a Marketing and Risk Management Decision Tool

AgRisk Example

This example is for a Northwest Ohio farm producing 1,000 acres of corn. Gross revenues are calculated using market prices on May 1, 1997, for two strategies:

Strategy 1—enter no marketing and insurance contracts.

Strategy 2—sell 100,000 bu of December corn futures contracts.

AgRisk shows gross revenues in the "view results" section. Strategy 1 has the following gross revenue:

		--------------Value At Risk--------------		
	Average	5%	10%	25%
Strategy 1	$390,311	$209,330	$242,760	$302,647

Strategy 1's average gross revenue is $390,311. A strategy with higher average gross revenue is likely to be more profitable than a strategy with lower average gross revenue.

The 5, 10, and 25 percent value at risks measure downside revenue potential. The 5 percent value at risk measure of $209,330 means that 5% of the time gross revenue will be below $209,330. Gross revenue will be below the 5% value at risk 1 in 20 years. Ten percent corresponds to a 1-in-10-year chance and 25% to a 1-in-4-year chance. A higher value at risk means lower risk.

Strategy 1 and 2 are compared below.

		--------------Value At Risk--------------		
	Average	5%	10%	25%
Strategy 1	$390,311	$209,330	$242,760	$302,647
Strategy 2	$387,810	$237,452	$264,574	$313,558

Strategy 1's average gross revenue of $390,311 is higher than strategy 2's average of $387,810. In an average year strategy 1 will be more profitable than strategy 2.

However, strategy 1 has more downside revenue potential than strategy 2. Five percent of the time, revenue will be below $209,330 under strategy 1 (see the 5% value at risk). Under strategy 2, revenue will be below $237,452 5% of the time, a $28,122 increase over strategy 1.

Source: Schnitkey, Gary, Mario Miranda, and Scott Irwin. AgRisk, Beta 1.0 software, The Ohio State University. 1998.

Note

1. Brock, Richard A., and James C. Hanson, *Greater Profits through Better Marketing,* Century Communications, Inc., 1990.

CHAPTER 8

Adjusting for Risk and Uncertainty

Introduction

Risk and uncertainty are facts of life for farm managers. Besides the annual production and market-price variability, agriculture has been characterized by boom-and-bust cycles. As agriculture enters the 21st century, will it be a boom or a bust? Agriculture started the 20th century coming out of the worst depression of the 19th century. A period of prosperity lasted to the end of World War I. From 1913 to 1920, farmland values increased by 70 percent for the country.[1] Then in 1920, farm prices began to fall. Farm prices declined by almost half within a year. Many farmers who borrowed to finance farmland purchases went bankrupt as their farms became insolvent. Hard times continued through the dust bowl years and the depression of the 1930s. Improvements came as export markets started to pickup with the onset of World War II and continued with a postwar boom. The 1950s and 1960s were characterized by stable prices and chronic excess agricultural capacity. In the early 1970s, led by increases in exports, agriculture experienced a boom, followed by the bust of the 1980s. A financial crisis of major proportions devastated a large proportion of farmers; by 1984, more than 50 percent of them did not have a positive cash flow.[2] Since the 1980s, farm financial conditions have improved. The years 1996 and 1997 saw records for net farm income, but 1998 was a different story. Hog prices fell as low as $10 per hundredweight. Corn and soybean prices were down by more than 20 percent from the previous years. Projections from Minnesota are estimating the first negative average accrual net farm income in the 59-year history of the Southwestern Minnesota Farm Business Management Association.[3] Market analysts now debate whether the current situation is short term or long term, but the issue for farm managers is how to manage risk and uncertainty.

Individual farmers may be able to reduce the variations in yields brought on by uncertain weather, disease, insects, and weeds; however, they can do little about market forces. But, as pointed out in Chapter 7, farmers can modify the prices they receive. Likewise, farmers can outlast a financial storm with good money management. Farming is a game of survival. The objective is to manage risks and not necessarily avoid them. Indeed, profit is a return for risk management. In a world of certainty a large master plan could be developed to schedule production, marketing, and finance so that all factors of production would receive a market return, but there would be no profit. Risk is not a negative concept. If properly managed, it may be a source of profit.

There is no one strategy for coping with risk and uncertainty, but rather a bag full of tricks that may work in individual situations. These strategies improve the chance for success and reduce the impact of failure. There are few guarantees, or even a best strategy. One who markets a "sure thing" probably should be avoided.

Economists and others studying risk and uncertainty often categorize events as being one or the other. Risk is not to have perfect knowledge, but rather events are considered to be probable occurrences. Risk may be a priori or statistical. Examples of probable occurrences include flipping coins or rolling dice. Statistical risk is based upon information gathered from past events. An egg producer may be able to predict with considerable accuracy the percentage of breakage on a daily basis from his or her records. Mortality tables are developed from death data as a means of setting life insurance premiums. However, uncertainty has no means for such probability distributions. It seems that no one is able to predict political happenings and world events. What may be a statistical probability to a nation, state, county, or large business could be uncertainty to an individual (e.g., the event of accident, sickness, or death may be uncertain to the business, but a probable happening to an insurance company).

Farm producers tend to think of risk and uncertainty as synonyms, even though there may be technical differences between them. Because the strategies may be the same for dealing with risk and uncertainty at the producer level, this text does not distinguish between the two terms.

Individual farm decision makers may find themselves in one of several knowledge situations as follows:

Subjective certainty—sufficiently confident in the completeness and accuracy of his or her knowledge to proceed with the decision.

Subjective risk—regards the decision sufficiently important to acquire additional knowledge so a decision can be made. The person is willing to decide and act when the chances for success seem adequate.

Forced action—does not have sufficient information to act willingly but the combination of circumstances is such that a decision must be made now. This condition is the most uncertain.

At the subjective risk level, farm decisions are made on the basis of subjective or personal probabilities. These probabilities are based upon the degree of belief or the strength of conviction that an individual has about the outcome of a particular decision. Personal probabilities are based upon many factors, including information about past events and outlook information. The total experiences of managers help formulate expectations. The final decision depends not only on the manager's personal probabilities but also on the manager's attitude toward risk and his or her ability to take risk. Some managers are willing or able to take greater chances than others.

The consequences of each decision are important to consider. A misreading of the states of nature, resulting in incorrect probability assignments, or a change of nature may cause a misreading of possible outcomes and a poor decision. Those who are forced into making a decision before they are ready to fully consider alternatives are in a particularly vulnerable situation. Thus, it is extremely important to do a thorough job of researching the alternatives with their associate probabilities and assess the possible gains and losses from each decision. A misjudgment on an important decision could cause the loss of the total business. The classifications or decisions introduced in Chapter 1 as to their importance, frequency, imminence, revocability, and available alternatives may be useful in this regard. The discussions that follow are meant to help a manager be a better decision maker by avoiding unnecessary and unwanted losses while receiving an acceptable return for his or her investments, labor, and management.

Sources of Risk

The causal factors of production, marketing, and financial risk were alluded to previously, but these factors are not the only sources of risk. Because the cure must fit the illness it is helpful to identify the source of the problem. The most important areas causing risk in farming are listed below.

Production variability—due to weather, disease, insects, mechanical failure, genetic improvement, and the art of farming are among the important explanations for the extremely variable yields sometimes observed. Even though some products exhibit more variance than others few are exempt from such forces. Not only are yields affected but also product quality.

Mechanical failure—has become an increasing risk factor as the share of inputs furnished by machines has increased. For example, the breakdown of a tractor or combine can be very costly. A 2- or 3-day delay in planting or harvesting can reduce yield several percentage points. A 2 percent reduction of corn yield can affect profits by more than 10 percent.

Prices—particularly those received, are a major source of income variability discussed and illustrated in Chapter 7. Supply and demand cause most of these changes, but imperfections in the market accentuate an already volatile situation. Although some price changes follow predictable patterns others do not. The expanded worldwide market for many products acts both to reduce price predictability and to level out the supply. Price making is no longer a national matter for the factors purchased or the products marketed.

Financing of operations and investments—are major concerns of most farmers. Few can operate without debt financing. Financing not only includes the borrowing of money but also the leasing of land and equipment. Financing could even include the merging of firms to join equity pools for its owners. Financial problems and their solutions are discussed in Chapters 9 and 10. As businesses have increased in size and more of the inputs are purchased commercially, the need for outside financing has increased dramatically. This increase has added to financing failures.

Technological change—most markedly observed during the last half of the 20th century, has completely changed the manner of producing and marketing many farm products. Adopting new methods and procedures may be very expensive, and machines that are not worn out are left sitting idle. Those who adopt new technology first may gain the advantage of a reduced cost or an increased yield, but they also run the risk that the change will not work or be accepted. Those who wait too long to adopt the new technology may lose the price advantage, and maybe their business.

Casualty loss—results from weather disturbances such as wind, hail, and floods; fires; wrecks and collisions; and thefts. Even though many of these losses can be insured against there are still a large number of casualty losses for which insurance is not available or is too expensive. For example, crop insurance and whole-farm insurance are available commercially but insurance to cover animal disease and death loss is difficult to obtain.

Government and legal risk—results from changing crop commodity programs, OSHA and EPA rules and regulations, export and import policies, and world disturbances. At the same time that agriculture was using more fertilizers and other chemicals, society was becoming increasingly aware of the effects these chemicals had upon human health and well-being. With an abundance of food, American society has been able to concern itself with environmental and health issues and the "good life." The preservation of resources and open spaces are important concerns. The consequences of these changing attitudes are increased laws and regulations.

Human risk—is brought on by death and illness, dishonesty, undependability, and unpredictability. The loss or illness of the manager or key employee can be very disruptive to the business. Most of the business affairs are done with people, who for the most part are highly reputable but on occasion can lie, cheat, and steal. Some firms with whom the farm does business even may fail.

It is true that some farmers seem to always be in the right place at the right time and become very successful even when they do not plan for risk. For example, the year one is born is important to success in farming. Thus, one may be tempted to ignore risk, but when the full story is told, there are many more who failed than succeeded because they did not plan for and incorporate risk strategies into their operations. Risk management is not free. Managing risk may require a sacrifice in yields, producing less profitable products, missing the highest prices, paying for insurance, and purchasing additional inputs. It also may require the cost of additional education or gathering additional information to position oneself to take advantage of future trends. But, as mentioned, increased long-run profit and survival are the rewards for risk management.

Selecting Goals for Managing Risk

Farmers and their families, like all people, have goals and objectives about what they want out of life. These were introduced in Chapter 1 and others are mentioned in Chapter 9. Even though goals and objectives are not the same for all individuals, it is likely that income and security will be among them. Beyond survival, personally and for the business, most people have hopes of increasing their standard of living. But, usually increased income and security are accompanied with increased risk. Thus, priorities need to be set to measure the trade-offs. Priorities are a means for expressing attitudes toward risk. Attitudes are dependent on farm and family goals, financial position, and potential gains and losses. Attitudes are formed over a lifetime and change with time.

An illustration demonstrates how attitudes about risk influence decisions. Consider how you would respond to each of the following hypothetical situations:

1. You are offered a wager in which you will gain $20 if a fair coin falls heads, or lose $10 if it falls tails.
2. You have accumulated a fortune of $5 million. You are now offered the opportunity to triple this amount if a fair coin falls heads, or lose it all if it falls tails.
3. You have accumulated a savings account of $5,000 for a much-earned vacation this month. You are offered the opportunity to double this amount, which will buy you a much better vacation if a fair coin falls heads, or lose it if it falls tails.
4. You have always wanted to take a fishing trip to Alaska. You have $5,000 but the trip would cost $10,000. You are offered the wager to double your $5,000 if a fair coin falls heads, or lose it if it falls tails.

It is not likely that all persons would make the same selections. It would seem that #1 and #4 are easier choices than #2 and #3. In #1 the loss is not great and the payoff is twice as large as the loss. In #4 the possible loss is a larger sum, but the whole fishing trip is at stake. In #2 the gain is double the possible loss as with #1, but the added utility of having $15 million may not be worth the utility loss of the $5 million. In #3 there is sufficient amount for a vacation and the trade-off is just for a more luxurious one.

Individuals can be placed in one of three classifications relative to their attitudes about risky decisions: risk takers, risk averse, and risk neutral. These classifications might be synonyms for optimistic or progressive, pessimistic, or conservative and expected. Risk takers are willing to gamble for the higher income possibilities. A risk taker selects the alternative with the higher income over a lower income with a greater probability of achievement. Risk-neutral persons select alternatives with the highest expected return. They are indifferent to the level of risk for outcomes having the same expected return. Risk-averse persons take the more secure position even though the outcome will probably be a smaller reward. It cannot be said that one classification is better than another. The position one takes may not be the same for each decision, and certainly will change with time. These attitudes about risk are further explored after the probabilities of the possible outcomes have been introduced.

Before delving more deeply into decision making under risk, it is helpful to consider the financial ability to sustain risk. Consider the balance sheets of the three individuals shown in Table 8.1. First, note the difference in the cash income requirement for Olsen, Jones, and Smith. Olsen needs $29,800 less cash than Jones and $35,600 less than Smith who has essentially the same asset base as Olsen. For Olsen the total cash need is only 11 percent of assets compared with 16 percent for Jones and 17 percent for Smith. Next, consider the leverage differences. Olsen has $1 of equity for each $0.07 of debt compared with $1.11 for Jones and $1.46 for Smith. For Smith a 41 percent reduction in asset values or a 68 percent increase in debts will make him or her insolvent. If grain prices decreased causing Olsen's grain assets to decline in value by $20,000, his or her equity would be reduced only by 3 percent. This same reduction would decrease Jones's equity by 18 percent and Smith's equity by 7 percent. Olsen can afford to take greater risks than either Jones or Smith. Even though Jones requires less cash than Smith, he or she has a smaller equity to fall back on. Jones and Smith are vulnerable in different ways and thus may change positions when confronted with different choices.

TABLE 8.1

Comparisons of Financial Ability to Take Risk Based on Net Worth, Debt-to-Asset Ratio and Cash-Flow Requirements

	Olsen (Owner)	Jones (Renter)	Smith (Buyer)
Assets:			
Cash	$ 10,000	$ 10,000	$ 10,000
Grain	100,000	100,000	100,000
Machinery	120,000	120,000	120,000
Land	480,000	0	480,000
Total	$710,000	$230,000	$710,000
Liabilities:			
Operating note	$ 45,000	$ 39,000	45,000
Machinery loan	0	41,800	41,800
Land rent	0	40,000	0
Land loan	0	0	334,700
Total	$ 45,000	$120,800	$421,500
Equity:	$665,000	$109,200	$288,500
Leverage (Debt/Equity)	0.07	1.11	1.46
Cash Requirements:			
Operating costs	$ 45,000	$ 39,000	$ 45,000
Machinery payment	0	13,200	13,200
Land payment	0	40,000	36,600
Capital and living	39,000	21,600	24,800
Total	$ 84,000	$113,800	$119,600

Failure to rationally analyze attitudes toward risk may affect not only financial well-being but also emotional health. Rational decision making requires quantifying the possible results for each alternative. The balance of this chapter concentrates upon rational approaches to risk.

Statistical Measures for Evaluating Planning Expectations

A priori risk is used to illustrate the concept of probability estimates. Consider the tossing of fair coins and the rolling of dice. If one coin is tossed, the probability of a head is equal to a tail, 50 percent each. Add two coins and the probabilities of two heads or two tails are 25 percent each, and for one head and one tail, 50 percent. If one die is rolled the probability of each of the six numbers coming up is 1/6th or 16.7 percent. Add one more die and the possibilities are six times six or 36 as depicted in Table 8.2 (first 3 columns).

These possibilities could be related to income deviations from some average level caused by production or price variability expectations as shown in Column 3 of Table 8.2. It is expected in this example that income will be within 10 percent of the mean 66.7 percent (11.1 + 13.9 + 16.7 + 13.9 + 11.1) of the time. Sixteen and two-thirds percent of the time profit would be 15 percent (8.3 + 5.6 + 2.8) or more below the mean and 16 2/3 percent of the time it would be 15 percent or more above the mean. The center outcomes (10 percent above to 10 percent below the mean) could be considered the expected outcome, with incomes 15 percent or greater above the mean optimistic and incomes 15 percent or

TABLE 8.2

Expected Outcomes from Rolling a Pair of Dice 36 Times

Outcome	Chances	Probability (%)	Profit or Loss (%)
2	1/36	2.8	–25
3	2/36	5.6	–20
4	3/36	8.3	–15
5	4/36	11.1	–10
6	5/36	13.9	–5
7	6/36	16.7	0
8	5/36	13.9	+5
9	4/36	11.1	+10
10	3/36	8.3	+15
11	2/36	5.6	+20
12	1/36	2.8	+25
Total	36/36	100.0	

less below the mean pessimistic. Farm managers set these expectations based upon personal experiences, historical information, outlook from others, and gut feelings.

Statistical risk is another means of assigning probabilities. Consider for example the yields of corn and soybeans in Illinois during the 1990s. Table 8.3 illustrates the tabulation of several statistical measures of variability. A discussion of each measure follows.

Mean (E)

The mean is the average of the observations and is depicted by "E." It is tabulated by summing the observations and dividing by the number of observations. The average outcome of past periods is a good estimator of what to expect (E) in the next period. The average yield for corn was 130 bushels and for soybeans 43.4 bushels. This expectation is reasonable if past observations have been consistent. The mean (E) statistic is nearly always a beginning tabulation and is useful for estimating yield, price, and income.

Not shown in these data are the mode and median. The mode is the value that occurs most frequently. With the few numbers shown in Table 8.3 it is difficult to observe a mode, but corn yields of 132 bushels were observed in three of the 10 years and soybeans had yields between 42 and 45 bushels in six of the 10 years. If the data were organized in order of magnitude the median would be the center, [(n – 1)/2], observation. For corn and soybeans this value would be 132 and 44.4 bushels, respectively.

Variance (V)

This statistic measures how observations are centered about the mean. It is measured by summing the squared deviations from the mean (i.e., subtract the mean from each observation, square the difference, and sum the squared amounts) and by dividing this total by the number of observations minus one. The smaller the variance, in relation to the size of the mean, the closer the observations are grouped about the mean. The reason that the denominator is reduced by one is because the variance thus tabulated is an estimate of a much larger number of observations (population). As N becomes smaller then the reliability is less. Reducing N by one removes the bias associated with sample size. The N – 1 is called the degrees of freedom. The probability distribution for many random variables can be approximated by the normal distribution. The shape of the

TABLE 8.3

Corn and Soybean Yields in Illinois with Statistical Measures

Year	Corn(bu) Yield	C_i-E_c	$(C_i-E_c)^2$	Soybeans(bu) Yield	S_i-E_s	$(S_i-E_s)^2$	$(C_i-E_c) \times (S_i-E_s)$
1988	77	–53	2,809	29	–14.4	207.4	763.2
1989	132	+2	4	45	+1.6	2.6	3.2
1990	132	+2	4	44	+0.6	0.4	1.2
1991	111	–19	361	42	–1.4	2.0	26.6
1992	154	+24	576	46	+2.6	6.8	62.4
1993	132	+2	4	45	+1.6	2.6	3.2
1994	162	+32	1,024	50	+6.6	43.6	211.2
1995	120	–10	100	42	–1.4	2.0	14.0
1996	143	+13	169	43	–0.4	0.2	–5.2
1997	137	+7	49	48	+4.6	21.2	32.2
Total	1,300		5,100	434		288.8	1,112.4

Mean(E) = $\Sigma Y_i/n$; $E_c = 1{,}300/10 = 130$; $E_s = 434/10 = 43.4$

Variance(V) = $\Sigma(Y_i - E)^2/(N-1)$; $V_c = 5{,}100/9 = 566.7$; $V_s = 289/9 = 32.1$

also $V_c = (\Sigma Y_i^2 - (\Sigma Y_i)^2/n)/(n-1) = (174{,}100 - (1{,}300^2/10))/9 = 566.7$

Standard deviation(s) = square root of V; $S_c = 23.8$; $S_s = 5.7$

Coefficient of variation (CV) = (S/E)100;

$CV_c = (23.8/130.0)100 = 18.3\%$; $CV_s = (5.7/43.4)100 = 13.1\%$

Range (G) = low to high; $G_c =$ 77–162; $G_s =$ 29–50

$$\text{Correlation}(r) = \sum_{i=1}^{N}(Y_{A_i} - E_A)(Y_{B_i} - E_B) / \sqrt{\sum_{i=1}^{N}(Y_{A_i} - E_A)^2 \sum_{i=1}(Y_{B_i} - E_B)^2}$$

$$r_{cs} = 1{,}112/\sqrt{(5{,}100)(288.4)} = 1{,}112/1{,}213 = 0.9$$

normal distribution is that of a bell and is depicted in Figure 8.1 as deviations from the mean. The continuing line is a cumulative distribution function (CDF). The CDF indicates the probability of observing outcomes equal to or less than a given value.

Standard Deviation (s)

The standard deviation is the square root of the variance. It is used to estimate how closely a future observation is expected to be near the mean value. It is a probability estimate as shown below the normal curve in Figure 8.1. In this figure the mean is assumed to be zero (0). Note that there are two probability distributions shown. The area underneath the bell-shaped line is referred to as a probability density function. Standard deviation lines are vertically depicted. The first standard deviation from the mean accounts for about 34 percent of the distribution of expected values (50.00% – 15.98%). Thus, one standard deviation on each side of the mean accounts for about two-thirds of the expected outcomes; two standard deviations account for about 95 percent of the expected outcomes. Referring to the corn and soybean yields shown in Table 8.3 it is estimated that two-thirds of the time corn yields are expected to be between 106.2 (130.0 – 23.8) and 153.8 (130.0 + 23.8) bushels per acre; comparable values for soybeans are 37.7 and 49.1 bushels. Please observe that products with smaller standard deviations have strong central tendencies and more predictable outcomes.

Coefficient of Variation (CV)

This statistic expresses the standard deviation as a percentage of the mean [i.e.,(S/E)100]. Whereas the mean and variance themselves may not be comparable

FIGURE 8.1

The Standard Normal Frequency Distribution and Cumulative Density Function

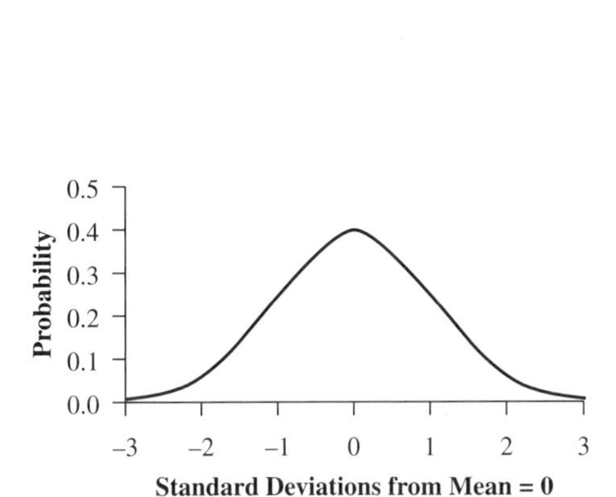

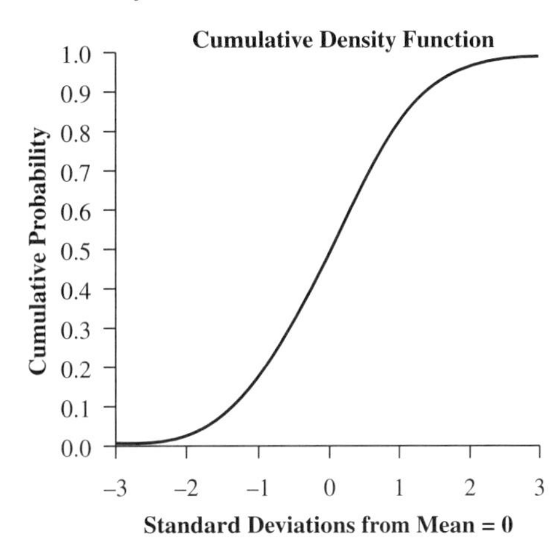

among products, CV percentages are comparable. This statistic is very useful when comparing the yield or price variability of one product against another. Soybean yield has 5 percent less variation than corn yield. If one were selecting products on the basis of variability soybeans would have the greatest promise of a less variable yield. If corn is the more profitable of the two products then the farmer might consider adding soybeans to reduce income variability caused by yield.

Range

A comparison of high and low values, or the range, is another measure of variability often used. It simply compares the lowest with the highest observation. For corn the range is from 77 to 162 bushels, and for soybeans from 29 to 50 bushels. The range provides the farm manager with lowest and highest expectations.

Correlation (r)

Farm managers often produce more than one product. Sometimes a second product is added specifically to reduce yield, price, or income variability. Correlation measures the degree to which different statistical series increase and decrease from their mean in the same direction and at the same time. It is useful to not only compare prices and yields among products but also to compare the yield against the price for the same product. The *r* value is a way for doing this. Consider for example the yields for corn and soybeans in Table 8.3. A comparison of $(C_i - E_c)$ and $(S_i - E_s)$ reveals that for all years except 1996 their yields moved above and below their means in the same years. This togetherness is revealed in the last column of Table 8.3. Now look at the correlation formula. The part of the equation before the division sign can be either positive or negative; the part after the division sign can be only positive. If all of the squared differences of the two means (last column) were positive then it is easy to see that the statistical series under consideration is highly correlated. But, if they were all negative then they differ from their respective means in the opposite direction and are highly negatively correlated. The correlation coefficient can vary from one to minus-one. If *r* is 0 then the two series are not correlated. In the example in Table 8.3, $r_{cs} = 0.9$ and we can conclude that the yields from these two crops are affected similarly by the same weather variables. Farm income variation caused by yield variation can be made less variable by diversification. Adding a second crop that has a smaller income variance or a correlation coefficient less than one can reduce income variability. The smaller the correlation coefficient the greater the income variation reduction.

Applications Using Subjective Probabilities

Farm planners must make projections about the future. These projections of yield, price, and income should be objectively studied as illustrated previously, but in the final analysis future expectations generally will be stated subjectively. These personal beliefs expressed as probability estimates have been called personal probabilities. Other considerations than past observations are included in arriving at a probability distribution (e.g., trends, world events, weather, substitution of other products, advertising, current conditions, and one's own gut feeling). All this information is synthesized by the manager and expressed in terms of a personal probability distribution. One method advocated for developing personal probabilities is to (1) list the range of all possible outcomes such as yield, prices, or income by selecting the lowest and highest possible outcome; (2) divide the range into increments; (3) weight each outcome or increment from 0 to 100 with the most likely outcome weighted the highest; and (3) divide the weight given to each possible outcome by the sum of the weights of all the possibilities. The conviction of each outcome is now a percentage, expressed as a decimal equivalent, of the total of 100 percent or 1.00. This procedure is illustrated in Table 8.4. An alternative procedure would have the individual select a probability distribution such as the normal distribution then give estimates of the mean and standard deviation. With these two estimates the individual could estimate the probability of any outcome. The individual could choose alternative distributions as reflecting one's degree of belief such as the uniform or triangular probability distribution. The important point is that for each possible outcome a probability is associated with it.

In Table 8.4, a manager evaluating the risk of a wheat enterprise considered the possible outcomes for gross margins to range between \$–150 and \$350. The manager then divided the range into intervals of \$100 and selected the midpoint of each interval to be the representative outcome. The manager then assigned weights to each outcome by using a scale from 0 to 100. A larger number indicated stronger belief in the likelihood of occurrence. The manager assigned a weight of 80 to the \$100 gross margin, the outcome believed to be most likely. The manager thought there was an equal

TABLE 8.4

Estimating Personal Probabilities for Gross Margins for Wheat Enterprise

Gross Margin (\$/acre)	Gross Margin Midpoint	Conviction That the Price Will Be in Given Range (Weights 0–100)	Conviction Converted to Personal Probability
–150 to –50	–100	10	0.05
–50 to 50	0	40	0.20
50 to 150	100	80	0.40
150 to 250	200	40	0.20
250 to 350	300	30	0.15
Totals		200	1.00

chance of the next most likely gross margin being $0 or $200. Each of these outcomes was assigned a 40. The $300 outcome was selected next most likely and assigned a weight of 30, and the $–100 outcome was selected least likely and assigned a 10. The weights were then totaled. The probabilities were calculated by dividing the individual weights by the total weights.

Subjective probability distribution functions for the wheat enterprise from Table 8.4 and a potato enterprise are illustrated in Table 8.5. From these data measures of return and risk are calculated. The expected value for gross margins is tabulated in much the same way as for the statistical series. Instead of a relatively large number (n) of historical observations each weighted by a probability (1/n), there are probability estimates which total to one (1), or 100 percent of the possibilities. The mean and variance are thus based upon the possible observations, but with variable expectancies. From Table 8.5, potatoes have a greater expected return ($270 per acre) than wheat ($120 per acre). The measures of risk, variance, standard deviation, and coefficient of variation all indicate that potatoes have greater income variability.

The measures of risk calculated in Table 8.5 are more accurate in assessing risk for a normal distribution. Note that neither of the distributions for wheat and potatoes is normal. The distribution for potatoes is more nearly a uniform distribution. The probability distribution for wheat shows some skewness. Suppose our manager is highly leveraged and needs at least a $100 gross margin per acre to meet his or her cash obligations. The cumulative probability distribution would indicate the risk of having a gross margin of $100 or less. Table 8.6 shows the cumulative probability distribution for each enterprise. There is a 65 percent probability of having $100 or less for wheat compared with a 40 percent chance for potatoes. From this perspective, wheat may have greater risk.

TABLE 8.5

Income (Gross Margin) Probabilities for Wheat and Potatoes

Gross Margin	Wheat			Potatoes		
($)/acre ($X_i$)	Subjective Probability	Expected (P_iX_i)	Squared ($P_iX_i^2$)	Subjective Probability	Expected (P_iX_i)	Squared ($P_iX_i^2$)
–$200	0.00			0.10	–20	4,000
–100	0.05	–5	500	0.10	–10	1,000
0	0.20			0.10	0	0
100	0.40	40	4,000	0.10	10	1,000
200	0.20	40	8,000	0.10	20	4,000
300	0.15	45	13,500	0.10	30	9,000
500	0.00			0.10	50	25,000
700	0.00			0.10	70	49,000
900	0.00			0.10	90	81,000
1100	0.00			0.10	110	121,000
Totals	1.00	120	26,000	1.00	350	295,000

Weighted Mean (E) = ΣP_iX_i ; E_w = $120/acre; E_p = $350/acre

Variance (V) = $\Sigma P_i(X_i - E)^2$; $V_w = 26{,}000 - (120)^2 = 11{,}600$; $V_p = 295{,}000 - (350)^2 = 172{,}500$

or $V = \Sigma P_iX_i^2 - E^2$;

Standard Deviation (s) = square root of V; S_w = $108; S_p = $415

Coefficient of Variation (CV) = (s/E)100; $CV_w = (108/120)100 = 90\%$ $CV_p = (415/350)100 = 119\%$

TABLE 8.6

Cumulative Probabilities for Wheat and Potatoes

	Wheat		Potatoes	
Gross Margin	Subjective Probability	Cumulative Probability	Subjective Probability	Cumulative Probability
–$200	0.00	0.00	0.10	0.10
–100	0.05	0.05	0.10	0.20
0	0.20	0.25	0.10	0.30
100	0.40	0.65	0.10	0.40
200	0.20	0.85	0.10	0.50

Sources of Biases in Subjective Probabilities

Although subjective probabilities appeal to the rational reasoning of agricultural economists, studies by psychologists and experimental economists suggest that eliciting subjective probabilities free of biases is a difficult task. Several sources of biases are briefly described in the hopes that managers can avoid these biases in their decision making.

Poor framing is when a manager allows a decision to be framed in the context of how the problem is presented rather than completely evaluating the problem. A manager who is told an investment has a 75 percent chance of success is more likely to accept the investment than a manager who is told there is a 25 percent chance of failure.

Availability bias is relying on the most readily available information such as the current market price or last year's yield without evaluating a longer historical picture of outcomes.

Anchoring is another example of being overly influenced by suggested values. A manager given the mean barley yield may place too much weight on the mean, resulting in a subjective probability distribution narrowly concentrated around the mean.

Representativeness or association bias occurs when a manager mistakenly associates one event with another event. Wet weather at the beginning of harvest may lead the manager to assume that a prolonged wet spell is beginning. This disregards the predictive accuracy of the associated event.

Misconception of chance is a misbelief that events are sequential rather than random. If the probability of a flood is one in five, a manager assumes since the past 4 years were flood free then this year must be a flood year. If floods are random then the chance of a flood this year is still one in five.

Decision-Making Strategies

Risky decisions involve choices, not only about expected income differences but also about the possibilities and consequences of being both above and below the mean or some critical income level. The statistical tools presented are useful for measuring mean differences and probabilities but there needs to be a framework for their use. These frameworks depend on the goals and objectives of the farm manager with regard to risk and uncertainty. This section examines alternative strategies for making decisions in an environment of risk and uncertainty.

Marginal Gain and Loss Analysis

Marginal analysis and diminishing returns were presented in Chapter 6, and in Chapter 5 partial budgets were discussed. It is useful in some analyses to budget the alternatives.

TABLE 8.7

Payoff Matrix for Corn Production under Stress[a]

Nitrogen Fertilizer (lb/acre)	Operating Costs ($/acre)	Low Stress (.3)		Typical (.5)		High Stress (.2)		Expected Value ($/acre)
		Yield (bu/acre)	Net Income ($/acre)	Yield (bu/acre)	Net Income ($/acre)	Yield (bu/acre)	Net Income ($/acre)	
0	125	106	140	78	70	50	0	77
20	130	117	162	88	89	58	15	96
40	136	126	179	96	104	65	27	111
60	143	134	192	102	112	70	32	120
80	151	141	201	108	118	74	34	126
100	160	146	205	111	119	76	30	127
120	169	150	206	114	116	77	24	125
140	177	152	203	114	108	76	13	118

Source: Voss, R. D., J. T. Pesek, and J. R. Webb, *Economics of Nitrogen Fertilization,* Department of Agronomy, Iowa State University, Ames.

[a]Net income is tabulated by multiplying the yield by $2.50 and subtracting operating costs. The expected value is the summation of the net incomes for each fertilizer level times the probabilities associated with each stress level as illustrated previously.

For most products risk increases as the optimal input level is approached. Consider for example the fertilization of irrigated barley, the return from the third 50 pounds of nitrogen fertilizer is less than for each of the first two 50-pound units. Feeder steers nearing their optimal market weight is another example. As they are fed to heavier weights the risk increases from holding them longer in the feedlot. The operator should invest where the potential gain from added inputs is clearly the greatest and consider the risk associated with alternative investments on the margin. A conceptual decision rule was provided by Young[4] in the equation $E(VMPx_1) = Px_1 + R{\cdot}I$ where $E(VMPx_1)$ is the expected value marginal product for input X_1, Px_1 is the pricc of input X_1, R is the managers risk preference coefficient, and I is the incremental increase in risk of additional input use. Together $R{\cdot}I$ is a risk premium. If the manager is risk neutral or indifferent to income variability, R is zero, and the manager would use an amount of X_1 in which $E(VMPx_1)$ equals the cost of the input. If the manager is risk averse, R is positive, and the manager would use less input. If the manager is a risk taker, R is negative, and the manager would use more input.

The Payoff Matrix

The payoff matrix is a good way to visualize the benefits and penalties of alternative actions. A payoff matrix is a table of possible outcomes caused by various states of nature. The data in Table 8.7 were developed from yield records at the Galva–Primghar Experimental Farm in Iowa and they show the effect of growing stress upon corn yields. Yields were separated into high-stress years and low-stress years, with probabilities for each. Stress might be caused by lack of moisture, high temperatures, early frost, or a combination.

First consider only the marginal values. Even for the high-stress (HS) level the added net income (NI) per added dollar of expense is $0.71 at 60 pounds of fertilizer [(32 – 27)/(143 – 136)]. At 80 pounds this expense narrows to $0.25. Investing at lower levels of marginal return could be compared with other alternative investments, particularly if operating capital is short. Without capital restraints fertilizer application is optimized in low-stress (LS) years at 120 pounds, in typical years (MS) at 100 pounds, and in high-stress (HS) years at 80 pounds. The mathematical expectancy [(NI for LS × 0.3) + (NI for MX × .5) + (NI for HS × 0.2)] is the highest at 100 pounds.

TABLE 8.8

Decision-Tree Illustration

Actions	Probability	Situation	Profit		Expected
Grow sorghum	0.20	Poor	$ 50		0.20 × 50 = $10
	0.50	Average	$ 60		0.50 × 60 = $30
	0.30	Good	$ 70		0.30 × 70 = $21
				Total	$61
Grow corn	0.20	Poor	$ 10		0.20 × 10 = $ 2
	0.50	Average	$ 60		0.50 × 60 = $30
	0.30	Good	$110		0.30 × 110 = $33
				Total	$65
Specialty crop	0.20	Poor	$ 0		0.20 × 0 = $ 0
	0.50	Average	$ 0		0.50 × 0 = $ 0
	0.30	Good	$160		0.30 × 160 = $48
				Total	$48

None of these considerations clearly indicates the fertilizer level to select, but the alternatives are more clearly spelled out. For this decision the penalties for applying too much or too little fertilizer are not very large. The final decision depends on an individual's risk preference. Individuals who are risk neutral would maximize expected net income and apply 100 pounds of nitrogen (N). Individuals who are risk averse would consider applying 80 pounds of N. Individual who are risk takers would possibly consider applying 120 pounds of N.

Decision-Tree Analysis

Decision trees are good ways to portray some alternatives when probabilities can be specified. Data like those in the payoff matrix shown in Table 8.6 could be used. But to visualize another application, consider the choice of which of three crops to grow, given three states of nature for each. The results for each consideration were budgeted and are shown in Table 8.8. This analysis would lead most individuals to select corn to grow because of its higher income expectancy. But not all individuals will make this selection because of the possibility of not receiving a higher income from producing the specialty crop or to avoid the possibility of income less than $10 by producing sorghum.

It would be helpful in selecting among the alternatives to know the utility function of the decision maker. The utility function would indicate whether the decision maker is risk averse, risk neutral, or a risk taker. The income utility preferences of these three types of persons could be conceived as illustrated in Figure 8.2.

The risk-neutral person would be indifferent and would receive the same added utility from each unit of added profit. The risk-averse person would worry about the possible losses associated with the added income and would discount the higher income levels. The risk taker would give higher utility to the higher income levels even though the penalties of being wrong are greater. The preference ratings of each individual from producing these three crops are shown in Table 8.9. Income levels and probabilities are from Table 8.8 and the utility measures from Figure 8.2. (In this illustration the utility functions are used to rate the different choices and do not measure pure utility.) The first figure in the tabulations is the expectancy and the second figure the utility.

FIGURE 8.2

Utility Preference Figure Representing Risk-Averse, Risk-Neutral, and Risk-Taking Individuals

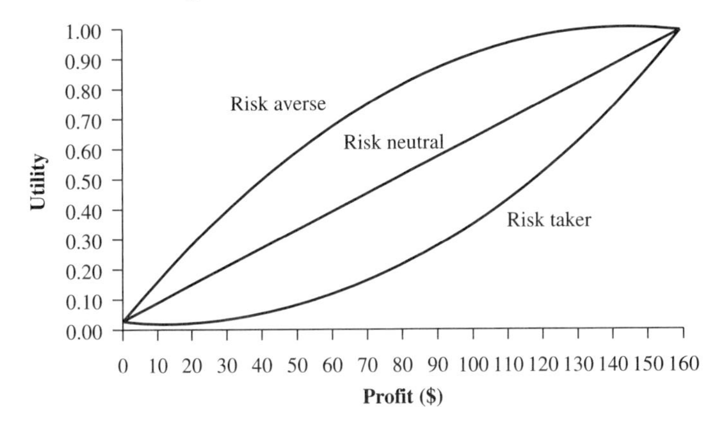

TABLE 8.9

Selected Decision Criteria Under Risk with Utility Known

		Expected Utility[a]		
Action	**Situation**	**Risk Averse**	**Risk Neutral**	**Risk Taker**
Grow sorghum	Poor	0.20 × 0.60 = 0.12	0.20 × 0.31 = 0.06	0.20 × 0.10 = 0.02
	Average	0.50 × 0.65 = 0.33	0.50 × 0.37 = 0.18	0.50 × 0.12 = 0.06
	Good	0.30 × 0.75 = 0.23	0.30 × 0.45 = 0.13	0.30 × 0.15 = 0.05
	Total	0.68	0.37	0.13
Grow corn	Poor	0.20 × 0.15 = 0.03	0.20 × 0.06 = 0.01	0.20 × 0.02 = 0.01
	Average	0.50 × 0.65 = 0.33	0.50 × 0.37 = 0.19	0.50 × 0.12 = 0.06
	Good	0.30 × 0.90 = 0.27	0.30 × 0.69 = 0.21	0.30 × 0.40 = 0.12
	Total	0.63	0.41	0.19
Grow specialty crop	Poor	0.20 × 0.00 = 0.00	0.20 × 0.00 = 0.00	0.20 × 0.00 = 0.00
	Average	0.50 × 0.00 = 0.00	0.50 × 0.00 = 0.00	0.50 × 0.00 = 0.00
	Good	0.30 × 10.0 = 0.30	0.30 × 10.0 = 0.30	0.30 × 10.0 = 0.30
	Total	0.30	0.30	0.30

[a]The expected crop incomes for each individual shown in the decision tree was associated with each person's utility function.

Under this criterion, the risk-averse person would select to grow sorghum. The risk-neutral person would select corn, whereas the risk seeker obtains the greatest utility from the specialty crop.

Stochastic Dominance

This procedure is used for viewing the cumulative probabilities associated with alternative courses of action. It rests upon the foundation that more is better than less. It is little more than a graphical depiction of income, or other common measure of output, plotted against respective cumulative probabilities. Consider the three lines A, B, and

FIGURE 8.3

Illustration of Stochastic Dominance

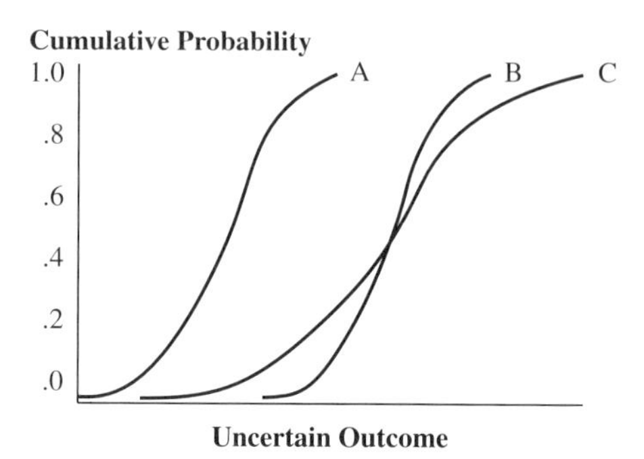

C in Figure 8.3. Note that the vertical axis measures cumulative probability and the horizontal axis measures the uncertain outcome, usually income.

Product line A lies to the left of both lines B and C at all outcome levels. Hence, it can be said that B and C dominate (is superior to) A to the first degree at all income levels. It is not so clear whether B is better than C because their outcome levels cross at probability levels greater than zero. Thus, it is necessary to consider the extent to which the one is better than the other. At the lower levels of probability B is better then C, but at the upper levels the reverse takes place. The area of measure between the lines gives some guidelines as to which is the best. If the area in which B dominates is greater than the area C dominates then the expected income of B is greater than the expected income of C. Individuals who are risk averse will prefer B over C. This consideration is called second-degree stochastic dominance and assumes that the decision maker is risk averse. If the opposite is true, the risk-averse individual would have to weigh the trade-off of higher income with C or less risk with B. In this last case, some risk-averse individuals may prefer B, whereas others may prefer C.

Stochastic dominance is illustrated in Figure 8.4 for two decisions: selecting fertilizer rates (Table 8.7) and selecting crops to produce (Table 8.8). In both cases the number of net income possibilities have been expanded to give continuous probability functions.

For fertilizer, 80 pounds dominates 60 pounds at all stress levels. One-hundred and twenty pounds dominates 80 pounds at the upper level of income, but 80 pounds dominates 120 pounds at the lower levels of income. The area that 80 pounds dominates is greater; therefore, risk-averse decision makers prefer 80 pounds because it has a higher expected income and less risk. Risk-neutral individuals also would prefer 80 pounds simply because it has the highest expected income of the three choices. The probability curves for the crop selection decision are more easily distinguished. No crop has first-degree dominance. The area [ige] is smaller than area [fgd], giving sorghum an income probability advantage over the specialty crop. Similarly, the area [chi] is smaller than [fha], giving corn an income advantage over the specialty crop. Both corn and sorghum dominate the specialty crop to the second degree. They both have greater expected returns and less risk than the specialty crop. The area [abd] is slightly smaller than [ebc], indicating corn has a higher expected return, but corn has greater probability of a lower income hence more risk. The risk-averse individual must weigh the trade-offs between higher expected return, (corn) or less risk, (sorghum). The risk-neutral person who is indifferent about risk prefers corn.

FIGURE 8.4

Stochastic Dominance Illustrations for Alternative N-Fertilizer Rate Decisions and Crop Selection Choices

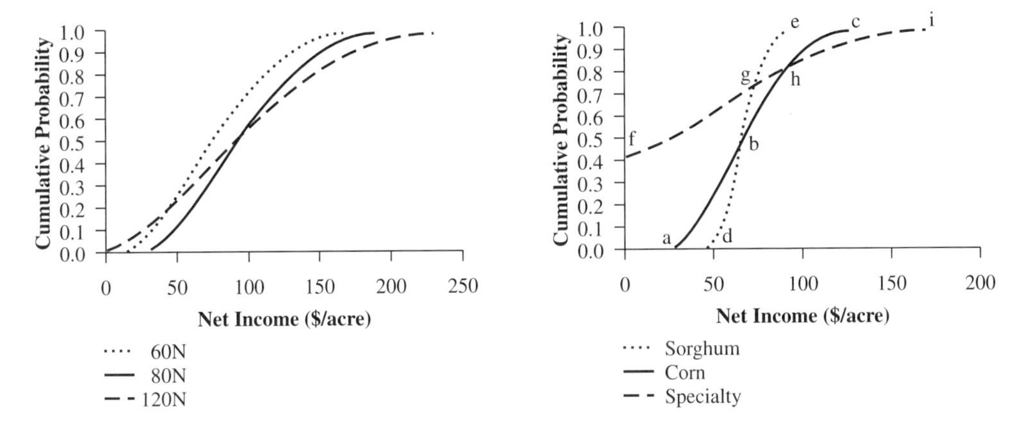

Game Theory Approaches

The previous examples used probability estimates to evaluate expected returns and risk. An individual's attitude toward risk or utility determines the appropriate choice of alternatives. There are other guidelines that do not require probability estimates for the decision-making process. These guidelines are perhaps appropriate in the classic definition of uncertainty in which individuals lack sufficient information to form subjective probabilities or are simply unwilling to do so. The decision maker still must decide to what extent each criterion is appropriate given his or her goals, the sizes of the associated gains and losses, and the financial ability to pay in the event of an adverse consequence. Three approaches are presented and illustrated—MAXIMIN, MAXIMAX, and INSUFFICIENT REASON. Each is defined as follows:

- MAXIMIN—the decision maker examines the worst outcome for each action and then selects the action that maximizes the minimum gain.
- MAXIMAX—the decision maker examines the best outcome for each action and then selects the action that has the best outcome.
- INSUFFICIENT REASON—the decision maker recognizes that it is very difficult to predict the future and thus considers all states of nature as equally likely of happening. Selection is on the basis of the highest average expected outcome.

The data from the decision tree in Table 8.8 are used to illustrate each of these methods in Table 8.10.

The manager who is pessimistic about the future or has strong financial commitments may consider one of the lesser outcomes. The MAXIMIN approach would select to produce sorghum because under a stressful growing year the income for sorghum would be higher than any of the other crop considerations. However, the optimist who is in a strong financial position might be drawn by the MAXIMAX approach to select the specialty crop. It has the highest income possibility. The risk-neutral person who does not wish to speculate about the future may select on the basis of the highest average, assuming each event is equally likely. The INSUFFICIENT REASON method would cause sorghum to be selected, with corn a close second. This last method is similar to using the highest mathematical expectancy as illustrated in Table 8.8, where the weighted incomes were $65 for corn, followed by sorghum at $61, and the specialty crop at $48.

TABLE 8.10

Decision Making by Using Game Theory

			MAXIMIN		MAXIMAX		INSUFFICIENT REASON
Action	**Situation**	**Profit**	**Worst**	**Select**	**Best**	**Select**	**Simple Average**
Sorghum	Poor	$ 50	$50	$50	–	–	0.33 × $50 = $17
	Average	$ 60	–	–	–	–	0.33 × $60 = $20
	Good	$ 70	–	–	$ 70	–	0.33 × $70 = $23
							Total $60
Corn	Poor	$ 10	$10	–	–	–	0.33 × $ 10 = $ 3
	Average	$ 60	–	–	–	–	0.33 × $ 60 = $20
	Good	$110	–	–	$110	–	0.33 × $110 = $36
							Total $59
Specialty crop	Poor	$ 0	$ 0		–	–	0.33 × $ 0 = $ 0
	Average	$ 0	–	–	–	–	0.33 × $ 0 = $ 0
	Good	$160	–	–	$160	$160	0.33 × $160 = $53
							Total $53

Reducing Risk Caused by Yield Variability

The best approach to reducing yield variability is to be technically competent and artfully efficient. No risk-preventative measure or gimmick can fill the void of not understanding the science of the products being raised and being wise in applying efficient and timely production practices. How can farmers select fertilizer applications if they do not understand their soils and the biology of their crops; select cattle that can withstand heat, cold, or drought if they do not understand breeds; select among crop varieties to reduce wind or hail loss if they do not understand genetics; limit losses from untimely operations if they do not understand crop seasonality; or develop a livestock breeding program if they do not understand animal anatomy and reproduction? These management skills impact not only upon the level of production and costs but also upon their variability. The more a farmer knows about what he or she is producing and marketing, the better equipped he or she will be in dealing with risk.

Persons who consistently have net incomes in the upper third of their farm classification do not suffer the severity of risk as those in the lower third. The benefits of high farming skills can be observed in Figure 8.5. Note that even in the low-return years the good producers generally make money. The solid line drawn through the middle of the figure approximates the break-even cost of production. The lower third of producers were below this line more often than above it, whereas the top producers were never below the line.

Selecting Genetically Stable Lines

Many agricultural experiment stations conduct crop variety trials. In addition to comparative yields, these trials provide information about freezing damage, lodging, insects, and diseases. Data are summarized and recommendations given. Selecting varieties that are resistant to yield and income-reducing problems will reduce production variability. However, the most resistant varieties may not always provide the highest yield. Livestock, too, can be selected to reduce production variability. Conception rates, birthing difficulty, environmental stress, and feed conversion are inheritable traits that are related to breed. It has been shown that cross-bred beef cattle not only convert feed to meat more efficiently but also are more healthy.

FIGURE 8.5

Livestock Returns per $100 Feeds Fed in Iowa, 1988–1997

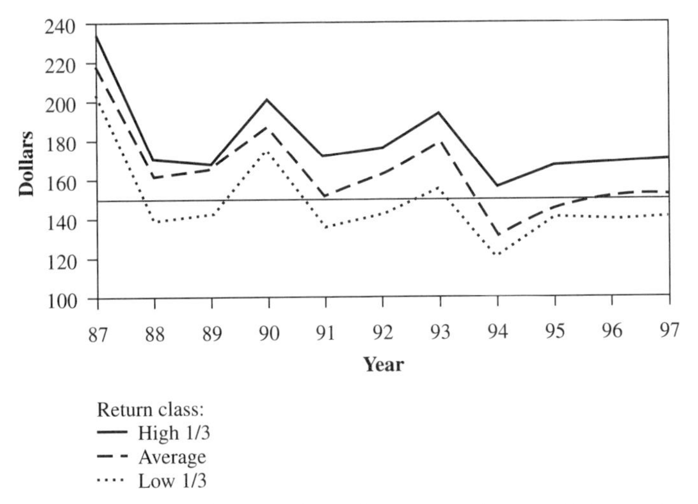

Source: *Iowa Costs and Returns,* Fm 1789, Iowa State University Extension, Ames.

Selecting Less-Variable Enterprises

Generally, income from field crops exhibits less variability than income from row crops, and grain crops are less variable than vegetable crops. Beef cows have a more variable income flow than dairy cows, and feeding yearling steers is more risky than feeding calves. A study completed in New Mexico several years ago showed income CV for peanuts of 10 percent, dry beans 13 percent, dry onions 27 percent, and spring lettuce 53 percent.[5] Profit was received in Iowa 71 percent of the years for farrow-to-finish hogs compared with only 62 percent for feeding yearling steers during a recent 10-year period. Variability tabulations of returns-per-$100-feeds-fed in Illinois between 1975 and 1987 showed CV for dairy of 12 percent, hogs 16 percent, fed cattle 21 percent, and beef cows 28 percent.[6]

Diversification

Diversification happens when enterprises are added to the farm product mix to reduce income variation. The principles of diversification were presented in Table 8.3 and measured with the correlation coefficient. It is difficult to reduce income variability by adding another enterprise if there are already three or four products being produced. In the first place income from crops tends to vary in the same directions. This is more true for yield than price. Furthermore, if there are already two or three enterprises, it is difficult to add another enterprise that does not correspond in its variability traits like one of the others. It is even possible to increase farm income variability. Nevertheless, diversification is a major means of reducing farm income variability as demonstrated by a Wyoming study. Income variability for several crops, with and without livestock, was studied for 11 years. The results are shown in Table 8.11. These data were then programmed to tabulate net income and income variability. Representative results of this study are shown in Table 8.12. Note that as variability decreased, as shown by the CV and probability range, so did the expected income. Also, note that when cattle were added income increased and variability decreased.

TABLE 8.11

Average Net Income and Income Variability for Alternative Enterprises

Enterprise	11-Year Average Net Income ($/acre)	Income Variability: 67% Probability Range Low ($/acre)	Income Variability: 67% Probability Range High ($/acre)	Income Variability: CV(%)
Alfalfa	115.27	77.23	153.31	0.33
Corn to silage	144.01	73.44	214.31	0.49
Corn to grain	136.84	45.16	228.52	0.67
Sugar beets	343.85	68.77	618.93	0.80
Dry beans	240.60	43.31	437.89	0.82
Oats (seeded to alfalfa)	10.94	–6.78	28.66	1.62
	($/head)	($/head)	($/head)	
Sell 725-lb steer	46.37	–25.97	118.70	1.56

Source: *Enterprise Combinations for Irrigated Farms in Torrington–Wheatland Area: Income and Risk.* B–771, Agricultural Experiment Station, University of Wyoming, 1982.

TABLE 8.12

Income and Risk Combinations and Corresponding Enterprise Mix

NET INCOME ($)	RISK: 67% Probability Range Low ($)	RISK: 67% Probability Range High ($)	RISK: CV (%)	ENTERPRISE MIX: Oat Grain (acres)	Sugar Beets (acres)	Corn Grain (acres)	Corn Silage (acres)	Alfalfa Hay (acres)	Dry Beans (acres)	Steers (head)
				crops only						
99,706	22,334	177,078	77.6	4	146	77	0	23	150	
72,000	51,336	92,664	58.9	23	63	118	23	150	21	
				crops and livestock						
109,847	54,814	164,880	50.1	3	135	86	21	20	135	289
72,000	51,336	92,664	28.7	25	63	118	23	150	21	308

Source: *Enterprise Combinations for Irrigated Farms in Torrington–Wheatland Area: Income and Risk.* B–771, Agricultural Experiment Station, University of Wyoming, 1982.

Controlling the Environment

Irrigation and the use of pesticides are so common that they are generally not considered risk-control devices. But anything that is done to modify or control the environment could be considered a risk-control measure. Examples of environmental control are orchard heaters and wind machines to modify the temperature, heated hog-farrowing houses, greenhouses for fruits and vegetables, and confinement egg and broiler production. Irrigation wells have been drilled and dams built to provide water to crops in rain-short years. Insecticides are used to reduce the chance of infestations as well as to control insects already present.

To economically justify new investments and increased annual operating costs, it is necessary to make cost–benefit analyses. Costs-and-returns budgets are made with various control devices inserted. When variable yields or prices are problems the use of averages does not seem adequate. It is important to consider the level of risk before and after the control device has been installed, as well as cost and income differentials.

Flexibility

Flexibility has to do with the costs of changing ones mind, or of using a practice or procedure that is not as currently efficient but will provide multiple use or allow for an earlier replacement. Examples include the selection of an annual crop instead of a perennial one, feeding calves for sale at different weights, having surplus machinery capacity to shorten planting or harvest time, buildings to house machinery and store crops, storing crops so that sales can spread out, and stockpiling feed for a dry year.

Comparative budgeting and programming are the most useful tools for studying flexibility. A long step is taken when the manager sets about to quantify the alternatives in an objective manner. Hunches and intuition are too often used because the variables are not known for certain. Payoff matrices provide some guidelines. Simulation (not illustrated in this book) may be a useful tool. The procedure is one of making decisions and then seeing what happens when random events are introduced. Although simulation procedures are beyond the scope of this book, the idea of considering the effects of different states of nature is well within the previous discussion.

Production or Yield Insurance

Insurance is a pure risk strategy. For a premium payment, the uncertain yield of the farmer is passed on to an agency where loss probabilities can be tabulated. This yield or income protection avenue is particularly relevant to many crops. Insurance is not generally available for livestock. There are two general sources of crop insurance: commercial and government-sponsored insurance. Commercial insurance is limited primarily to hail and wind damage of grain crops. All-risk government insurance is more broadly based, both in terms of the crops protected and the causal factors. Furthermore, it is income based, which includes price.

Crop hail insurance protects the insured against loss from crop damage due to hail, and generally wind damage. Most companies include fire and lightening damage in their policies. Annual costs per acre of coverage vary from 2 to 15 percent of the insured value of the crop. The chance of damage depends on crop maturity, season of the year, and the probability of storms in the area. Insurance providers maintain very strict weather records of small geographic areas to tabulate local probability estimates. Hence, rates vary by area and by crop. The amount for which a crop can be insured is controlled by the provider. Operating costs are a common amount carried. For one crop insurance company in Iowa, the cost per $1,000 of coverage was $120 in the most costly counties, compared with $29 in the least costly county for soybeans. Similar costs for corn were $52 and $15, respectively. Claims for losses are calculated by multiplying the amount of the insurance by the percentage of damage. Because there are some reasons for misrepresentation, damage is assessed by an unbiased regulated adjuster rather than by an employee of the insurance company.

The Office of Risk Management of the Farm Service Agency (FSA), offers several crop insurance programs, including Multiple Peril Crop Insurance (MPCI), Catastrophic Insurance (CAT), and Group Risk Plan (GRP). They are considered to be the prime sources of federal disaster assistance. Forty-four crops were listed in 1990 for MPCI. Risks covered include drought, excess moisture, insects, disease, wind, hail, fire, and all hazards of nature. Quality as well as yield is considered. FSA insurance is financed from premiums and a subsidy from the Federal Government. Thus, a significant part of government crop insurance is born by taxpayers. Agents may be private insurers. Because these insurance premiums are based upon actuarials relevant to each farm, every premium is as cost-effective over time as the other. Thus, the selected levels are

TABLE 8.13

Gross Revenue from 500 Acres of Corn at Nine Market Price–Yield Scenarios with Two MPCI Options and a Self-Insured Option

Scenario	Yield (bu/acre)	Price ($/bu)	APH 65/100	APH 75/100	Self-Insured
1	163	$3.35	$270,275	$268,775	$273,025
2	163	$2.75	$221,375	$219,875	$224,125
3	163	$2.25	$180,625	$179,125	$183,375
4	130	$3.35	$215,000	$213,500	$217,750
5	130	$2.75	$176,000	$174,500	$178,750
6	130	$2.25	$143,500	$142,000	$146,250
7	64	$3.35	$135,200	$153,200	$107,200
8	64	$2.75	$116,000	$134,000	$ 88,000
9	64	$2.25	$100,000	$118,000	$ 72,000

Source: *Managing Change—Managing Risk: A Primer for Agriculture,* Iowa State University Extension, Ames.

based upon the expectation and consequence for each insurer. Premiums are based on a two-part selection by the insured. First, there is a historical yield guarantee—the farmer's yield or the nearby community yield. The farmer may select an insured yield from 50% to 75% of the historical expected yield. Second, there is a price selection. Price elections are from 60% to 100% of FSA projected market prices. Premiums vary by yield and price election. To illustrate the MPCI consider 500 acres of corn with an expected yield of 130 bushels and an FSA price of $3. Table 8.13 shows gross revenue (GR) for three choices at 65% and 75% yield guarantee election and self-insured. Gross revenue also is shown for three market prices and three yield outcomes. GR is calculated as the product of (yield · price + insurance receipt – insurance premium). APH 65/100 means 65% of the expected yield and 100% of the expected price. It should be recognized that no insurance receipt is received until yield drops below 85.5 bushels at the 65% level and 97.5 at 75% level. The insurance receipt is the difference between the actual yield and these levels times the price selected. In the table only the 64-bushel yield receives an insurance payment. Thus, at yields 130 and 163 bushels the only difference in GR is the insurance premiums. It is clear that the 75 percent yield at the highest price selection provides the greatest income protection, but it costs $4,250 in premiums per year over no insurance.

Reducing Price Variability[7]

Means for reducing market and price risk were discussed in Chapter 7. Thus, there is no need for a lengthy discussion here. Only very brief mention is made to complete the discussion of risk and uncertainty.

Spreading Sales

Rather than selling all of the crop or livestock at the same time of year, sales are spread throughout the year. This approach may not be feasible for some products, such as those that are nonstorable and can only be produced once per year. Grain crops, beans, forages, peanuts, and cotton are examples of crops that can be stored and sold in different months of the year. It is more difficult to spread the sale of livestock. Market livestock can be fed to gain weight at different rates. Beef calf producers can calve at different seasons of the year.

Hedging and Options

The commodity futures market can be used to narrow the sales price range for those commodities traded. The crop price is hedged by selling a futures contract at the beginning or during the growing season, thus locking in a price for all or part of the crop. When the crop is harvested, the contract is filled with a commodity purchase on the board of trade and the product is sold locally. The purchase of a put option buys the right to sell a standard futures contract at a fixed price on the same exchange. Options are sold at a fixed price and require no additional margin payments. A put option establishes a floor or minimum price for the expected production. Before beginning commodity trading, the farm manager should study the commodity market and become familiar with the basis of the product traded. The basis is the price differential between the board of trade and the local market. This, too, is a source of price variability.

Contract Sales

Some products may be sold in advance of harvest at a fixed price. Most vegetable crops are produced under contract with the purchaser furnishing some of the inputs. Grain crops and beef calves are sometimes contracted for by an elevator or feed yard. These contracts, while reducing price uncertainty, introduce a quantity uncertainty. Contracts usually specify a certain kind, quality, and quantity of product. If the harvest is short of their contractual items, penalties may be assessed.

Participation in Government Programs

If the farmer produces one of the crops covered by a federal government support program such as feed grains, wheat, cotton, sugar, milk, wool, and tobacco the price received may be protected by a loan deficiency payment or a loan payment, or by an acreage diversion or set-a-side payment. These programs place a floor under the price received or supplement the market-determined price.

Controlling Financial Risk

Financial management is the third major area of risk control. Controlling financial risk requires an understanding of financial accounting (Chapters 2–4), particularly the balance sheet, and the financing of business ownership and operations (Chapter 10). Broadly based financial difficulties have plagued farmers in the past (e.g., the financial crisis of the 1980s), and such difficulties are likely to happen in the future. And, in any 1 year a large percentage of farmers have financial concerns. Even though it is more difficult to avoid the financial difficulties associated with widespread economic failure, most of the common financial problems can be reduced, if not eliminated, by planning and safeguarding. The major means for controlling financial risk are as follows:

- Pacing of investments and asset sales.
- Adjusting consumption and personal investment.
- Income tax management (see Chapter 13).
- Control the proportion of debt in relation to total investment.
- Having sufficient liquidity on hand as a reserve.
- Discounting of returns on long-term investments.
- Insuring against sickness, injury, and death.

Many investments are made without adequate preparation. Purchases are made before assessing profitability and feasibility. Feasibility has to do with financial risk. It is possible for a new investment to show high return prospects and still be difficult to finance—borrow and payback. Payback includes both interest and principal. Annual cash needs include business operations and fixed expenses; purchases, or down payments on new investments; principal and interest payments on old loans; family living expenses; and investments and transfers out of the farm business. Money to meet these costs comes from sales of produce and capital investments, nonfarm income, transfers into the business, and new loans. These monetary sources all must be considered, not only for the coming year but also over the life of all secured debt. Pacing of new investments to match payment ability, with a measure of error, is necessary to avoid possible

FIGURE 8.6

Operator Net Farm Income by Interest Paid as a Percent of Gross Farm Returns, 1992–1996

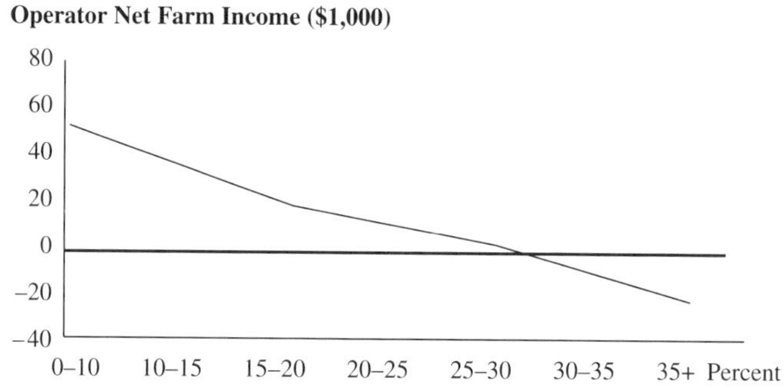

Source: *Farm Income and Production Cost Summary from Illinois Farm Business Records,* ARE-4566, Cooperative Extension Service, University of Illinois.

TABLE 8.14

Financial Leverage for Returns or Failure

Item:	**Level 1**	**Level 2**	**Level 3**
Leverage:			
Equity	$100,000	$100,000	$100,000
Debt	0	100,000	200,000
Assets	$100,000	$200,000	$300,000
Situation A: Rate of returns (i_a) is greater than debt costs (i_d). Assume $i_a = 15\%$; $i_d = 10\%$.			
Return to assets	$ 15,000	$ 30,000	$ 45,000
Cost of debt	0	10,000	20,000
Net to equity	$ 15,000	$ 20,000	$ 25,000
Return to equity (%)	15	20	25
Situation B: Rate of returns is less than the rate paid on debt. Assume $i_a = 5\%$; $i_d = 10\%$.			
Return to assets	$ 5,000	$ 10,000	$ 15,000
Cost of debt	0	10,000	20,000
Net to equity	$ 5,000	$ 0	$ 5,000
Return to equity (%)	5	0	–5

failure. Many business difficulties are encountered because of the extravagances of the owners. Controlling personal consumption may be necessary during stressful periods.

Controlling Debt–Leverage (D/E)

Debt is required in most farm operations to acquire assets, and for operations, but too much debt can greatly increase financial risk. Interest as a percentage of gross income and leverage are the statistics usually measured to consider the effects of too much debt. Interest effect of debt can be seen in Figure 8.6 where net income is plotted against interest payments as a percentage of gross returns for some Illinois farms. Note that when interest rose above 30 percent the prospects for a positive net income were dim. Leverage is the ratio of debt to equity. The smaller the ratio the less the business is leveraged. For an illustration of how leverage is tabulated see Chapter 4. How leverage can be used to increase business return, as well as being a source of risk, is illustrated in Table 8.14. As leverage increases from level one to level three, the rate of return to equity increases if the rate of return to assets is greater than the interest rate. The opposite trend occurs if the rate of return to assets is less than the interest rate.

Maintaining Liquidity

Liquidity has to do with buying power. Buying power is measured by the amount of cash on hand plus saleable inventories and other current assets minus current obligations and credit reserves. The tabulation and significance of the current asset ratio was discussed in Chapter 4 and is not repeated here. Financial planning and loan justification are discussed in Chapter 10, including the risk of loan failure and the marginal efficiency of capital. Maintaining a credit reserve is not adequately covered elsewhere.

The credit reserve is borrowing power. It is tabulated by subtracting present debts from the total debts that could be obtained under normal and reasonable circumstances. Liabilities should be partitioned as shown in the balance sheet. Additional borrowing power is tabulated by first listing all of the assets. Next, list the maximum loan value (the amount that could be borrowed by using each of the assets as collateral) of these assets. The difference between the maximum amount that could be bor-

TABLE 8.15

Tabulation of a Credit Reserve from Assets and Production

Asset	Asset Value		Maximum D/A	Credit Capacity	Present Debt	Credit Reserve
Asset credit:						
Personal property	$250,000		0.5	$125,000	$ 35,000	$ 90,000
Real property	600,000		0.6	360,000	150,000	210,000
Total	$680,000			$485,000	$185,000	$300,000
	Value/ unit	**No. of units**	**Credit rate**			
Income credit:						
Crops	250/acre	400	0.60	$ 60,000	$ 25,000	$ 35,000
Livestock	550/head	350	0.75	144,000	45,000	99,000
Total				$204,000	$ 70,000	$134,000
Total credit reserve						$434,000

rowed and the current debt shows the additional borrowing power available in case of unforeseen circumstances and new investment opportunity. Borrowing power is illustrated in Table 8.15.

Discounting of Future Returns

Discounting future returns is a safety-first approach. Examples include conservative income projections and relatively high discount rates for tabulating present values. Discounting procedures are explained in Chapter 10. Present value analysis is used to evaluate the profitability of new business investments. The pessimistic evaluation should be made even if the decision is made on a more optimistic outlook.

Farm and Farmer Insurance

Farmers need to consider insurance for two major reasons: to protect the future income during the lifetime of the owner and to provide income for family members in case of a premature death. Two common types of farm insurance coverage are liability and property insurance. Common types of personal insurance are health and life.

Liability insurance gives protection to the insured against legal liability for death or injury to another person, or damage to the property of another. It pays any court and judgement costs up to the limits of the policy that might be obtained by someone who brings suit because of an accident or other cause. Judgments can be very expensive and can wipe out total business assets. Liability insurance is of three general types: (1) comprehensive personal liability insurance for injury to the person or property of others, (2) employer's liability insurance for personal injury of employees, and (3) motor vehicle insurance. Most states require vehicle insurance before licensing. As costly as this insurance might be, few, if any, farmers can afford the risk of being without it.

Fire and extended coverage provides protection from losses caused by fire, lightening, and hazards that can cause serious financial loss. Coverage is available for dwellings, household contents, barns and outbuildings, farm products, livestock, machinery, and other items of personal property. Often this insurance is required before outside financing can be obtained. Crop insurance was discussed previously.

Health and life insurance are more personal but important to the farm business. Some farm loans require the borrower to maintain personal insurance. Health insurance, in particular, can be very expensive, even when limited to major medical coverage. The amount of life insurance to carry is thought by most to be a guessing game, but there are means for improving the guess. The amount of life insurance to buy is the difference between the discounted amount needed to care for the survivors and the current financial equity. Discounting procedures are discussed in Chapter 10. Many reputable insurance sales persons have programs for tabulating this amount. It pays to shop around.

Managing Other Forms of Risk

Two other means of controlling risk, which did not fit into the categories previously discussed, are land lease arrangements and backup labor and management. A brief discussion of each follows.

FIGURE 8.7A

Cash Rent vs. Crop Share for the Tenant with Corn–Soybeans Rotation

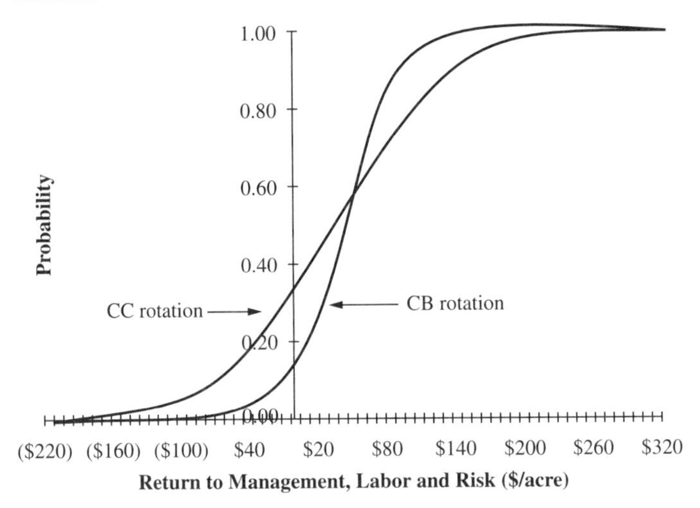

FIGURE 8.7B

Cash Rent vs. Crop Share for the Landlord with Corn–Soybeans Rotation

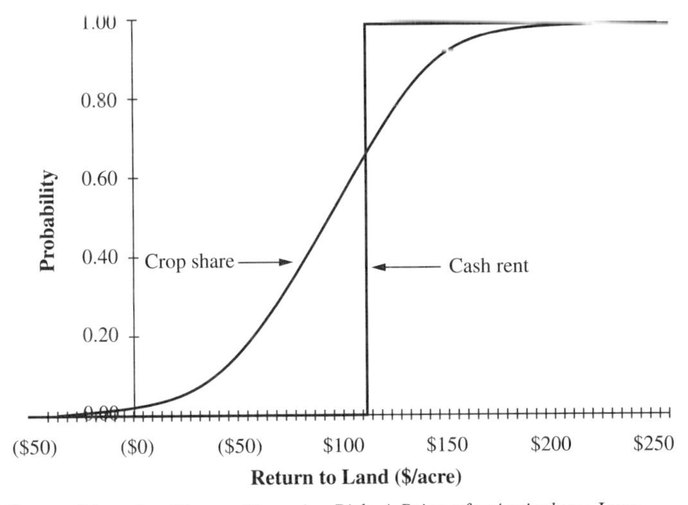

Source: *Managing Change–Managing Risk: A Primer for Agriculture,* Iowa State University Extension, Ames.

Land Lease Arrangements

Leasing arrangements are discussed in Chapter 9. Cash and share leases are covered here. Under pure cash leases the operator pays a fixed cash rent for the use of the property and receives all of the product. Thus, the lessee assumes all of the risk, but there are means for reducing this impact. The lease could specify that the payment is to be made in kind, thus sharing the price risk. Another version has the payment based upon a fixed price and a flexible yield. Both of these types were illustrated. A product-share lease divides risk between the lessee and lessor. The differences in risk associated with cash versus crop-share rents are illustrated in the probability curves shown in Figures 8.7a and b. The tenant will be better off with the cash lease when yields and prices are high and worse off when they are low. There is a trade-off that favors the crop share lease as a protection against poor yields and low prices. The landlord's probability distribution is different from the tenant's. The areas between the two curves favor the cash rent as a protection against low gross returns. The preferences of the tenant and landlord may lead to an adjustment of the crop shares or an adjustment of the cash rental rate.

Backup Labor and Management

Backup labor and management is partly an estate-planning matter and partly a means of sharing farm tasks and decision making. Proving for backup labor and management relates to the form of business organization and ownership. Partnerships and corporations bring more people into these responsibilities. (See Chapter 10.) Many agricultural cooperative extension services have legal and management specialists who can provide helpful suggestions for organizing the farm business. Some attorneys specialize in estate planning. It is extremely important that the manager not be pushed in directions that are not compatible with the family goals and objectives. It is appropriate to raise "what if" questions about health and life possibilities to determine their effects upon the business, its owners, and managers. It may require the service of experts to help solve estate problems. Having backup labor and management requires the involvement of others.

Notes

1. Cochrane, Willard. *The Development of American Agriculture.* Chapter 6, 2nd ed. University of Minnesota Press, 1993.
2. Melichar, Emanuel. "A Financial Perspective on Agriculture," *Federal Reserve Bulletin,* January 1984.
3. Olsen, Kent. "Net Farm Income Estimated to Drop Precipitously in 1998," Staff Paper Series P98–6, Department of Applied Economics, University of Minnesota, St. Paul, MN, August 1998.
4. Young, Douglas. "Risk Preferences of Agricultural Producers: their use in extension and research." *American J. of Agricultural Economics.* 61(Dec. 1979):1063–1070.
5. *Production and Marketing Considerations for New Mexico Vegetables.* Bull. 482, Agricultural Experiment Station, New Mexico, 1963.
6. Tabulations were made from annual farm business summaries and costs and returns publications issued by Cooperative Extension Services.
7. A more complete discussion of managing market risk can be found in *Managing Change—Managing Risk: A Primer for Agriculture,* Iowa State University Extension.

CHAPTER 9

Organization and Ownership of the Farm Business

Relationship of the Farm Business to Its Owners

In the United States there are approximately 2.1 million farms encompassing about 1 billion acres of land. In 1992, 58 percent of these farms were owned outright with no rented acres. Thirty-one percent were part owners and 11 percent were fully tenant operators. Most farms were operated as sole proprietorships (86 percent). The balance was operated as partnerships (10 percent), family corporations (3 percent), and nonfamily corporations and others (1 percent). Considering the 1 billion acres of farmland, 57 percent were owned by its operators and 43 percent were rented land. (Source: 1992 Census of Agriculture.) Farm organizational patterns by size of farm are shown in Figure 9.1 and the amount of investment by farm size is shown in Figure 9.2. These statistics reveal that most farms are family operated by sole proprietorships, particularly at the medium and smaller size levels. As farm size gets larger the percentages of partnerships and corporations increase. At the large size level nearly one-half of the farms were operated as partnerships (22 percent) and corporations (23 percent). Ownership and operational patterns are highly diversified by geographic location and product marketed. Farms producing cash market crops are much more likely to be tenant operated than livestock farms, where most of the crops produced are marketed through livestock. Farms with high investment requirements in machinery and equipment or livestock are more likely to be operated by partnerships and corporations. Livestock business arrangements are difficult to define for tenants, and their operations are not very stable. Thus, cash-and-share rental

---

Some agricultural extension services have issued publications outlining forms of farm ownership and organization. Examples are: Harl, Neil E., *Organizing the Farm Business,* Pm-878, IA St U; Calkner, Richard W. and Gayle S. Willett, *Estate Planning Basics for Washington Farm and Ranch Families,* EB1231, WA St U. *Farm Estate and Business Planning,* by Neil E. Harl, Century Communications Corp., provides a more complete discussion on this topic.

FIGURE 9.1

Farm Organization by Size of Farm

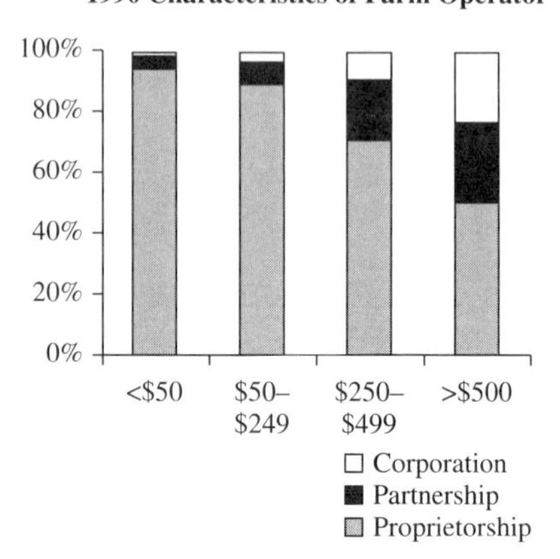

Source: USDA–ERS.

FIGURE 9.2

Farm Investment by Size of Farm

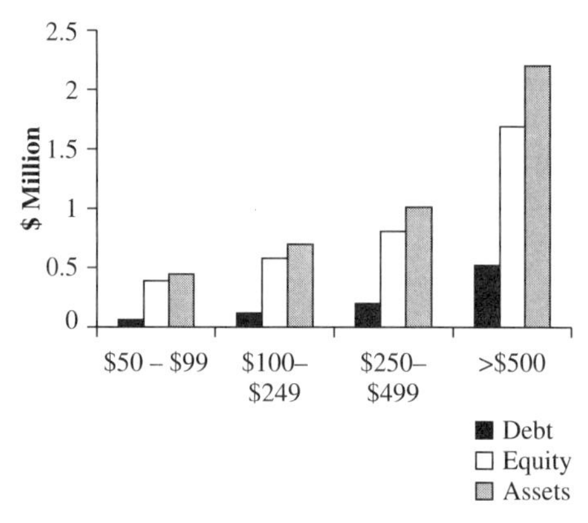

Source: USDA–ERS.

agreements are much more prevalent in the Midwest and southern United States where more cash crops are grown than in the Rocky Mountain west.

Most farmers would have among their goals the following:

1. Own their farm business. The business might include rented property but the base of operation would be owned. Farmers still believe farming offers the most business and personal freedom of any other occupation. Ownership is a form of saving for security, retirement, and estate building.
2. Be debt free. Few people like debt, even though it is often the source for expanded opportunities and growth. Most farmers borrow to finance operations and investments. A $1-million farm that was 60 percent financed at 7 percent

FIGURE 9.3

Average Size of Farm in the United States

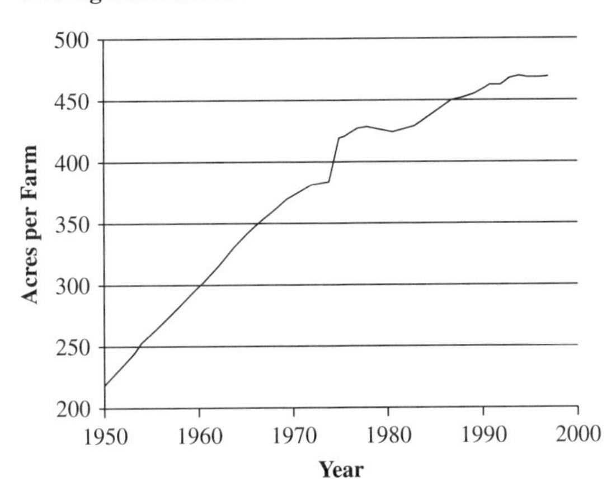

Source: USDA–ERS.

FIGURE 9.4

Number of Farms in the United States

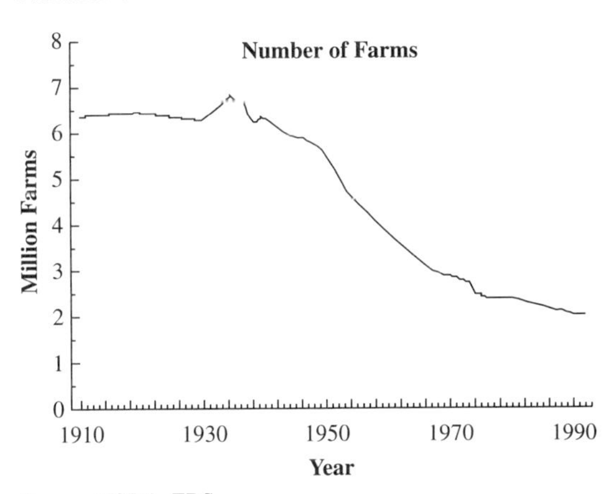

Source: USDA–ERS.

interest for 25 years would have annual loan payments of more than $51,483. Over the loan period, these principal and interest payments exceed the property value by 29 percent—a large drain on the cash flow of the business. Even though there will be a savings (equity) of $600,000 at the end of 25 years, the annual cash available for family living costs and enjoyment will be greatly reduced. Thus, debt is often viewed as a plague, even though it may be required and profitable.

3. Generate sufficient income to support the family at an acceptable level. Being economically and technically efficient is not sufficient to guarantee cash to support the needs of the farm family. Either the farm business needs to be large enough to generate enough family income or family members must find employment off the farm. In reality farms have been getting larger as the number of farms declines (Figures 9.3 and 9.4) and more nonfarm income is being earned

FIGURE 9.5

Distribution of Farms and Gross Cash Income by Size of Farm

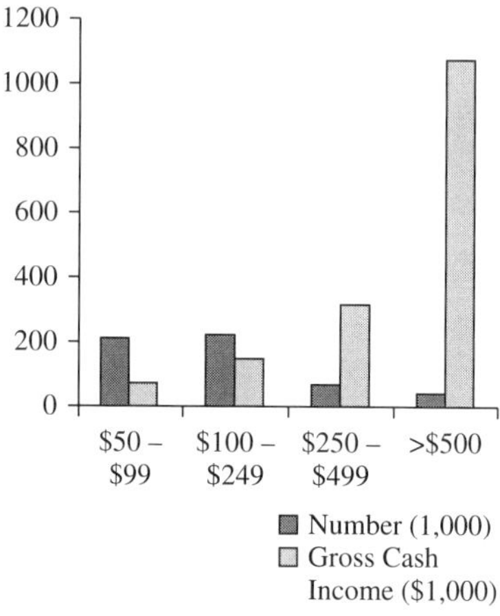

Source: USDA–ERS.

FIGURE 9.6

Total Farm Operator Income and Off-Farm Income by Size of Farm

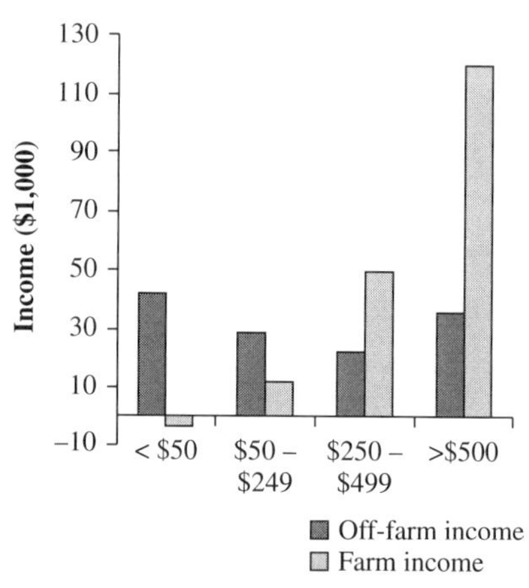

Source: USDA–ERS.

by farm families (Figures 9.5, 9.6, and 9.7). The purchase of additional land, equipment, livestock, and other productive capital is one means for increasing farm size, but there are other means. Expanding farm size by renting additional land and other real property is common. Machinery and equipment are likewise acquired by rental and lease agreements.

The form of business organization also is important when seeking to increase the income-producing capacity of the farm business. Partnerships are used to combine the capital assets of two or more individuals into a single business. Cor-

FIGURE 9.7

Income per Farm in the United States by All Farms and by Commercial Farms

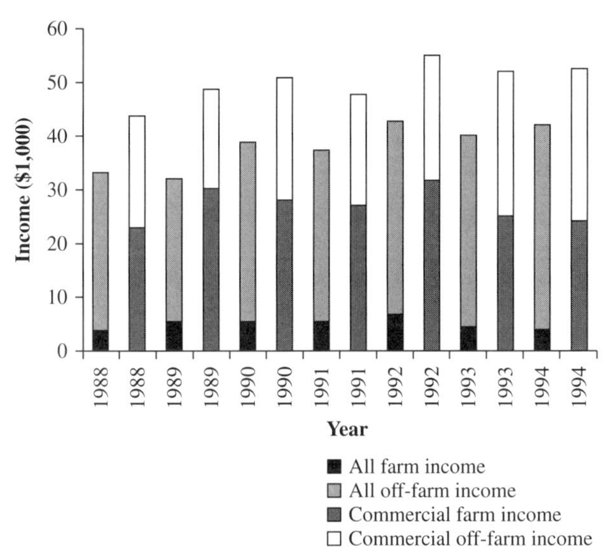

porations have the structure for consolidating capital resources from even a broader number of individuals, including family members.

4. Have a family business. Many farms are considered to be family businesses where parents and children work together. Family togetherness is proclaimed to be one of the real benefits of farming. Youth from the farm are recognized for having a strong work ethic, showing initiative and being honest and dependable. This benefit to society is used by politicians to promote one farm policy over another. Farm organizational structures and ownership patterns are means for bringing family members into the farm business. Many farm partnerships are between the parents and one or more of their children. Most farm corporations are closely held by members of the same family. Labor and operational agreements are used to bring family members into the business.
5. Maintain the farm business in the family through inter-generational transfer. There are several forces behind this goal. Most families would like to maintain a tight structure, particularly in farming where the children have worked with the parents in the business. The investment requirement to own and operate a farm business, even when part of the business is rented, is beyond the capacity of most beginning persons. The average asset value per farm in 1996 was nearly $508,000 for all farms. This investment is probably insufficient to generate a satisfactory level of family income. Farms having enough resources to support two families may need to manage more than $1,000,000 of resources (Figures 9.2, 9.5, and 9.6). Gifts during life, preferential sales, and inheritance transfers are important means for financing the next generation of farmers. Money flowing off the farm by asset sales and inheritance to off-farm children can create a real burden to those who wish to remain on the family farm. Farms can pass through an estate income-tax free, but if sold the capital gain will be taxed.
6. Reduce the amount of family wealth going to federal and state governments through income and estate taxes. Taxable estates valued in excess of $625,000 can

have considerable estate taxes to pay. Taxes on a $1-million estate in 1998 would be $143,750 with a marginal tax rate of 41 percent. Farm properties sold during life may have sizeable capital gains taxes to pay. If a farm with a tax basis of $200,000 is sold for $500,000 the capital gains income tax would be nearly $60,000 federal, plus another $20,000 to the state for a total of more than $80,000. Most families would like to avoid this tax drain and keep these dollars in the family. But, this approach is not always in the best interest of the parents in their retiring years. There are other means for reducing these taxes than through gifts and inheritance. Farm business organization and ownership are important considerations in whatever means are used to secure the retirement of the farm owners and transfer of farm resources to succeeding family generations.

The material that follows is important in achieving success in any one, or a combination, of these goals. How this can happen should be apparent to the reader in the following discussion. The previous discussion should help to make the presentation that follows more meaningful and understandable. Business organization and ownership are the topics discussed. The focus is not on the tax consequences of each option.

It is sometimes supposed that ownership of property should specify its organization. The fact that it may not could be confusing. Even though the nature of the ownership may give an indication of the type of business operation, the ownership could be separate from the organization. A husband and wife may own property jointly but operate the farm business as a single proprietorship. A husband and wife may even own property separately and still operate as a single proprietorship. However, they could operate as a formal partnership. Partners in a partnership can co-own property. But, a partnership can exist and not own any property. In many states a partnership may own property outright. A corporation is owned by its stockholders and the corporation may have title to property. Thus, the form of ownership could indicate the business organization, but it does not follow that it always does. It is even possible for a business to be declared a legal partnership when the owners thought they had a single proprietorship or a rental agreement.

Forms of Business Organization

Five forms of business organization are discussed: sole proprietorships, partnerships (general and limited), corporations, trusts, and limited liability companies. Trusts may not be considered a form of business organization, but they are important means of having farm property managed by a trusted individual, a trustee, on behalf of self or someone else. In recent years limited liability companies have come into being. These companies are organized like a corporation but taxed like a partnership. Business organizations are not only means for holding resources but also define patterns for managing the business. There are many important reasons for selecting one particular organization pattern over the alternatives.

There are significant differences in how businesses are taxed. These differences cause individuals to have ownership and organizational preferences. Tax planning is important and can make a difference, not only on an annual basis but also at retirement when the principal owners are ready to dispose of the property or pass it to the children. The form of business organization is important, even in death.

There are rules surrounding federal farm program participation and payments that affect the selection of a business arrangement. These rules have caused problems for some farm rental arrangements. Because of single payment limitations some farms have been divided to qualify for a greater total payment.

One form of business may be more suitable in estate planning than another. Whether it is keeping the business in the family, reducing estate and income taxes, providing an equitable distribution of family wealth, or assuring the retirees of cash flow, the form of business organization is important. The organization that meets the goals of the family may involve single proprietorships, partnerships, corporations, leasing arrangements and trusts, or some combination.

Sole Proprietorship

The sole proprietorship is the most common (86 percent) of all forms of farm business organization. It is the least complicated organizational structure. The simplest of all organizations is where a sole proprietor owns all the land, machinery, livestock, and other resources, borrows no money, and hires no labor. The sole proprietor receives all of the income and is liable for all of the taxes. However, sole proprietorships are usually more complex than outlined. The resources are often financed through one or more credit agents, land and machinery may be rented, and there may be several employees. In fact this person may be part of another business organization as a partner and own stock in separate business ventures.

Sole proprietorships are characterized by the single individual who acquires the resources, manages the business, is responsible for all of the liabilities, assumes all of the risk, and most importantly receives all of the profit and bears all losses. If any legal judgement is awarded personally or against the business, his or her belongings, future income, and business property may all be subject to attachment and sale. The resources commonly are individually owned or co-owned with a spouse.

The sole proprietor must pay all property and other business taxes and is liable for personal income taxes, including farm business income. Any capital gains increase is also subject to taxation when the property is sold.

The sole proprietorship is established simply by obtaining resources by purchase or lease and beginning to do business. No legal permits, licenses, or other authorization are required unless the particular business activity requires it. There may be zoning regulations, Environmental Protection Agency (EPA) restrictions, or other legal and permit requirements for particular activities. There are no restrictions on the size of business as to resources owned or controlled, or the volume of inputs used or products sold. The real limitations are with the acquisition of resources and the individual's managerial talents. Resource limitations and managerial ability may be the real handicaps of sole proprietorships.

The advantages and disadvantages of the sole proprietorship are reviewed in the following list.

Advantages:

- Personal freedom to act alone. There are fewer arguments over business decisions.
- Simplest form of business structure. It is easily formed and operated in a legal manner.
- The business can be easily expanded or contracted and enterprises can be changed without the interference from others.
- The operator is his or her own timekeeper and as such can begin and end the day at any time and can take a vacation when desired.
- All business profits belong to the operator.

Disadvantages:

- Specialization in a multiple-enterprise operation may be more difficult.
- Finances to purchase and operate the business may be more difficult to obtain because there is only one contributor. Obtaining adequate financing may limit the size of the business so as to reduce the volume to an inefficient level.
- Business and personal liabilities for debts and illegal acts can result in the attachment of not only business assets but also personal assets.
- The business is a fragile one, being based upon the health, well-being, and life of the operator. It is as easily dissolved as it is created. It is difficult to bring children into the business except as employees. At death the business may need to be sold and reorganized. If the assets are sold during life the capital gains tax could be large. The result could be a smaller inheritance for the surviving heirs. Estate planning may be more limited.
- The sole proprietor cannot be an employee and thus benefit from any employee benefits, some of which have important tax implications. Lost employee benefits may include workers compensation insurance, life and health insurance, retirement plans, and counting some limited personal costs as business expenses.

Proprietorships can be modified with various types of contractual agreements. Most are related to bringing a family member into the business. One of these agreements is the profit-sharing plan under which the employee can be rewarded for certain measurable performances with bonuses or a share in the gross income. Examples of profit sharing include rewards for the number of livestock births saved, high crop yields, and the amount of gross or net income of the business. Usually, a guaranteed base salary is paid. Profit sharing gives the employee an incentive to perform well. The benefits could include a program for gaining ownership in parts of the business such as the livestock.

Another arrangement is an operating agreement. The employee operates the total business in the absence of the owner, or part of the business in the presence of the owner. The employee receives a salary and also participates in an incentive plan. There may be provisions for the employee to begin buying into the business and for furnishing some operating capital such as machinery or livestock. The operating agreement could call for the major responsibility of the hog operation or the corn crop.

It is important to keep these agreements out of a partnership definition, which includes participation in losses as well as profits. Having agreements in writing should help. Some of the things to avoid are listed in the discussion of partnerships. Employee agreements are particularly useful for testing out a relationship before committing it to joint ownership and operation.

Partnerships

General partnership. The general partnership is a business organization where two or more persons have joined themselves to operate a business for profit. Note that ownership of property is not mentioned. Likewise, there is no mention about any legal procedure or percentage allocation of income. The partnership may be formal (i.e., written specifications of ownership, name, and responsibilities) or informal. A partnership may be distinguished by any of the following characteristics: a partnership name; a joint or partnership bank account; a joint set of business accounts; a division of management responsibilities; and the sharing of net income, particularly in the losses of the business. Partnerships ought to be in writing but many are just verbal agreements. In some cases it is the Internal Revenue Service (IRS) who declares the business to be a partnership in some tax dispute. A Uniform Partnership Act (UPA), which has been adopted by most states, specifies rules and requirements to be followed. But there remains much latitude for differences and uniqueness.

The formal written partnership (the preferred way) should contain the following elements:

- The name and location of the business.
- The names and addresses of the partners.
- Description of the business in general terms, including its purpose.
- Duration of the partnership, even if there is no specific termination date.
- The capitalization of the partnership, including the property each partner is contributing, how contributions are to be paid, whether the partnership will own property, etc.
- The amount of labor each partner will provide, the division of responsibilities, and the settlement of disputes.
- The division of income and when and how payments are to be made. Retained earnings are important in this decision.
- The accounting system to be used and who will keep the books. The method of tabulating taxable income as well as determining farm profitability needs to be specified.
- How disputes are to be resolved.
- Procedures to be followed in case of dissolution, death, withdrawal, expulsion, and injury should be specified.

Individual partners may furnish assets to the business in the form of real and personal property, and may be paid for their use. The partners may co-own the property, and in some states the partnership may own property outright as "tenancy in partnership." Forms of co-ownership are discussed later. The partnership also may rent or lease assets such as land and machinery. This arrangement makes it possible for a son or daughter to enter the business as a partner without owning much property. Assets can be transferred gradually through gifts, preferential sales, and joint ownership of new purchases.

Thus, it is not only important to define the ownership of assets and how they will be compensated but also how the profits will be divided. The right to a share of the income is in itself a property right and is recognized in the UPA. The division of income does not need to be equal for all of the partners. An equitable distribution is arrived at by considering the contributions of each party, aside from property furnished that is paid for out of joint income. An example of a division of income by the contribution of each party is shown in Table 9.1. Even though this illustration shows an uneven share of profits the common procedure is to share profits (and losses) evenly after compensating each partner for any uneven contribution of assets to the business.

Typically the partners share equally in management responsibilities; thus, each partner has the same voting power. Usually the business is divided into major income-producing activities and each partner has major responsibilities for its day-to-day activities, and only major management decisions are discussed. Each partner is considered an agent of the partnership and may legally bind the partnership, so long as they act within the scope of the business. An exception is where the partner is not authorized to act and the person with whom he or she is doing business knows that he or she lacks authority. If a partner acts in an authorized manner to acquire property by purchase or contract, or his or her actions cause a loss or injury, the partnership is liable for any damages or losses.

The partnership pays property taxes and sales taxes but it does not pay income taxes. The partnership files an information return (Form 1065) that sets forth the individual's income from the partnership. The individual then adds this income to other income that he or she might have, subtracts deductions, and pays the tax. Partnership income is reported on Schedule E and is a line item on Form 1040. This is also true for

TABLE 9.1

Sharing Partnership Income According to Individual Partner Contributions

Contribution:	Partner #1	Partner #2	Total
Land	$500,000	$ 0	$500,000
Machinery	80,000	50,000	130,000
Livestock	30,000	30,000	60,000
Labor	9 months	12 months	21 months
Allocation:			
Land @ 4%	$20,000	$ 0	$20,000
Machinery @ 10% + depreciation	16,000	10,000	26,000
Livestock @10%	3,000	3,000	6,000
Labor @ $1,500	13,500	18,000	31,500
Total	$52,500	$31,000	$83,500
Percent of total	63	37	100

social security tax. The partners are considered to be self-employed and thus they must file Schedule SE. Children of partners working for the partnership are not treated like children of sole proprietorships in that social security taxes are required to be withheld from wages and paid. If salaries are paid to individual partners these are expenses to the partnership and taxable income to the receivers. The method of accounting to determine taxable income may be either cash or accrual.

One of the problems of partnerships is with the personal liability for partnership obligations. This problem was alluded to earlier. The order of liability is as follows: First, partnership assets are used to secure partnership liabilities. Second, individual partner's personal assets are used to satisfy their personal obligations to creditors and others for accidents or wrongful acts. Third, the personal assets of partners may be used to satisfy partnership creditors or other obligations of the partnership. Fourth, partnership assets, to the extent of the individual partner's share, may be used to meet the credit and other personal obligations of individual partners. This liability can be partially alleviated by having adequate liability, and in some cases credit insurance. It is important to know well those others in the partnership.

Another problem of the partnership is with its instability, although it may be more stable than the sole proprietorship in certain aspects. Any time a partner wants out, or is forced out through death or other reason, the partnership may cease to exist. The larger the number of partners the greater this problem is, but there are means for reducing its effects. In the case of death, a new partnership may be formed by the remaining tenants, or an heir can be brought into the business. The remaining partner(s) may form a new business. The partnership agreement may be worded so as to bind the surviving spouse of the deceased person. Many partnerships contain buy-sell agreements that provide the remaining partners the first opportunity to purchase the leaving partner's share.

The advantages and disadvantages of the general partnership have been put forth but are reviewed below.

Advantages:

- There is a pooling of capital, which may provide a larger and more efficient business.
- There can be a better-trained management core to handle the specialized needs of the business.
- The partnership agreement is relatively simple and easy to change as experience is gained and circumstances change.

- Business continuity may be enhanced over that of the sole proprietorship.
- Transferring of family resources to succeeding generations through gifts and preferential sales is made easier.

Disadvantages:

- Decision making may be more difficult with more persons opinions to contend with.
- Disagreements over contributed shares and income distribution may arise.
- The parents' retirement may be disrupted (i.e., it may be difficult to collect social security payments because of active participation in the business and the younger partners may wish to expand the business when the parents wish to slow down and reduce debt and other involvements.
- The amount of book keeping may be increased. This may be an advantage if the records are used for management purposes.
- The liability of the individual partners is increased do to the obligations of the partnership. This can be improved with insurance.
- The partnership is by its nature relatively unstable. Its stability can be improved through the agreement.

Limited partnership. Limited partnerships have one or more general partners and one or more limited partners. The liability of the limited partners is their investment in the partnership and they do not participate in management. General partners have their usual features of general partner status, including unlimited liability and the right to participate in the management of the limited partnership.

Limited partnerships have been used in farming to consolidate capital and as an investment opportunity. In times past the limited partnership was often a tax-loss operation with the limited partners using the farm loss to offset nonfarm income, with the hope that there would be better times ahead. This tax opportunity no longer exists. More recently the limited partnership has been used to hold the land, which is then rented to the general partnership, sole proprietorship, or corporation. The off-farm heirs could be limited partners and receive a payment for their share of the ownership in the land. Other investment opportunities exist.

The limited partnership is more formal than the general partnership and must not only be written but also filed for public information with a state agency. It is important to distinguish it from the corporation to be income taxed as a partnership and not as a corporation. There are distinguishing characteristics for tax purposes:

- Continuity of life—power of any partner to dissolve the partnership.
- Centralized management—only general partners have management control. Centralized management generally exists only when the general partners own more than 20 percent of the interests in partnership capital or profits.
- Free transferability of interests—if substitution of a successor partner is dependent upon the consent of the general partners, the limited partnership does not have free transferability of interests.
- Limited liability—despite the unlimited liability of general partners, a limited partnership is considered to have limited liability if the general partners have no substantial nonpartnership assets reachable by creditors and the general partner(s) are merely acting as agents of the limited partners.

Limited partnerships offer flexibility in organizing the farm business for the transfer of family wealth to succeeding generations. Those interested in farming can operate the farm as general partners, or other organizational structure, while the family resources are held in a limited partnership to benefit the retiring farmer and the off-farm children. The organizational structure is less formal and involved than the corporation.

Corporations

A corporation is an artificial being created under state law. Thus, there may be minor differences among states, although laws are enough alike to allow for doing business across state boundaries. The major characteristic of the corporation is the sharp line of distinction between the business and its owners. The corporation is a separate legal "person" and may be a separate income taxpayer. It is a separate business entity, distinct from its owners, the stockholders. Corporations may conduct business, own property, make contracts, pay taxes, be sued, and sue, much the same as any person.

Corporations can be "publicly owned" or "closely held." A publicly owned corporation may have thousands of stockholders, whereas closely held corporations have only a relatively few stockholders, often family members. Most farm corporations fall into this latter category and are used to keep the farm business in the family. The pooling of capital increases the size of business and enhances growth to provide income for more than one family.

State law usually specifies that one or more persons may act as incorporators. These persons file necessary papers with the proper state public agency, often the secretary of state. The articles of incorporation usually detail the following:

- Name of the corporation
- Purpose(s) of the corporation
- Number of shares of stock authorized to issue
- Kind of stock issued and any special restrictions regarding voting rights, etc.
- Whether the stock is closely held or public
- Registered office or agent
- Duration of the corporation
- Number of initial directors, their names and addresses

Bylaws also are developed at the time the corporation is formed. They specify the internal governing rules and are less formal than the articles. Included are procedures governing the board of directors, officers, and stockholders. The time and place of shareholder and board meetings, quorum requirements, and duties of officers are covered. The shareholders elect directors and the directors elect officers who run the day-to-day activities of the corporation.

One of the most important steps in the formation of the corporation is its financial structure. Corporations are financed by stock and debt securities. The stock is termed equity capital. It includes common stock and may include preferred stock. The value of common stock, although initially set, fluctuates over time on the basis of dividends paid, the growth and stability of the company, and other investment opportunities of the stockholder. Preferred stock values are more stable because the dividend rate is set somewhat like interest at the time of purchase. Sometimes preferred stock does not carry voting rights. Debt capital comes from notes, bonds, or debentures. Bonds are long-term obligations and are secured by corporate assets. Debentures are long-term unsecured obligations. Debt securities do not carry the right to participate in management, whereas stockholders do participate in management. Holders of common stock receive income in the form of dividends as declared by the board of directors. Dividends are paid out of business profits after paying interest on debt securities. Interest on debt has priority over dividends.

Corporation stock is highly transferable, such that in a closely held corporation restrictions are placed on who may own it. There may be requirements that sales must first be offered to remaining shareholders. Buy–sell agreements may require that in the case of the death of a shareholder the other shareholders, or the corporation must buy the stock. Some of these restrictions may become unwieldy with the passing years.

The initial stock of a farm corporation is usually contributed by family members. It consists of the equity in farm assets of the business organization in existence prior to the incorporation. Individuals trade their interests in real and personal capital (i.e., land, machinery, livestock, and inventories) for shares of stock. Debt on the assets traded may be transferred to the corporation. Shares of stock also can be purchased for cash to supply operating capital. The corporation can draw upon cash reserves and use corporation assets as collateral to borrow money for operations and investment. The corporation can lease property and become a shareholder in other organizations to gain resources for an efficient operation. Assets can be transferred into the corporation in a tax-free exchange. The assets carry with them the basis they held in the prior organization and no capital gain is taxed at this time.

In addition to receiving stock for assets transferred into the corporation, bonds and debentures can be issued. They carry fixed interest obligations, and thus may be used to secure the retirement income of parents in the family corporation. Bonds and debentures, along with preferred stock if authorized, can be held (through purchase or gift) by nonfarming members of the family.

Voting in the corporation is according to the shares of stock held. It is easy to recognize that any one holding 51 percent of the stock controls the voting, can be on the board of directors and serve as an officer. All of the stock necessary to control the corporation is a majority share. The management control is easily retained by the parents in a newly formed family corporation. This process of funding the corporation makes it possible for parents to transfer stock to their other children through annual gifts and favorable purchase. It is easy for others to buy into the business through stock purchase.

An example may be helpful in understanding how the capitalization of a family corporation may take place. Assume a family composed of a father and mother with four children, two of which are interested in farming. Table 9.2 sets forth the assets before incorporating and the shares of stock and loans after incorporating.

TABLE 9.2

Asset Conversions When Organizing a Farm Corporation

Assets before incorporating:		
Father and mother	$600,000 farm, $400,000 equity	
Child #1	$ 40,000 cash or equity in assets	
Child #2	$ 60,000 cash or equity in assets	
Child #3	no interest in farming	
Child #4	no interest in farming	
Assets after incorporating: (Stock shares are valued at $1,000 each.)		
	Stock shares	Debentures ($1,000 each)
Father	100	100
Mother	100	100
Child #1	40	
Child #2	60	—
Total	300	200

The parents can make gifts, during life and at death, of stock to children #1 and #2 and of debentures to children #3 and #4. The interest on debentures also may provide cash flow to the parents in their retirement.

Farm corporations can be regularly income taxed as a business or elect to be taxed as a partnership; that is, the corporation pays no tax but the shareholders are taxed for the dividends and other income received from the corporation. The net income it makes is subject to income taxation, and dividends may again be taxed to the shareholder. The first is known as a regular, or Subchapter C corporation. The second is known as a tax-option, or Subchapter S corporation. "C" corporations pay taxes at rates between 15 and 34 percent (1997) on taxable income. "S" corporations file an information return (Form 1120-S) and pay no taxes. Tax-option corporations were created to benefit small businesses, including farmers.

Subchapter S corporations must meet certain criteria to qualify for preferential tax treatment. The following are included:

- Only common stock can be issued.
- There can be no more than 35 stockholders.
- Stockholders must be individuals, estates, or grantor trusts, and not partnerships or other corporations.
- All shareholders must consent to the election.
- There can be no nonresident alien stockholders.
- No more that 20 percent of the corporation's gross receipts can come from rents, royalties, dividends, interest, or annuities.
- The farm must be oriented toward making a profit.

Corporations may have employee benefits that are not accessible to sole proprietorships and partnerships. Because stockholders in most farm corporations are also employees they may participate in employee benefit plans. These benefits include health and life insurance and retirement plans. In addition to favorable group rates, the premiums, with some restrictions, are business deductible. In a few cases where the corporation owns housing its cost may be deductible, and in even fewer cases meals may be deductible. Social security tax rates are the same for the corporation as for the sole proprietor, but the corporation can include the employer share as a business expense. Children of stockholders are treated as other employees relative to social security taxes and other tax withholdings.

There are disadvantages to corporations, even though this discussion may sound very positive. The corporate form of business is more complex, making possible more chances for problems to arise. There needs to be annual stockholders meetings and the board should get together more often. The amount of bookwork is increased and accounting becomes more complex. Dividends must be determined and paid. An annual report must be filed with a state agency and an annual fee must be paid. Because of the legal complexities lawyer services will be increased.

One of the real advantages of the corporation is with its limited liability. Shareholders and owners are legally responsible only to the extent of the capital they have invested. Personal assets of the shareholders cannot be attached by creditors to meet the obligations of the corporation.

The three forms of organizing a farm business (sole proprietorship, partnership, and corporation) are compared in Table 9.3.

TABLE 9.3

Comparison of Farm Business Organizations[a]

	Sole Proprietor	Partnership	Corporation
Nature of entity	Single individual	Aggregate of two or more individuals	Legal person separate from shareholder-owners
Life of business	Terminates on death	Agreed term; terminates at death of a partner	Perpetual or fixed term of years
Liability	Personally liable	Each partner liable for all partnership obligations	Shareholders not liable for corporate obligations
Source of capital	Personal investment; loans	Partners' contributions; loans	Contribution of shareholders for stock; sale of stock; bonds, and other loans
Management decisions	Proprietor	Agreement of partners	Shareholders elect directors who manage business through officers selected by directors.
Limits on business activity	Proprietor's discretion	Partnership agreement	Articles of incorporation and state corporation law
Transfer of interest	Terminates proprietorship	Dissolves partnership; new partnership may be formed if all agree	Transfer of stock does not affect continuity of business—may be transferred to outsiders if no restrictions.
Effect of death	Liquidation	Liquidation or sale to surviving partners	No effect on corporation. Stock passes by will or inheritance.
Income taxes	Income taxed to individual	Partnership files an information return but no tax. Each partner reports share of income or loss, capital gains and losses as an individual.	**Regular Corporation** Corporation files a tax return and pays tax on income; salaries to shareholder employees deductible. **Tax-Option Corporation** Corporation files an information return but pays no tax. Each shareholder reports share of income, deductions, losses, and credits of the corporation.

[a]Adapted from: *The Farm Corporation,* NCR 11, Iowa State University Extension, Ames.

Trusts

Trusts are the least used, but one of the most flexible instruments in estate planning. Because trusts do not own property, they are different from the other three forms of business organization previously discussed. A trust is composed of a trustee, the property to be held in trust, and the beneficiaries. As strange as it may seem the trustee can be the one owning the property and also the beneficiary. A retiring person may put in trust property to benefit a surviving spouse, or a married couple may place property in trust to benefit a surviving minor child. A trust can be used to set up a sequence of distributions triggered by certain events. The one setting up the trust (grantor or settler) should know exactly what is to be accomplished.

Trusts can own property, outright or in combination with others, and this property can be managed by a trustee. Instructions can be given in the trust instrument of how trust property is to be managed and how the income and principal are to be distributed. Trusts can be the receiver of capital from outside sources such as insurance policies, retirement programs, and wills.

There is a variety of trusts available for business management and transfer to beneficiaries. Living trusts are set up during the grantor's life with property transferred to the trust. They may be either revocable or irrevocable. Revocable trusts can be changed, amended, and even canceled. At the death of the grantor or other specified event, they may become irrevocable. Irrevocable trusts cannot be changed and may be used to provide income to a surviving spouse. After the death of that spouse, the trust properties may be passed to designated heirs or to some favorite purpose.

Trusts may be used to reduce estate settlement costs. The trustee can be empowered to pay off creditors, pay the costs of death, and distribute the residual as directed. Cost of probate and other legal fees, the hiring of an administrator, and other costs can be reduced. Trusts can be used to reduce federal estate taxes with advanced planning. For example, the properties of a business can be divided between the husband and wife with each leaving their share in an irrevocable living trust for the life of the survivor, after which the residual passes to the deceased person's heirs. Thus, the survivor can have the income of the total property for life, but only the survivor's trust property is subject to estate taxes at the death of the last survivor.

Testamentary trusts are often set in personal wills. These may not save taxes but they could save estate settlement costs. Their major purpose is to manage property for a beneficiary. These persons could be minors, handicapped, or in other situations for which it is desired that the benefactor's share is to be received over a period of time.

It is the remainder persons or beneficiaries that discipline the management of trusts, particularly after the death of the grantor. Unless contested the trustee conceivably could make decisions not authorized in the trust document. Thus, it is important that communication lines be kept open with regular reporting to all who might be affected.

Limited Liability Companies

The limited liability company (LLC) is a hybrid form with the limited liability of the corporation, which is taxed like a partnership. Most states have recognized them. There must be at least two members of any entity type. Ownership interest can be of a single type or multitiered. Management is vested in its members and they may define how their LLC is to be operated and member rights.

Members of an LLC may need to pay self-employment tax (FICA). Members are treated as limited partners if they are not managers. Managers pay FICA but others do not. But if the LLC has no designated managers then all members are managers, and hence pay the tax. The cash method of accounting may be used unless the operation is a farming syndicate, in which case the accrual method must be used.

Whether an LLC is treated as a partnership or a corporation depends upon its characteristics. Corporations have continuity of life, centralized management, and free transferability of interests, which partnerships do not have. LLCs do have limited liability.

It can be easily understood from this brief explanation that LLCs can be very complicated and involved. Their formation needs to take place under the guidance of well-informed legal council. Furthermore, its members must be able to understand and operate under this form of organization. Saving taxes is an insufficient reason for selecting any form of organization.

Farm Business Ownership

The most common method of obtaining resources to farm with is through purchase. Thus, it is important to understand property ownership rights, particularly for real property. Even though these rights in the United States are the broadest and freest of any place in the world, restrictions still apply. The rights in property are often referred to as a bundle. Ownership may not have unlimited use of all of the rights in the bundle. The rights retained by government in real property include the following:

- Condemnation or eminent domain. This is authority to take private property for public use. Public use includes acquisition for highways, dams, utility lines, and buildings such as schools. The owner is paid a fair price for the land and the creditor is protected.
- Police power. Government retains the right to regulate the use of the property to protect the health, safety, and well-being of the public, including the owner. Zoning laws, building codes, and EPA regulations are among those things enforced.
- Taxation. Not only can the property be taxed outright to support government and community services but also can be used as a bonding base for capital improvements such as schools and municipal buildings. These taxes all tend to reduce the value of the owner's property.
- Escheat. If an owner dies intestate and has no legal heirs at death, then the property reverts to the government.

The rights in ownership are those of fee simple, commonly abbreviated "fee." These rights include uses for business and pleasure and are not limited in time. Thus, property can be pledged as security for a loan, perhaps to finance its purchase, and can be sold or traded. But, even with fee, it is important to know the extent of ownership. In addition to surface rights, there are aboveground and underground rights to consider. Also, the title may specify certain use restrictions.

Historically it has been assumed that the landowner controlled the rights to the air above and the minerals and water below. These rights may have been partitioned away and retained or sold and thus are not part of that bundle. Prior owners may have placed restrictions that limit the use for prescribed purposes such as schools and churches. Land with these restrictions may be referred to as determinable fee estates. It is important that the purchaser have the title thoroughly researched and have an abstract prepared to know exactly what he or she is getting before signing any purchase contracts.

The title to real property generally passes from one owner to another by deed. If a seller provides a quit claim deed, all rights this person had are transferred to the owner, but the title is not warranted against other claims. To be sure the title is free from any and all outside claims, the buyer should request a warranted deed from an abstract or title company.

Ownership in property can be by an individual or by a group. For a single owner or proprietorship, the title is in one name only; that person has all of the rights of ownership. However, a spouse, particularly the wife, may be required to sign the papers because of dower interests.

Co-ownership of property takes two forms, tenancy-in-common and joint-tenancy. Under both forms the owners hold undivided interests and there are few differences during life. Each co-owner's interest can be sold or pledged as security for a loan and each can partition to get out, as in the case of dissolutions of marriage. The big difference comes at death. The deceased shareholder's interest in a joint-tenancy passes directly to the surviving tenants, and not the deceased person's heirs. A will to the contrary has only secondary power. Under tenancy-in-common ownership the deceased's shares pass to his or her heirs, or as otherwise specified in the will. This distinction is of great importance to partnership co-owners. It has happened that a spouse has lost his or her inheritance to the deceased's partners co-tenant. Tenants-by-the-entirety is similar to joint-tenancy and is used in several states for husbands and wives.

Several states have community property laws, meaning that all property acquired after marriage, except by gift, devise, or inheritance, belongs equally to both spouses. This rule supercedes any particular title. Property brought into the marriage or received by gift, devise, or inheritance is classified as either the husband's or wife's separate property. There is some latitude in how shares of community property may be distributed at death but if there is no will then one-half is distributed according to state statutes, and the survivor's share is protected.

There are other rights in property than ownership. Two important rights associated with ownership are life estates (similar to estates for years) and remainder interests. As mentioned earlier in this chapter, life estates can be created to secure the well-being of a surviving spouse, and an estate for years for a minor child. These rights are usually accomplished through trusts. The holder may have the right to manage the property for its income but usually not to diminish the principal value. Husbandman conditions may be specified. The property could even be sold and the principal converted into another form. Remainder interests, sometimes called reversionary, are those remaining after some other person is finished using the property, or certain conditions are met. Remainder persons could be the children after the death of the last parent. Thus, remainder persons discipline the provisions of wills, life estates, and trusts to protect their own interests.

Leasehold estates are another form of holding interests in property for business purposes. In this case the user does not own the property but pays for its use over some period of time. The owner receives a cash payment or shares in the products. The period of the lease may be specified or be for an indefinite time with notification of termination. Some states have passed legislation that requires lessees to have 90 or more days of notice. Fraud laws may govern certain aspects of leases.

Farm Lease Arrangements

As was mentioned at the beginning of this chapter, a substantial part of the land in farms is not owned by its operators. In some farming communities more than two-thirds of the land is owned by absentee landlords. In the Midwest, south, and many other parts of the United States where cash crops predominate, farm management firms thrive by managing farm property for absentee landowners. Renting is often used to pass the operation of a farm from the parents to their children. The parents may even rent the land to a family partnership or corporation. Some nonfarm business persons own farm land for an investment, similarly as they would stocks and bonds. Farmland has been considered a good hedge against inflation. For whatever reason, farm leasing is a major means of obtaining land, and in some cases personal property, including machinery and livestock, for farm operation.

There are two basic forms of leasing arrangements: cash and share. Under cash, the lessor (owner of the resources) receives a cash payment from the lessee (operator of the farm). Share arrangements divide the costs of production and the produce between the lessor and the lessee according to some prearranged formula. The prevalence of these two forms varies by community and over time. Most retiring farmers prefer a stable cash payment to a variable share of the product. Other owners want to be more involved in management and share in the benefits from risk. In all leases the details and payments are arrived at by bargaining, and can be best worked out over time; it is a game of which both can be winners. Following community standards can cause lease failure by not recognizing the specifics of each arrangement.

Leases may be written or simply verbal. Verbal leases leave more room for disagreements. Written leases not only have the advantage of improving people's memories but also cause the contracting parties to be more thorough in developing important details of agreement. A legally drawn written lease should provide the following:

- An accurate description of the property
- Period of time over which the lease extends and renewal dates
- Price or allocation of receipts and expenses
- Names of both parties
- Signatures

These legal minimums are insufficient to ensure a continuing lessor–lessee relationship. The lease needs to provide for a profitable system of farming with a fair division of income and expenses. The yearly income needs to be reasonably stable without unexpected fluctuations. There needs to be security to both parties. Being initially careful to work out an economically efficient lease, and then having regular meetings to communicate difficulties can help ensure a long-time relationship.

Factors affecting leasing terms include the following:

- Value and kind of contribution each contracting party is making.
- Alternative investment and income opportunities for each party.
- Bargaining position and ability of each party.
- Prevailing rental arrangements in the community.
- Improvements, facilities, location, and size.
- Crop production opportunities and livestock considerations.
- Productivity of the land and other resources.
- Economic climate.

Cash Leases

Cash leases are comparatively simple, but arriving at a fair payment may not be easily achieved. The tenant pays a set sum for the farm and in return receives all of the income and pays all of the expenses of production. The lessor usually pays all real property taxes and property insurance and maintains the buildings and improvements. He or she wants a fair return on investments. Table 9.4 illustrates how the landlord might approach an asking price.

TABLE 9.4

Landlord Approach to an Equitable Cash Rental Rate

Interest on investment in land, buildings, and other resources furnished:	
320 acres × $1,200 × 4%	$15,360
Repairs on buildings @ historical average	650
Insurance	3,000
Depreciation:	
Buildings, $60,000/30 years	$ 2,000
Fences, $8,000/20 years	400
Water system, $5,000/20 years	250
Total	2,650
Management expenses	700
Total	$22,360
Landlord's per acre rental fee	$ 70

The tenant, however, approaches the amount he or she can pay on the basis of the expected net return to labor and management. The specific crops to be grown, yields, prices, inputs and their costs, hired labor, interest, and depreciation on machinery are important. The items shown in Table 9.5 may be on the cash tenants list.

A review of Tables 9.4 and 9.5 shows that the landlord's proposed rate is $5 per acre more than that developed by the tenant. This difference is reasonably close and an agreement should not be difficult to reach. There are times when prices are good and landowners have the advantage in the bargaining process. The bargaining power depends on the number of lessees and lessors there are actively looking to make a deal. It should be remembered that in the long run, given an active market, rental rates will move toward a common price.

There are advantages and disadvantages with the cash lease, and these are not the same for the landlord and tenant. Some of these advantages and disadvantages are as follows:

TABLE 9.5

Tenants Approach to Arriving at a Fair Cash Rent

Gross value of crops produced:					
Crop	Acres	Yield	Production	Price	Value
Corn	150	130	19,500 bu	$ 2.75	$53,625
Soybeans	50	38	1,900 bu	6.25	11,875
Alfalfa	30	6	180 ton	65.00	11,700
Small grain	15	65	975 bu	3.60	3,513
Government	35				4,500
Total					$85,213
Operating costs of production:			Unit	Cost	
Corn	150		acre	$125	$18,750
Soybeans	50		acre	85	4,250
Alfalfa	30		acre	145	4,350
Small grain	15		acre	95	1,425
Government	35		acre	25	875
Total					$29,650
Fixed costs:					
Depreciation on machinery					$15,000
Interest on average investment @ 9%					6,500
Insurance					2,500
Total					$24,000
Overhead costs:					
Operator labor on production and maintenance					$ 8,500
Management @ 3% of gross					2,500
Total					$11,000
Net income to total farm					$20,563
Tenant's net income per acre					$ 65

Advantages:

To the landlord:

- It provides a steady flow of cash income.
- Close supervision and involvement is not required.
- Leasing terms are relatively simple and there is less room for controversy.

To the tenant:

- Independence in the farm operation.
- Receives all of the benefits from his or her management.

Disadvantages:

To the landlord:

- The tenant may tend to exploit the farm. This tendency can be reduced with longer contracts and protection specifications in the leasing terms.
- The cash payment, in the long run, is probably less than for a share lease because the tenant assumes nearly all of the price and yield uncertainty.

To the tenant:

- The landlord is reluctant to furnish improvements. Because he or she does not participate in the increased benefits there is little incentive to do so. This could be partially solved with a higher cash payment.
- Cash rents do not automatically adjust for price and production changes.
- There is greater income uncertainty.

Flexible Cash Leases

Because the tenant assumes nearly all of the income uncertainty, there are pressures to modify the strictness of the fixed rent payment by developing flexible payments. These are tied to the price or yield of the principal crop, or both. Three examples are illustrated below.

1. Price flexible cash rent: $(Y \times P \times F) + (Z \times I) =$ cash rent

where Y = base yield of major crop,
P = base price of crop per yield unit,
F = a fixed percent such that when multiplied by Y and P gives cash rent which is near the community standard,
Z = some dollar amount that is related to the change in net income as I increases or decreases by one unit, and
I = the number of unit changes the current price is above or below the base price. The unit is a price change, set at some percent of the base price (i.e., 4 percent).

Assume the crop is corn with Y = 125 bu, P = \$2.50, F = 25%, Z = \$4.00, I = 3 (the unit is \$0.10 [0.04×2.50] and the current price is \$2.80):

$$(125 \times \$2.50 \times 0.25) + (\$4.00 \times 3) = \$90 \text{ per acre}$$

2. Yield flexible cash rent: $(Q \times G \times P) =$ cash rent

where Q = actual yield per acre,
G = percentage of product that will estimate the fixed cash rent under typical yields and prices, and
P = base price per yield unit.

Assume the above example and Q = 125, G = 25%, P = \$2.50:

$(125 \times 0.25 \times \$2.50) = \$78$ per acre cash rent

3. Yield and price flexible cash rent: $(Q \times G \times R)$ = cash rent

where R = actual price per unit of product.

Assume: Q = 130, G = 25%, R = \$2.80:

$(130 \times .25 \times 2.80) = \$91$ per acre cash rent

Cash rents can be made so flexible that they lose their intended purpose. Cash rents should not be used to accomplish what a share lease can do better.

Share Leases

Share leases are of two kinds: crop share and livestock share. This discussion focuses on the crop share lease because it is the most common. Livestock share leases follow the same principles. The crop share lease is often called the crop share–cash lease because a combination of payments is used. Under a crop share lease the landlord furnishes the land and pays the related ownership fees. The tenant furnishes the machinery and the operating labor, and pays those costs not related to production. The tenant and landlord share in the costs of production. They also share in the crop production before sale. Because the pasture and forage crops are not easily divided a fixed cash payment is usually paid for this land.

There is a simple tried-and-proven economic principle that guides the division of expenses and production. It is that expenses are shared on the same percentage rate as

FIGURE 9.8

Economics of Sharing Costs and Returns in Share Renting

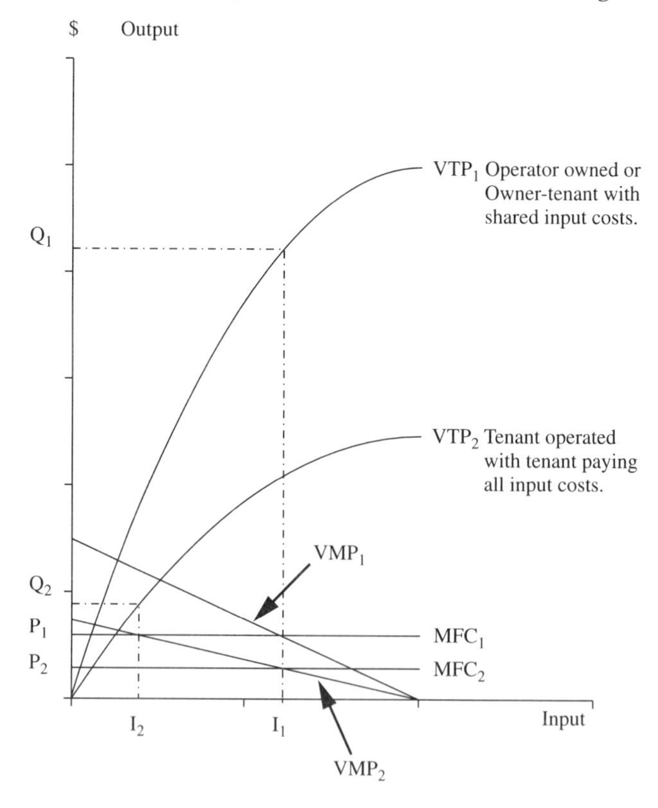

is the production. Consider Figure 9.8. If this were an owner-operated farm, the operator would maximize income by applying input I1 and producing Q1. If this is a tenant-operated farm and he or she furnishes all of the input and receives only one-half of the product, the optimal input drops to I2 and the total product drops to Q2. The input price is P1 in both of these situations. In this tenant situation, both the tenant and landlord suffer by not getting as high of an income as they could by sharing the costs and returns equally. If the landlord shares in the expenses, then the input price drops to P2 and the optimal production level returns to Q1. The landlord should participate in all input costs that affect yield and where the tenant may reduce the input to save money if the landlord does not share in their costs. Beyond these inputs, the sharing of costs is a bargaining affair. Harvesting costs are a case in point. Will these affect the yield? Probably not, but landlords have been participating in their costs anyway.

There are two approaches for arriving at a fair division of expenses and receipts (Table 9.6). The first, fixed contribution approach, considers only fixed inputs. The second, fixed and operating share approach, considers both fixed inputs and operating costs.

Including operating expenses in the tabulation could change "fixed contribution" percentages slightly. The cooperators may begin by sharing operating expenses equally. Under the fixed contribution approach, the landlord would pay 40 percent of the operating costs and receive 40 percent of the share-rent products, whereas under the fixed and operating share approach, the landlord's share of production increases slightly from 40 to 43 percent.

TABLE 9.6

Fixed and Operating Share Approaches to Leased Land Income

	Whole farm	Landlord	Tenant
Fixed contribution approach			
Land			
Interest ($384,000 @ 4 percent)	$ 15,360	$15,360	$
Taxes ($384,000 @ 0.3%)	1,152	1,152	
Buildings and improvements			
Depreciation ($60,000/25 yr)	2,400	2,400	
Repair ($60,000 @ 5%)	3,000	3,000	
Insurance ($30,000 @ 0.4%)	120	120	
Power and machinery			
Interest ($50,000 @ 8%)	4,000		4,000
Depreciation ($110,000/12)			9,167
Repair ($110,000 @ 4%)	4,400		4,400
Labor and management	11,000	1,000	10,000
Other			
Cash rent	00	–2,500	2,500
Liability insurance	200	100	100
Miscellaneous	1,200	400	800
Total	$51,999	$21,032	$30,967
Percent	100	40	60
Fixed and operating share approach			
Fixed contribution totals:	$51,999	$21,032	$30,967
Operating expenses:			
Corn production	18,750	9,375	9,375
Soybean production	4,250	2,125	2,125
Small grain production	1,425	712	713
Total	$76,424	$33,244	$43,180
Percent	100	43	57

These approaches are first approximations for the division of contributions and products and can be fine-tuned after one or more year's experience. Good records and candid communication is the key to a long-term association.

With livestock share arrangements, most of the income is from livestock sales. It is better adapted to lands not suited to cash crops. The major difference is that the landlord and tenant co-own the livestock. The landlord furnishes the land, buildings, and other real property. The tenant furnishes the machinery and equipment and most of the labor and management. Because of the livestock ownership the livestock-share lease is sometimes interpreted to be a partnership. Thus, it is important to have leasing terms specifically defined.

As with cash leases, there are advantages and disadvantages to share leases. Some of these are as follows:

Advantages:

To the landlord:

- His or her share of profits is generally larger than with cash.
- There is greater opportunity to participate in management.
- There is a more direct return from added investments made.

To the tenant:

- There is less risk than with cash rent.
- There is less cash and capital needed to operate the business.
- The landlord is more willing to make new investments.

Disadvantages:

To the landlord:

- Greater cooperation is required.
- The cash flow of income may be more variable.
- There is a possibility that social security benefits may be interrupted.

To the tenant:

- The landlord may be overly interested in high yields.
- The landlord may be unwilling to furnish livestock improvements if it is a crop-share arrangement.
- There is a considerable loss of freedom.

Family and Farm Estate Planning

Traditionally, farms were born, reached a peak of efficiency, matured, and died, much like their owners. This resulted in a great loss of efficiency and a transfer of wealth out of the family. Estate planning is a tool for maintaining both a high level of efficiency and keeping wealth within the family. Various organizational structures have been presented for achieving this goal. But there needs to be a testing and transition period.

The spin-off model has been proposed as a means of looking at the growth of farming alternatives for the child. This model is presented in Figure 9.9. The parents create the opportunity for the child to test the farming way of life through a wage agreement. This approach can have income incentives built into it. If the parents and child can work together then management responsibility in the form of enterprise agreements could be added. The child now shares in the income of the business. Eventually the child becomes a full partner, or a corporation is organized. There is opportunity during this testing and development period for the child to drop out and move away from farming or to farm by him or herself. If the child wishes to farm, and the parents and child are compatible, then a series of gifts and sales is negotiated to help the child farm.

FIGURE 9.9

Road Map of Farm Entry Transfer Process or Spin-off Model

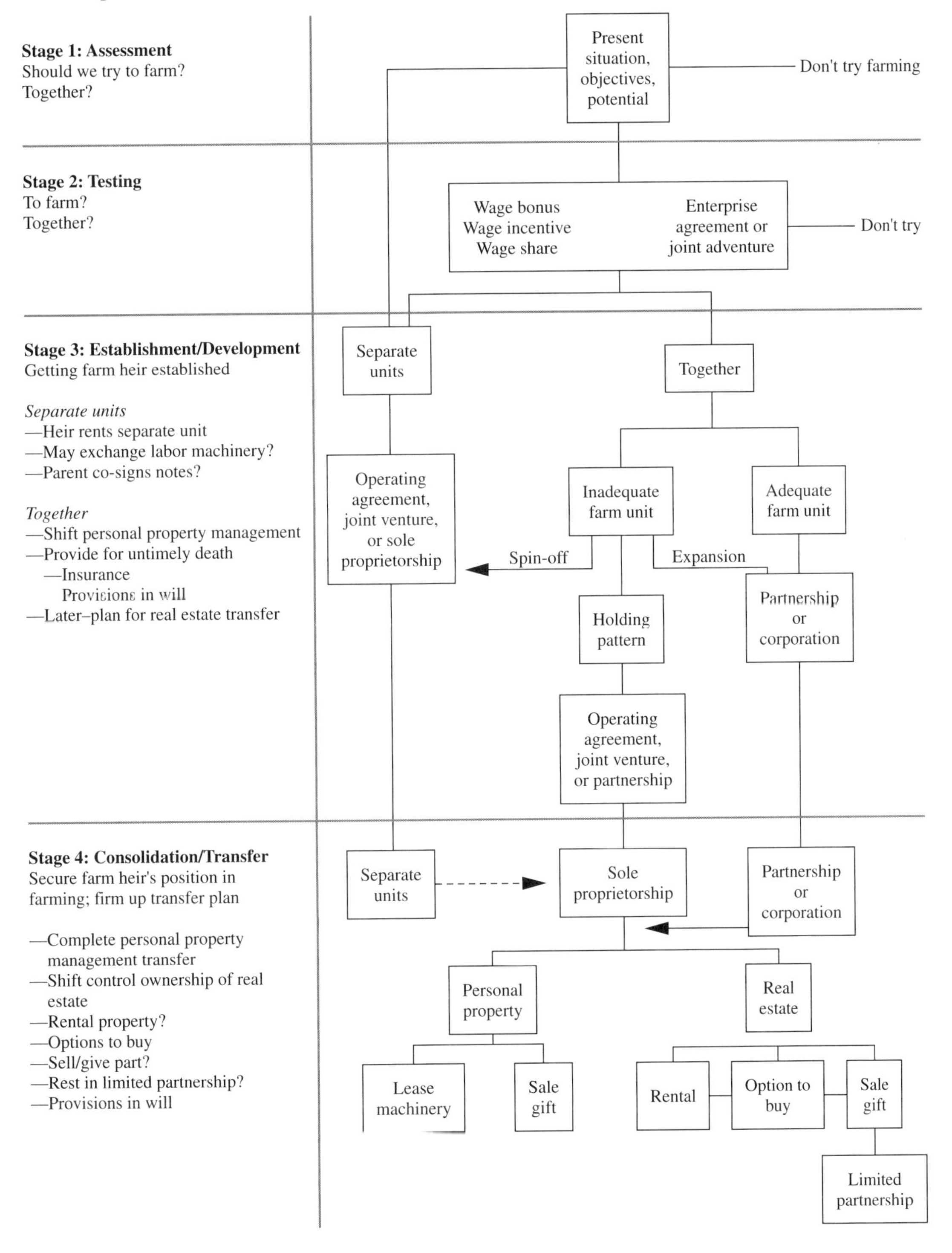

Source: Thomas, Kenneth H. and Michael D. Boehje, *Farm Business Arrangements: Which One for You?* Extension Bulletin 401, Agricultural Extension Service, University of Minnesota.

It is a good idea to consider the compatibility of the persons to be farming together before they move in with each other. The compatibility test needs to include spouses.

If children join the farm business, the principles and organizational structures presented in this chapter can be used to develop workable solutions for passing the farm business and family wealth to the heirs, while protecting the retirement of the parents. Through gifts and sales the value retained in the parent's estate can be reduced to avoid high estate taxes. Children can enter a farm business, which otherwise might be prohibitively expensive to do so. Capital gains income tax can be avoided by passing the farm to the heirs and by not selling it to outsiders. However, selling the property might be in the best interests of those involved. Even though paying taxes is repulsive, selling the farm may in some cases be the best solution in estate planning. The important thing is to plan for the welfare of family members, particularly the parents, and the transfer of family wealth, rather than letting time dictate unwanted choices.

Life Insurance

Life insurance is an important part of many estate plans. It can be used to build an estate, provide liquidity to meet estate settlement costs, satisfy the rights of nonbusiness heirs, and furnish retirements benefits. Some advantages and disadvantages of life insurance follow:

Advantages:

- Proceeds can go directly to beneficiaries.
- The owner's estate is immediately increased, providing financial security.
- Estate taxes may be avoided if incidents of ownership have been transferred to a survivor during life.
- Proceeds received are not subject to income tax.

Disadvantages:

- Premiums are a drain on current cash and may cause problems of meeting business expenses.
- Beneficiaries often are not changed as situations change.
- The proceeds could be included in the gross estate if the deceased retained the incidents of ownership.
- Insurance may not return as much for the investment as other alternatives, and could be less flexible.

Federal Estate Taxes

This tax is an excise tax levied on the transfer of property upon the owner's death. It is based upon the gross estate being transferred less allowed deductions. The process begins with the balance sheet as outlined in Chapter 2. The adjusted gross estate is roughly the decedent's equity, using the market approach, less estate settlement costs. Form 9.1 provides a review of the items generally found in a farmer's estate. Gifts to charities, other qualified receivers, and the marital deduction are deducted from the adjusted gross estate to arrive at the taxable estate. This amount is taxed using a graduated scale much like the income tax. The estate tax is then adjusted by the uniform credit, which reduced the amount considerably. Furthermore, if the state also wants a share of the estate, there is a state credit before the final federal estate tax is levied. Some of these elements need further explanation.

It should be understood that before the estate is taxed the IRS income tax must first be satisfied. The deceased may have unpaid taxable income in the form of harvested and unharvested crops and livestock produced prior to death. However, if taxes have been paid, there could be a refund due because tax rates are on an annual basis.

Estate settlement costs can vary based upon how property is owned, whether an estate plan is in place and if there is a will. If an administrator must be appointed and there is a probate proceeding, the cost can be high. Note that 10 percent of the gross estate is used in Form 9.1. For a $1-million estate this would be $100,000. Estate planning is worth the effort.

FORM 9.1

Estate Inventory and Settlement Costs

	Separate Property	Community Property	Total	
A. Property owned				
1. Cash and savings	$________	$________	$________	
2. Stocks and bonds	$________	$________	$________	
3. Life insurance	$________	$________	$________	
4. Grain inventory	$________	$________	$________	
5. Livestock inventory	$________	$________	$________	
6. Land and buildings	$________	$________	$________	
7. Machinery and equipment	$________	$________	$________	
8. Nonfarm property	$________	$________	$________	
9. Personal property	$________	$________	$________	
10. Other property	$________	$________	$________	
GROSS ESTATE (Total of above items)	$________	$________	$________	(A)

B. Less debts and settlement expenses
1. Debts . . . . $________
2. Settlement expenses (10% of item A) . . . . $________
TOTAL . . . . $________ (B)

C. Adjusted gross estate (item A–item B) . . . . $________ (C)

D. Less exempt gifts and marital deduction
1. Exempt gifts . . . . $________
2. Marital deduction . . . . $________
TOTAL . . . . $________ (D)

E. Taxable estate (item C–item D) . . . . $________ (E)

F. Tentative federal tax (use E above and Table 1) . . . . $________ (F)

G. Estate tax credit
1. Federal unified credit . . . . $________
2. Estate tax on prior transfers . . . . $________
3. State credit (see Appendix Table 1) . . . . . . $________
TOTAL CREDITS . . . . $________ (G)

H. Federal tax due (item F–item G) . . . . $________ (H)

I. Liquidity needs
1. Federal estate taxes (H) . . . . $________
2. Washington estate taxes (G3) . . . . $________
3. Settlement expense (B) . . . . $________
4. Debts (B) . . . . $________
LIQUIDITY NEEDED (total) . . . . $________ (I)

J. Liquidity available (total of items A1–A3) . . . . $________ (J)

K. Additional liquidity (item J–item I) . . . . $________ (K)

Source: *Estate Planning Basics for Washington Farm and Ranch Families*, EB1231, Washington State University.

The marital deduction reduces to zero that portion of the gross estate passing to the surviving spouse. Thus, no tax is due. But, the size of the survivor's estate is augmented, and a larger estate tax may be due when he or she dies and there is no marital deduction.

The unified credit reduces the estate tax liability. The tax is first tabulated and then the credit is subtracted to determine the tax due. Table 9.7 shows the federal estate and gift tax schedule. Note that the rate increases from 18 percent to 55 percent on estates valued at $0 to $3,000,000. The unified credit is $650,000 in 1998 and increases to $1,000,000 in year 2006. Interpreted this deletes $213,000 of taxes in 1998. But note that the tax rate at this level is 37 percent and any estate value above this amount will be taxed at an even higher level.

Gifts during life can be used to reduce the size of the estate. Each year, each spouse can give to any one person, without limits on the number of persons, amounts up to $10,000 without gift tax liability. Larger amounts may be taxed. The adjustment is made in the estate of the donor. There is no limit on the amount of gifts to a spouse. Most gifts within 3 years of death are included in the gross estate. Amounts above this will be used to reduce the unified credit. There are a few exceptions. In addition to reducing the estate, gifts can be used to transfer income tax to one who is in a lower tax bracket.

Various schemes have been formulated to reduce estate taxes by changing the way property is owned, managed, and willed to take advantage of the marital deduction and unified credit, etc., to reduce estate taxes. The tax savings, including settlement costs, can be large. This is another place where deathbed repentance may not be too helpful. It is the young and fearless that may need estate planning the most. The incidence of death may be lower but the consequences on the survivors could be at its highest level.

TABLE 9.7

Federal Estate and Gift Tax Rate Schedule

Taxable Estate and Lifetime Gifts From: ($)	To: ($)	Tax ($)	+	%	Of Excess Over: ($)
0	10,000	0		18	0
10,000	20,000	1,800		20	10,000
20,000	40,000	3,800		22	20,000
40,000	60,000	8,200		24	40,000
60,000	80,000	13,000		26	60,000
80,000	100,000	18,200		28	80,000
100,000	150,000	23,800		30	100,000
150,000	250,000	38,800		32	150,000
250,000	500,000	70,800		34	250,000
500,000	750,000	155,800		37	500,000
750,000	1,000,000	248,300		39	750,000
1,000,000	1,250,000	345,800		41	1,000,000
1,250,000	1,500,000	448,400		43	1,250,000
1,500,000	2,000,000	555,800		45	1,500,000
2,000,000	2,500,000	780,800		49	2,000,000
2,500,000	3,000,000	1,025,000		53	2,500,000
3,000,000	—	2,290,800		55	3,000,000

Much more could be written about income and estate tax planning. That has not been the purpose of this chapter. A much longer treatise would be required. There are so many rules and regulations, which seem to be continually changing, that it is difficult to discuss them all. The footnote on the chapter-opening page lists some sources that provide useful study before consulting a professional estate planner. Personal experience has shown that not all professional consultants are equally qualified to both execute an acceptable plan and educate the participants. The latter may be as important as the former. Avoid those who move too quickly without gaining a complete understanding of the goals and objectives of the participants. There probably are structures that will leave all participants better off except for state and federal governments.

CHAPTER 10

Financial Planning for Ownership and Operation

Introduction

Five farm- and family-type goals were presented at the beginning of Chapter 9. All of them had to do with financing the farm business. There are only four sources of funds to finance the investments and operations of a business. If funds are not available from equity resources, they must be either borrowed, rented, or transferred from gifts or other investments. The size of the business required for efficient operations and to provide full-time employment are such that few, if any, can operate without the use of borrowed funds.

To visualize investment and operating-fund requirements, consider the amounts shown in Table 10.1 for a sampling of commercial U.S. farms. External financing is very apparent.

Sources of Funds for Investment and Operations

With farm numbers declining and farm size increasing, there is not a shortage of people to farm, but this scenario indicates a financial

TABLE 10.1

Financial Characteristics of United States Commercial Farms by Gross Sales, 1997[a]

Item	Gross Sales Class				Corn	Beef
	$50,000–$99,999	$100,000–$249,999	$250,000–$999,999	$500,000 or more	Farm Businesses	Farm Businesses
Balance sheet						
Assets	$ 571,149	$ 734,351	$1,045,230	$2,703,504	$ 786,744	$1,303,407
Liabilities	75,655	112,312	208,858	539,513	135,699	148,817
Equity	$ 495,494	$ 622,039	$ 836,372	$2,163,991	$ 651,045	$1,154,590
Income statement						
Cash:						
Sales and miscellaneous income	$ 76,997	$ 157,742	$ 326,145	$1,380,626	$ 181,502	$ 250,956
Variable expenses	47,222	92,662	194,075	910,336	90,734	169,886
Fixed expenses	16,971	30,683	57,792	151,163	44,180	33,809
Net cash	$ 12,805	$ 34,396	$ 74,278	$ 319,127	$ 46,588	$ 47,261
Noncash:						
Depreciation(–)	$ 8,910	$ 16,317	$ 29,882	$ 71,440	$ 18,822	$ 16,572
Inventory change	5,687	10,541	29,864	39,413	7,888	9,284
Other	4,053	4,186	3,906	2,103	3,968	4,247
Net to equity and unpaid labor	$ 13,634	$ 32,806	$ 78,166	$ 209,222	$ 39,592	$ 44,220

[a] Source: "1993–95 Farm Cost and Returns Study" and "1996–97 Agricultural Resources Management Study" ERS,USDA.

problem of acquiring larger-sized units. Financial means of acquiring farms include the following:

- Using retained earnings (equity and savings).
- Receiving gifts and inheritance.
- Pooling of equity capital with others to form partnerships and corporations.
- Leasing and renting of real and personal property.
- Using custom services to perform farm operations.
- Producing contract crops where the contractor furnishes some or all of the inputs such as seed, fertilizer, and harvesting.
- Borrowing to finance purchases.

Equity Financing

Items purchased with equity alone are purchased with cash or traded for using assets, which are debt free. Cash financing of investments and operations is very important and necessary, but debt financing is more often used, even if only for short time periods. Consider for example an investment of $400,000 required to purchase a small farm. Just to accumulate a 25 percent down payment with savings of $10,000 per year at 10 percent interest would require more than 7 years; at 5 percent it would require 9 years. If land appreciated in value at 5 percent per year, in 11 years the land would cost $684,000 and a $171,000 down payment would be required. Currently, the equity requirement to purchase land is nearer to 40 percent, making the savings requirement even more difficult. Down payments are necessary for most all debt financing. Without adequate equity the farmer will not have the necessary resources to farm with, and principal and interest payments will be a financial drain on the business.

Balance sheet ratios were presented in Chapter 4 and are of relevance here. Liquidity, solvency, and leverage are all used in evaluating equity and debt financing.

Gifts and Inheritance

Gifts and inheritance have played and will continue to play important roles in financing agriculture, particularly the acquisition of real property. Family farms received following the death of a parent or grandparent may be tax free and debt free. Intergenerational transfers are goals of many farm families. Gifts during life also are used to transfer family wealth between generations. Each parent can give up to $10,000 (plus inflationary increases) per year to each beneficiary without triggering any gift tax. But there are problems with these types of acquisitions. Just a few of them are listed below:

- The receiver may be near retirement when the gift is received if the donor lives even to the average life expectancy.
- Inheritance often splits the property among the heirs.
- Gifts during life may leave the giver without adequate retirement funds.
- Property at death often has limited access and use may be limited by estate settlements, life estates, and trusts.
- Some property acquisitions could have major tax liabilities to be paid. Estates valued in excess of $625,000 (1998) are subject to estate taxes that are high at the margin.

Thus, estate planning is very important and needs to begin early in life when there is still time to incorporate interested family members into the business, change the organization of the business, and give gifts when they are most beneficial to all parties.

Pooling of Equity Capital

Two or more parties can join their resources into a single farm unit to achieve greater efficiency and profitability: parents often want to bring a child into the family business; brothers and sisters may want to farm together; or unrelated friends sometimes decide to start up a new business together. In Chapter 9 partnerships and corporations were discussed. A review of that material provides the guidelines of how partnerships and corporations can be formed to combine the finances of two or more persons. The principal advantages are to accumulate financial resources, increase farm size efficiency, and share risk. Distributed management could improve the efficiency of operations.

Leasing and Renting

About 45 percent of all U.S. farms have land rental arrangements. In addition to land, many farmers lease and rent machinery and equipment. Leasing and renting provide use rights without ownership. Leasing and renting should benefit both the lessor and lessee. The lessee obtains the use of the item without the requirement of having to come up with the money to purchase it. Details of farm leasing were discussed in Chapter 9.

Contracting

Contracting takes many forms. Fruit and vegetable processing firms contracting to purchase farm products may furnish inputs such as seed, fertilizers, and chemicals and often do the harvesting. Thus, contracting reduces the financial requirements of the producer. Many farms use custom operators to harvest their crops and perform other functions. Many broiler and some hog farmers produce on contracts where they provide labor and facilities, with all other inputs provided by the integrator. These operations reduce the producer's need for ownership of equipment and releases funds for other uses.

Borrowing

Borrowing is last mentioned but not the least used. The combination of debt and equity financing is a typical means for obtaining resources to farm with and paying operating costs. Considerations of using a combination of equity and debt to finance farm investment and pay operating costs is the subject of the balance of this chapter.

Financial Planning and Loan Justification

There are three levels of accountability when using debt financing: profitability, feasibility, and risk of failure. Often most of the energy is spent in worrying about interest rates, loan terms, lending agency, and other things the borrower can do little about. Although these levels of acountability are important, they seldom are the primary cause of loan failure. Interest ought to be considered when making the decision to invest, but the borrower has the greatest control over profitability after adjusting for risk.

Profitability

The determination of profitability is not a financing matter. Financing is part of the implementation phase of management, whereas the determination of expected profits is part of production planning. Budgeting is the principal tool used to test the profitability of alternative farm plans. Linear programming and economic principles are also means for testing profitability. If you are unacquainted with these tools, refer to Chapters 5 and 6. Whether the activity requires debt financing is irrelevant to this first and necessary step in planning. But, if it is necessary to borrow funds to finance the activity under consideration, then it is even more important that the expected returns are sufficiently large to meet contractual payments before going ahead with the project. Point of emphasis: profitability determination, so far as possible, precedes all financing with either equity or debt financing. Indeed, profitability assessment is required by any reputable credit agency.

Feasibility

Some production activities may show a profit, but the cash flow therefrom may be insufficient to meet principal and interest payments over the period of the loan. The length of the loan is often shorter than the life of the asset being financed. Interest rates and loan terms now become very important. Determining feasibility is like going to an after-Christmas sale and finding a real bargain only to discover that pockets and wallet are both empty and the store will not accept your credit card or check.

The major tool for determining feasibility is the cash-flow statement. It was defined and illustrated in Chapter 3 and applied in Chapter 5. Your attention is called to Table 3.7, Example 5.4, and Table 5.4c where cash-flow statements are illustrated. Pay attention to the structure of this statement. Note that cash income is from sales of produce, capital asset sales, and off-farm sources. Expenditures go to pay operating expenses, ownership expenses, and principal and interest payments on old loans; make down payments on new capital purchases; purchase livestock; and pay family living costs. To the extent that expenditures are greater than receipts this difference must be paid from savings (equity) or be financed with new debt. A line of credit can be determined. Not only are needs accessed but also paybacks are projected. All reputable lending agencies will want to see this statement before making a new loan.

Risk of Loan Failure

Financial risk was discussed in Chapter 8. It generally can be classified under two headings, security and income. If income is predictable and steady, there is no need for backup security, but, if income falls below some planned level, it may be necessary to make up the difference by reducing equity, sometimes referred to as balance sheet financing. All reputable credit agencies will want to analyze historical balance sheets and income statements, not only to look at their levels but also to study their variability. Financial accounts should be studied as part of financial planning.

The credit agency is likely to attach certain property for security of new loans. Security instruments are discussed later in this chapter. Security requirements were presented in Chapter 4 where the concepts of solvency, liquidity, and leverage were illustrated.

The attachment of the financed property, along with other property, is protection for the lending agency but is of no particular value to the borrower, except as a means for obtaining the loan. The real security to the borrower, and the lender, is in the income statement. No one borrows to finance business activities thought to be unprofitable. Furthermore, most farm producers use a variety of measures to control production and price risk as means of securing stable and profitable levels of income, thus being able to meet their credit obligations.

To understand more clearly the borrowers approach to credit, consider the marginal efficiency of capital illustrated in Figure 10.1. Output is measured in dollars and is shown on the vertical axis. The horizontal axis measures the amount of investment

FIGURE 10.1

Marginal Efficiency of Capital and the Effects of Risk

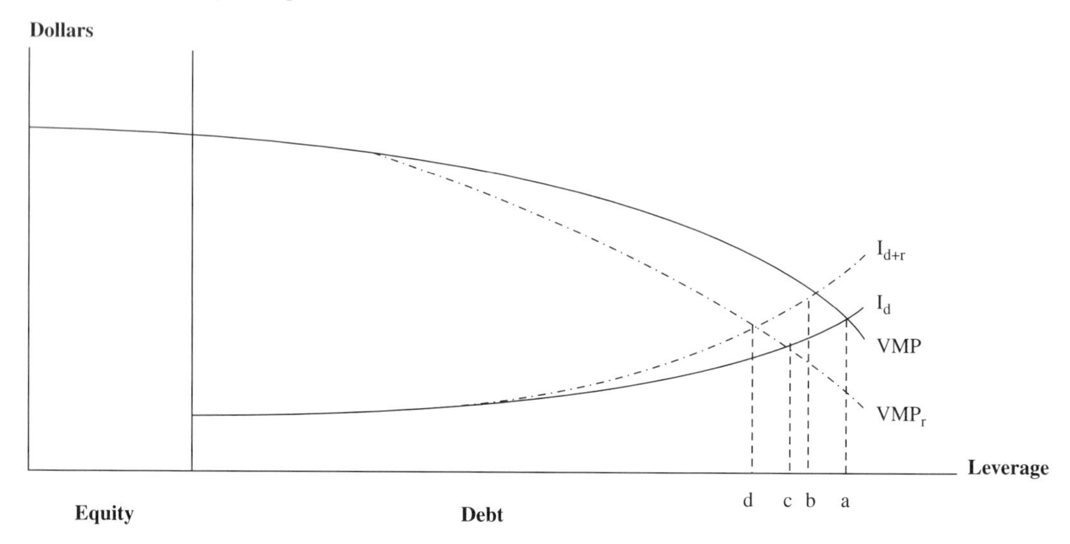

and increases from left to right. The value marginal product curve (VMP) is typical in that returns from additional investments decline at the margin. Because of increased risk with higher investments the marginal returns may be risk discounted, causing the VMP curve to decline even faster (VMP_r).

Note that equity and debt financing are shown. As debt increases, so does the debt-to-equity leverage. The lower the proportion that is debt financed the lower is the ratio and the more secure the investment. As the ratio increases credit agencies become less willing to risk their funds and charge higher interest rates (I_d). Furthermore, as debt increases the borrower becomes less willing to borrow because his or her liquidity is slipping away (i.e., cash for family living, making new investments, and meeting emergencies) and he or she wants a larger expected income, as expressed in the higher interest rate, before investing (I_{d+r}). As a risk precaution, the borrower also may subtract a safety margin from the expected rate of return on the investment. The variables in Figure 10.1 are defined as follows:

I_d = interest charged on new debt
I_{d+r} = interest charged on debt plus the borrower's liquidity preference
VMP = value of the marginal return on new and expanded investments
VMP_r = VMP adjusted for risk. Risk increases as leverage increases
a = level of investment without risk
b = level of investment adjusted only for liquidity preference
c = level of investment adjusted only for income risk
d = level of investment adjusted for finance and income risk

Notice that risk reduces the level of production and interest efficiency by moving the optimum amount of debt to the left from "a" to "d."

The Time Value of Money

Before discussing business loans and their repayment plans, it is important to understand the time dimensions of money. Most everyone knows that if they deposit money in a bank savings account it draws interest and becomes more valuable with time. Furthermore, if the bank loans money there is an interest cost. Many do not understand the arithmetic of how money grows or how principal and interest payments on loans are tabulated. Compounding and discounting may sound like technical financial terms that only bankers understand, but borrowers also need to understand them. Thus, the principles and terms used in saving, borrowing, and investing of money are presented in this section. Following this discussion applications are made to borrowing and investing.

Compounding and Discounting a Single Payment

Compounding means that the interest received at the end of a period of time is added to the principal amount and draws interest in the following period. Discounting is the reverse. The present value of an amount to be received in the future, say 1 year from now, is the amount that if deposited in a savings account at some rate of return would grow to be worth the future value. It is the "bird in the hand, worth two in the bush" idea. A single payment means that there is only one deposit or receipt amount compared with the situation where there may be several payments made or received. Compounding and discounting are illustrated here for the single-dollar amount without elaboration. The examples demonstrate the application of the principles.

Compounding, or the future value of a present amount plus interest (see Appendix Table 10.1.)

Formula: $V_n = V_0 (1 + r)^n$ (See footnote below for formula verification)

where: V_n = future value of the amount invested
V_0 = present value, or the amount invested
n = the number of compounding periods (Each period can be 1 day, 1 month, or 1 year in length.)
r = the rate of return during each of the time periods, expressed as a decimal (The rate of return and the interest rate "i" are often used interchangeably. Interest rates and rates of return are usually expressed as annual percentages and must be converted to the time period under consideration.)

EXAMPLE 10.1

$1,000 is placed in savings for 1 year with interest compounded annually at the rate of 8 percent. What is the future value?

$$V_1 = \$1,000\ (1 + 0.08)^1 = \$1,080.00$$

Verification of the formula. $V_n = V_0(1 + r)^n$ (Assume n = 2)

$$V_2 = V_0 + rV_0 + r(V_0 + rV_0)$$
$$= V_0 + rV_0 + rV_0 + r^2V_0)$$
$$= V_0 + 2(rV_0) + r^2V_0$$
$$= V_0(1 + 2r + r^2)$$
$$= V_0(1 + r)(1 + r)$$
$$= V_0(1 + r)^2$$

Note that the first equation tabulates the value of a single deposit interest bearing investment at the end of two years: the amount invested (V_0), the interest added the first year (rV_0), and the interest for the second year $r(V_0 + rV_0)$. The proof just simplifies this equation into the general form: $V_n = V_0(1 + r)^n$.

EXAMPLE 10.2

$1,000 is placed in savings for 1 year, interest is compounded semiannually at 8 percent annual rate of return. What is the future value?

$r = (0.08)\ 6/12 = 0.04$

$V_2 = \$1,000\ (1 + 0.04)^2 = \$1,081.60$

Discounting, or the present value of a future income (See Appendix Table 10.2.)

Formula: $V_0 = \dfrac{V_n}{(1+r)^n}$

Note: This formula is derived from the one above by solving for V_0.

EXAMPLE 10.3

What is the present value of $1,000 received after a waiting period of 1 year when the discount rate of interest is 8 percent.

$$V_0 = \frac{\$1,000}{(1+0.08)^1} = \$925.93$$

EXAMPLE 10.4

What is the present value of $1,000 received after a waiting period of 2 years when the discount rate is 8 percent.

$$V_0 = \frac{\$1,000}{(1+0.08)^2} = \$857.34$$

Compounding and Discounting a Flow of Funds

The formulas that follow illustrate procedures to tabulate the future value of a series of payments, such as with a payroll savings plan, and the present value of a series of receipts, such as periodic income payment to be received over several future years. This is how many payments are made and income is received. The formulas are little more than extrapolations of the formulas previously given. Again, few explanations are given with the formulas. Study carefully the illustrations.

Future and present values when payments or receipts are not of uniform amounts

1. Compounding, or the future value of a series of uneven payments:

Formula: $V_n = A_1 (1 + r)^{n-1} + A_2 (1 + r)^{n-2} + A_3 (1 + r)^{n-3} + \ldots + A_n (1 + r)^{n-n}$

$$= \sum_{t=1}^{n} A_t (1+r)^{n-t}$$

where A_t = the payment received at the end of period t*
r = the discount rate over the compounding period
n = the number of payment periods over the life of the investment

* If payments are received at the beginning of the periods then corresponding adjustments need to be made in the formula.

EXAMPLE 10.5

What is the value of the following payments received at the end of each of the years if r = 9%: year 1, $1,000; year 2, $2,000; year 3, $1,500; year 4, $2,500?

$$V_4 = \$1{,}000\,(1.09)^3 + \$2{,}000\,(1.09)^2 + \$1{,}500\,(1.09)^1 + \$2{,}500$$
$$= \$1{,}295 + \$2{,}376 + \$1{,}635 + \$2{,}500 = \$7{,}806$$

2. Discounting, or the present value of a series of uneven receipts:

Formula: $V_0 = \dfrac{A_1}{(1+r)^1} + \dfrac{A_2}{(1+r)^2} + \dfrac{A_3}{(1+r)^3} + \ldots + \dfrac{A_n}{(1+r)^n}$

$$= \sum_{t=1}^{n} \frac{A_t}{(1+r)^t}$$

where A_t, r, and n are as defined above.

EXAMPLE 10.6

What is the present value of the following payments to be received at the end of each of the years shown if r = 9 percent: year 1, $1,000; year 2, $2,000; year 3, $1,500; year 4, $2,500?

$$V_0 = \frac{\$1{,}000}{(1.09)^1} + \frac{\$2{,}000}{(1.09)^2} + \frac{\$1{,}500}{(1.09)^3} + \frac{\$2{,}500}{(1.09)^4}$$
$$= \$917 + \$1{,}683 + \$1{,}158 + \$1{,}771 = \$5{,}530$$

Future and present values when the amount paid in or received is of a constant amount (uniform series of payments)

1. Compounding of a finite uniform series ($n \neq \infty$):

$$\text{Formula: } V_n = A\left[\frac{(1+r)^n - 1}{r}\right] = A(USFV)_{r,n}$$

where A = a uniform (constant) series of future incomes
USFV = a table value for a uniform series of future values (See Appendix Table 10.3.) (The rate of return and the number of payments are shown as subscripts.) Note: The proof for this formula is similar to the one shown for discounting a uniform series below.

EXAMPLE 10.7

What is the future value of $1,000 received in each of the next 5 years at an interest rate of 8 percent?

$$V_0 = 1000\left[\frac{(1.08)^5 - 1}{.08}\right] = 1000\left[\frac{1.469 - 1}{.08}\right] = \$5867$$

or $= \$1,000\ (USFV)_{.08,5} = \$1,000\ (5.867) = \$5,867$

2. Discounting of a finite uniform series ($n \neq \infty$):

$$\text{Formula: } V_0 = A\left[\frac{1-(1+r)^{-n}}{r}\right] = A\ (USPV)_{r,n}$$

where USPV = a table value for a uniform series of present values. (See footnote below for formula development; see Appendix Table 10.4.)

Formula development:
Given: A = a uniform series of net after tax income flows received over some future period of time.
r = the present value discount rate for A.
n = some future time period.
V = the discounted value of A.

If A is received over an infinite period of time beginning now: $V_0 = \frac{A}{r}$

If A is received over an infinite period beginning in year "n": $V_n = \frac{A}{r}$

If A is received over some finite period, beginning now and continuing through year "n" then the finite V_0 equals the infinite V_0 minus the discounted infinite V_n. Thus:

$$V_0 = \frac{A}{r} - \frac{A/r}{(1+r)^n} = \frac{A}{r} - \frac{A}{r}\left[\frac{1}{(1+r)^n}\right] = \frac{A}{r} - \frac{A(1+r)^{-n}}{r} = A\left[\frac{1-(1+r)^{-n}}{r}\right]$$

EXAMPLE 10.8

What is the present value of a uniform series (annuity) of future receipts of \$1,000 per year over the next 5 years with a discount rate of 8 percent?

$$V_0 = 1000\left[\frac{1-\frac{1}{(1.08)^5}}{.08}\right] = 1000\left[\frac{1-\frac{1}{1.469}}{.08}\right]$$

$$= 1000\left(\frac{1-.68}{.08}\right) = 1000\left(\frac{.319}{.08}\right)$$

$$= \$1,000\ (3.993) = \$3,993$$

or

$$V_0 = A\ (USPV)_{.08,5} = \$1,000\ (3.993) = \$3,993$$

---

3. Discounting of an infinite uniform series ($n = \infty$). Land is an example of this kind of an asset. Ownership in fee simple is without bounds. The current owner will not live forever but the land can be sold or given away. It retains its value through time.

   Formula: $V_0 = \frac{A}{r}$

   (See footnote below for formula development.)

EXAMPLE 10.9

Assume a uniform series of \$1,000 over an infinite period at a discount rate of 5 percent. What is its present value?

$$V_0 = \frac{\$1,000}{.05} = \$20,000$$

Formula development:

$$V_0 = \frac{A}{(1+r)} + \frac{A}{(1+r)^2} + \ldots + \frac{A}{(1+r)^\infty}$$

Multiply both sides by (1 + r) and simplify:

$$(1+r)V_0 = A + \frac{A}{(1+r)} + \frac{A}{(1+r)^2} + \ldots + \frac{A}{(1+r)^\infty}$$

Subtract the second formula from the first and simplify:

$$V_0 - (1+r)V_0 = A + \frac{A}{(1+r)^\infty}$$

$$V_0(1-(1+r)) = A\left[1+\frac{1}{(1+r)^\infty}\right] = A$$

$$V_0 = \frac{A}{r}$$

Loan Costs and Repayment Terms

Understanding the time value of money is necessary for comprehending how interests costs are arrived at and repayment terms set. The costs of financing a loan is discussed, followed by a presentation of several methods of charging interest. Because of the different methods of charging interest and various added charges and penalties the efficiency of loan payments is assessed.

Lender's Cost of Loanable Funds

The borrower is often critical of the lender's interest rates and costs of processing a new loan application. There may be some justification for this cost, but it is still good practice to shop for borrowed funds. The reasons why credit institutions charge higher interest rates than the rates they pay on savings and checking accounts are as follow:

- *Rates paid out on savings accounts and checking accounts.* These are the primary sources for loan funds.
- *Reserve requirements.* A minimum percentage of the funds received in deposits must be held in cash or near-cash reserves to ensure liquidity. These percentages are set by federal and state laws. Although some money can be held as balances in Federal Reserve Banks (FED) that pay small interest premiums, the largest percentage must be held as vault cash with no interest. Although variously applied the cash requirement can amount to as much as 15 percent or 20 percent and changes occasionally.
- *Rates paid for funds borrowed.* Credit institutions borrow from other banks and the Federal Reserve Banks. The local credit institution may use the security received from the farm borrower as security for a loan from another larger credit institution or the FED. The FED discount rate becomes important for this source of money. Some banks are correspondents of larger banks who loan money to them and service the large loans.
- *Operational costs.* Credit institutions have buildings and equipment to maintain, employees to pay, utility payments, and other costs of doing business.
- *Risk and inflation.* Some loans default and in most of these cases the credit institution loses money. Furthermore, interest rates are variable in the market and thus the institution can loan money at rates that are below their future costs. Inflation favors the borrower.
- *Profit.* Profit is a credit institutions primary purpose for being in business. Competition holds profit levels down.

Borrower's Noninterest Costs of Credit

- *Legal fees.* Particularly in the case of real property an abstract or title search will be made to determine if the title is clear of other claims and conditions. Certificate of title, license fees, and mortgage fees may need to be paid. Some credit institutions even charge for signature notarization.
- *Appraisal fees.* Some property needs to be appraised to determine its market value before the credit institution will make the loan. This service can cost hundreds or thousands of dollars, depending upon the size and complexity of the property being financed.

- *Credit searches.* If the lender is unacquainted with the borrower a credit-worthiness check probably will be required. This search is done by an outside agency that charges for its service.
- *Initiation fees.* Some credit institutions charge a fee to process a new loan. This requires that each loan pays for its own processing. Some loans are processed before the lender learns that the borrower does not want the loan. This fee protects the lender and assures that the borrower is serious.
- *Insurance and taxes.* Some lenders require that the borrower have life and health insurance to secure payment in case of the unforeseen death or illness of the borrower. Property taxes are seldom paid up to date. The new purchaser may need to pay the seller for prepaid taxes and other such items.
- *Points.* Points are a prepaid interest. They sometimes are used to bring a quoted interest rate in line with the market. One point is 1 percent. The point value is the present value of the discounted payment difference from using the two rates over the life of the loan. (The point value when added to the loan amount will give the same schedule of payments as would the market rate of interest.) Points most often are used by the Federal Housing Association (FHA) and other federal lending agencies whose rates are fixed. The point amount is paid up front but may be added to the loan amount.

These added costs may be tax deductible, but the borrower may be required to distribute them over the life of the loan rather than deducting them all in the year paid. This is particularly so when these costs are added to the amount borrowed.

Methods for Tabulating Interest Costs

There are two methods currently in use for tabulating borrower interest costs: remaining-balance or simple interest method and the add-on method.

Interest rate efficiency is tabulated by comparing other methods with the simple method of tabulating interest. Hence, it is sometimes called the true interest rate or annual percentage rate (APR). In 1969 the Consumer Credit Protection Act took effect. Title I is known as the Truth in Lending Act and requires lenders to reveal the costs of credit to borrowers, including interest, initiation fees, and other charges. This act was revised in 1980 under the Truth in Lending Simplification and Reform Act. Thus, if the simple interest rate is not provided it should be requested. The differences in assessing interest will become apparent after methods are presented.

Remaining balance method (simple interest or APR)

- Borrower receives full amount of loan.
- Interest is paid after the money has been used.
- Interest is paid at the end of each payment period and not less than once per year.

Formula: $I_n = RB_t\,(i_e)\,(y)$

where I_n = interest payment at the end of each scheduled payment period

RB_t = the beginning amount of the loan, or the balance remaining following the last payment (This is sometimes referred to as the principal.)

$$RB_t = v_0 - \sum_{t=1}^{n-1} P_t$$

V_0 = original amount borrowed
i_c = contractual interest rate expressed on an annual basis (normalized), i_c expressed as a decimal
y = percentage of year since last payment of interest, expressed as a decimal
n = total number of payments

EXAMPLE 10.10

Given a $1,200 loan, 9 percent interest,and total principal and interest due at the end of 6 months. What is the interest charge?

$$I = \$1,200 \times 0.09 \times 6/12 = \$54$$

Note: The borrower receives $1,200 in the beginning and pays $1,200 back at the end of 6 months plus interest of $54.

EXAMPLE 10.11

Given a $1,200 loan, 9 percent interest, principal payments of $100 per month plus interest on the unpaid balance.

Month	Principal	Interest Tabulation	Payment Total	Unpaid Balance
0	$ —	$ —	$ —	$1,200.00
1	100.00	1,200 × 0.09 × 1/12 = 9.00	109.00	1,100.00
2	100.00	1,100 × 0.09 × 1/12 = 8.25	108.25	1,000.00
3	100.00	1,000 × 0.09 × 1/12 = 7.50	107.50	900.00
⋮				
12	100.00	100 × 0.09 × 1/12 = 0.75	100.75	0.00
Total	$1,200.00	$58.50	$1,260.00	

$$\text{Average amount borrowed} = \frac{1200 + 100}{2} = \$650$$

$$\text{Annual interest rate } \frac{58.50}{650} = 0.09 \text{ or } 9\%$$

Add-on method

- Borrower receives the full amount of the loan.
- Interest is tabulated at the beginning of the period and added to the amount borrowed.
- The payment is constant and is arrived at by dividing the amount borrowed plus the interest amount by the number of payments.
- It is useful only for multiple payments.

Formulas: $I = (L \times i_c)(y)(n)$
$A = (L + I)/n$

where L = the amount of the loan
A = principal plus interest payments
y = percentage of year between payments, expressed as a decimal
n = number of payments

EXAMPLE 10.12

Given a \$1,200 loan, 9 percent interest, 12 monthly payments. What are the principal and interest payments?

$$I = (\$1{,}200 \times 0.09)\ (1/12)\ 12 = \$108$$
$$A = (\$1{,}200 + 108)/n = \$1{,}308/12 = \$109$$

EXAMPLE 10.13

Given a \$3,000 loan, 10 percent interest, three annual payments.

$$I = (\$3{,}000 \times 0.10)(1)(3) = \$900$$
$$A = (\$3{,}000 + 900)/3 = \$1{,}300$$

Amortized Payments

The word *amortize* has its roots in death—to kill by degrees. In loans it means to liquidate or extinguish by periodic payments. The two common methods for doing this are the constant principal payment method (sometimes referred to as the Springfield plan) and the constant total payment method (often referred to as the standard amortized loan). Both methods use the simple or APR method of assessing interest where interest is paid after an interest bearing period. They are illustrated here.

Constant principal payments (Springfield)

Principal payments are the same over the life of the loan. Principal payments are tabulated by dividing the borrowed amount by the number of payments. Interest is tabulated for each payment by multiplying the unpaid balance by the interest rate. This payment method is illustrated in Figure 10.2.

FIGURE 10.2

Principal and Interest Payments with Constant Principal Payments

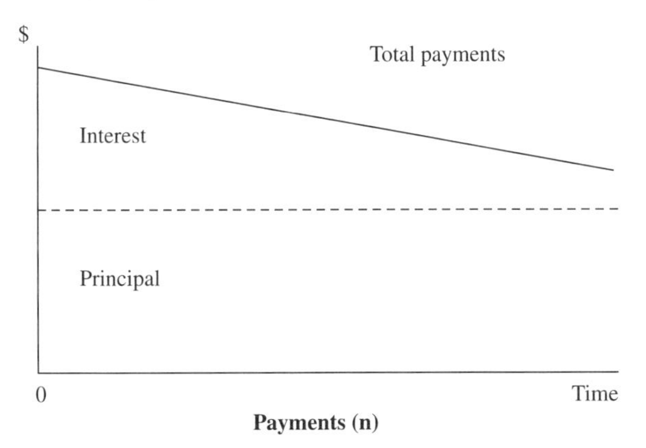

Constant principal-payment tabulations were shown in Example 10.11. This method of payment is most often used by noncommercial lenders such as a private individual.

EXAMPLE 10.14

Loan of \$10,000 for 5 years with equal annual principal payments and interest at 11 percent on the unpaid balance

Year	Principal	Interest	Total	Unpaid Balance
0	\$ —	\$ —	\$ —	\$10,000
1	2,000	1,100	3,100	8,000
2	2,000	880	2,880	6,000
3	2,000	660	2,660	4,000
4	2,000	440	2,440	2,000
5	2,000	220	2,220	0
Total	\$10,000	\$3,300	\$13,300	

Constant total payments (Standard)

This method is the standard method used by commercial lenders. For a discussion and illustration of the principle see "Discounting of a finite uniform series" presented earlier in this chapter. Its payments are illustrated in Figure 10.3: This is an application of discounting in which the discounted sum of the future payments is just equal to the amount of the loan. This assumes that the discount rate is equal to the interest rate. Note that the payments are constant as with a uniform series. The interest rate is sometimes referred to as the actuarial rate of interest.

FIGURE 10.3

Principal and Interest Payments Using a Constant Total Payments

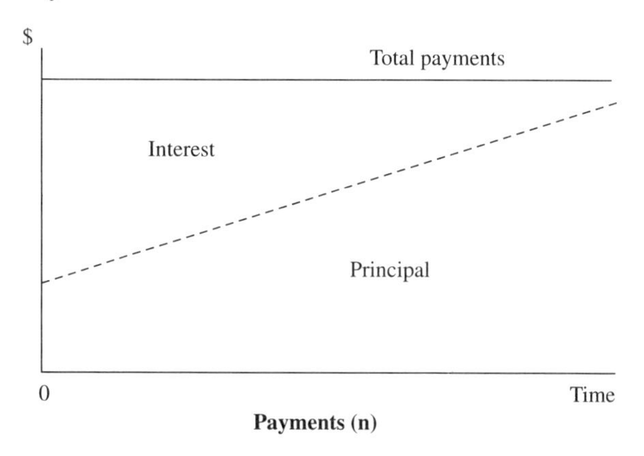

Tabulation formula: $0 = V_0 - P(USPV)_{1,n}$. (See footnote below for explanation of P.)

Explanation of P: The principal and interest payment (P) is obtained by dividing the amount borrowed (V_0) by USPV. Appendix Table 10.4 assumes $V_0 = \$1$ and shows P for various interest rates and time periods.

where V_0 = the amount borrowed
P = the uniform payment of principal and interest
$USPV_{1,n}$ = uniform series of present value. These values may be tabulated by financial calculations or found in a table for various i and n. See Appendix Table 10.4.
i* = the contractual interest rate
n* = number of payments to retire the principal of the loan

*Note: "i" and "n" must be of the same reference period; that is, if payments are monthly then i is a monthly rate, if payments are annual then i is an annual rate.

EXAMPLE 10.15

Assume a loan of $10,000 with 40 monthly payments and a normalized (annual) interest rate of 11 percent. What are the monthly payments?

$$0 = \$10{,}000 - P\,(USPV)_{.12/12,40}$$

Note: USPV is a value found in Appendix Table 10.4 and was derived by using the discount formula previously developed in this chapter.

$$0 = \$10{,}000 - P\,(32.84)$$
$$P = \$10{,}000/32.84 = \$304.51$$

Balloon Payments

Loan principal and interest payments are tabulated in a normal way, but a time is set for the balance of the principal to be paid off in advance of the time used in tabulating regular loan payments. This type of loan allows the borrower to refinance the unpaid balance at a more favorable time period or interest rate. Also, the lender is not required to tie his or her resources up so long before closing out the loan.

EXAMPLE 10.16

A $10,000, 5-year loan with a balloon payment at an annual interest rate of 10 percent with payments tabulated over 20 years.

Year	Payment	Principal	Interest	Unpaid Balance
0	$ —	$ —	$ —	$10,000
1	1,175	175	1,000	9,825
2	1,175	192	982	9,632
3	1,175	212	963	9,420
4	1,175	233	942	9,187
5	10,099	9,187	912	0
Totals		$ 10,000	$4,799	

Interest Rate Efficiency

The interest paid on any particular loan is compared with the interest that would be required to finance a loan using the unpaid balance or simple rate method. The most accurate method is to make the comparison to the actuarial rate. Several methods are shown below.

Average amount borrowed method. This method gives a quick estimate of the simple interest rate. It is tabulated by first calculating the average amount borrowed.

$$AB = \frac{BL + LP}{2}$$
$$i_n \approx \frac{I}{AB} \times 100$$
$$i_N = i_n T$$

where AB = the average amount borrowed
BL = the beginning amount of the loan
LP = the amount of the last principal payment
I = the total amount of interest paid
i_n = the percentage of interest paid based upon the simple rate method during the time borrowed
i_N = the annual rate of interest
T = the number of times the loan is turned over per year

EXAMPLE 10.17

Assume an add-on interest loan for \$6,000 for 3 years with monthly payments and 10 percent interest.

$$I = \$6,000(.10)(3) = \$1,800$$
$$\text{Payment} = (\$6,000 + \$1,800)/(3(12)) = \$216.67$$
$$AB = \frac{\$6,000 + \$200}{2} = \$3,100$$
$$i_n \approx \left(\frac{\$1,800}{\$3,100}\right)100 = 58\%$$
$$i_N = 58/3 = 19.4\% \text{ annual rate}$$

EXAMPLE 10.18

Assume you purchased gasoline with your credit card at a 2 cents/gal added cost. There is a 30-day lag between the time you purchase the gas and when you pay your credit card bills. Gasoline costs \$1 per gallon. What is the interest rate on this transaction?

$$AB = \frac{\$1 + \$1}{2} = 1 \qquad \text{Annually } AB = \$1/12 = \$0.08/\text{month}$$
$$i_n = \$0.02 \text{ per } \$1.00 \text{ borrowed}$$
$$T = 365/30 = 12.2 \text{ assuming 30 days between the purchase and payment}$$
$$i_N = i_n T = 0.02\ (12.2)\ 100 = 24.4\%$$

The Stelson method. This is a relatively accurate method of tabulating true interest rates. It was developed by Hugh E. Stelson in *The Mathematics of Finance.* (Princeton, N.J.: Van Norstad, 1963, p. 76.)

Formula: $i = 2\ Im/[n\ (B + a)] = 2\ I/[t\ (B + a)]$

where i = approximate nominal rate of interest
I = the total amount of interest paid
m = the number of regular payments made during the year
n = total number of payments to repay the loan
t = term of the loan in years
B = the beginning principal amount
a = the amount of each periodic payment

EXAMPLE 10.19

Assume an add-on loan of $1,200 to be paid back in 12 equal monthly installments and a rate of interest of 8 percent.

$I = \$1{,}200(0.08) = \$96$
$a = (\$1{,}200 + \$96)/12 = 108$
$i_n = [2(96)]/[1(\$1{,}200 + 108)]$

The actuarial rate method. This is the same as the standard amortized loan rate and the USPV rate discussed earlier. Except in this case the payments and the time periods are known and one solves for the interest rate that equates these to the beginning loan amount, similar to the internal rate of return (IRR) discussed later.

Formula: $0 = V_0 - A\ (USPV)_{i,n}$

where all variables are known except "i"; $USPV_{i,n} = V_0/A$; and "i" is calculated using the USPV formula and is read from a table, or is solved on a financial calculator.

EXAMPLE 10.20

Assume the same variables as given in Example 10.19.

$$0 = \$1{,}200 - \left[\frac{(\$1{,}200)(1+0.08)}{12}\right] USPV_{i,12}$$

$0 = \$1{,}200 - \$108\ (USPV)_{i,12}$
$USPV_{i,12} = \$1{,}200/\$108 = (11.11)_{i,12}$
$i_n = 1.20\% / \text{month}$
$i_N = (1.20)(12) = 14.4\%$

Nominal and Effective Rates of Interest

The nominal rate is the periodic rate converted to an annual basis; 1 percent/month is 12 percent per year, 3 percent per quarter is also 12 percent per year. The rate of interest actually earned in a year is the effective annual rate. For example, a savings account that compounds monthly at 1 percent has an annual effective rate of 12.68 percent [$i_e = (1.01)^{12} = 1.1268$].

Effect of Interest Rate and Time upon Debt Servicing Capacity

You may have realized that the rate of interest and the number of payment periods have much to do with the size of payments. An amortized standard loan of $100,000 at 10 percent interest with monthly payments for 10 years will have payments of $1,321.50. If interest were 12 percent, the payments would increase to $1,434.71. If the time period were stretched to 20 years at 10 percent interest the payments would be reduced to $965.02. These considerations should be factored into any loan payment negotiations.

Appendix Table 10.6 shows the total principal and interest paid over the life of the loan for various interest rates and time periods. This table is based upon a loan of $1. Appendix Table 10.7 illustrates the size of a purchase (V_0) that can be made from a periodic payment of $1 at various interest rates and time periods. These table values may be expanded by the uniform amount available for investment.

Asset Preferences in Loan Making

The ability of the borrower to meet payments has a great deal to do with the assets being financed. The following figure, which illustrates the lenders asset preference when making a loan, may help visualize this point:

	Self liquidating	Nonself liquidating
Asset generating	High	Intermediate
Nonasset generating	Intermediate	Low

Assets that become more valuable with time create repayment ability. Feeder cattle are an example of this kind of asset. Also, the feeder cattle loan is paid off from gross income. Assets that must be paid off with net income are more difficult to pay than those paid out of gross income. Land principal payments are an example of an asset paid out of net income. Interest is paid out of gross income. Land is an intermediate type. To the extent that the life of an asset exceeds the loan period, equity is created with each payment. This is a forced savings similar to a land payment. Loans to finance consumables are the least desirable type and in the "low" quadrant.

Legal Considerations When Financing Asset Purchases

The rights in property and the forms of ownership were presented in Chapter 9. That information is relevant here and should be reviewed if unclear. The legal requirements are somewhat different for real property and personal property. Each is discussed separately.

Financing Real Property

Two types of purchase are common, contract and mortgage or trust deed. Under the contract sale the purchaser does not receive full title to the property until after the terms of the contract have been met. Normally the seller, or a third party, holds the title. The buyer receives the property and has near full rights of ownership so long as contractual conditions are met, primarily that principal and interest payments are made. If the buyer does not make these payments then the property may revert back to the seller with past payments considered as rent. Generally, the seller finances the sale. Typically, the amount of down payment is much smaller than under a mortgage. The buyer pays the property taxes and receives all of the profits and losses. The contract may call for a balloon payment, after which it is anticipated that the purchase will be alternatively financed.

The mortgage sale is generally associated with third-party financing (commercial credit agency). The buyer receives title to the property and gives the creditor a security certificate called a mortgage or trust deed. If the owner of the property defaults in payment there is a legal court process for causing the owner to make payment or sell the property, and the holder of the mortgage normally does not receive the property back. If the property is sold then the loan company receives its rightful portion after legal fees and any unpaid taxes have been paid. It is an expensive process and most often there are no winners. For this reason down payments for this kind of a loan are higher than for the contract—ranging between 25 percent and 40 percent minimum.

Financing Personal Property

Personal property is secured under the Uniform Commercial Code—Article 9 covers most farm property. Lenders secure payment with a security interest agreement. This document lists the secured property and credit terms. The security interest is the right or claim that a lender has in the secured property in case of default in payment. In case of default the property can be reclaimed. A financing statement is generally used to describe the secured property, particularly when it is common practice to finance the purchase using credit. The creditor receives the financing statement and files it in a public office, such as the Secretary of State, so that other lending agencies can determine if said property is already pledged as collateral to someone else. Secured loans have priority over unsecured loans to the extent of the secured property in case of defaults in payments. Unsecured loans must be worked through the court system for collection. Promissory notes are used to spell out the terms of the loan such as contracting parties, interest rate, and length and payments particulars.

Bankruptcy

Bankruptcy is used by the borrower when loan payments cannot be made and all avenues have been exhausted for finding means of payment. The purpose of bankruptcy is to allow individuals to start over. The first bankruptcy act was passed in 1898. Current laws are generally described in the Bankruptcy Act of 1978. A substantial change affecting farmers was passed in 1986, following the financial problems farmers had in the 1980s.

There are four separate approaches to bankruptcy available to farmers; only two are in common use. Although farmers cannot be forced into bankruptcy directly by law there may be no other alternative left when all operating capital is gone. There is a special bankruptcy court established to handle these matters. Each method of bankruptcy is briefly described below.

- *Chapter 7.* This is the straight bankruptcy act in which the individual gives up all property except that which is legal to keep. The debtor's assets are liquidated and the creditors are paid so far as possible. Student loans and alimony are exempt and cannot be written off. Federal exempt property includes $7,500 in a home, $1,200 in a vehicle, and other listed property with values of about $4,000. This is not much, but it leaves a positive equity rather than a negative one.
- *Chapter 11.* This plan offers rehabilitation as an alternative to liquidation. The debtor may enter into an agreement with creditors under which all or part of the business continues and debts are restructured. This can only benefit creditors and debtors if the "going concern" value of the property is greater than the liquidation value. Interest is usually stopped on unsecured loans, but runs on secured loans to the extent of the secured property. This is the act used by big business firms.
- *Chapter 13.* This act is for wage earners and small businesses. The objective is to develop a plan under which debts can be paid from future earnings.
- *Chapter 12.* This act was passed in 1986 specifically for farmers, but congress failed to extend it beyond April 1, 1999. It was similar to Chapters 11 and 13, but these had not worked well for farmers. Under this plan a trustee was appointed to act as an advisor, and a reorganization plan was developed within 90 days of filing for bankruptcy and approved within 45 days by the bankruptcy court. The new secured indebtedness could not exceed the market value of the property. The interest rate was generally reduced. Any unsecured debt was paid only after secured debt and living expenses had been met. At the end of the reorganization period any unsecured debt was erased.

It is easy to understand that it is preferable to stay out of bankruptcy if at all possible. Even serious loan problems most often can be settled better out of court by reasonable parties. Out-of-court settlements often benefit both the lender and borrower, primarily because of the high legal and court costs.

Capital Investment Analysis

Many expenditures that farm producers make are justified on the basis of expected income flows. Decisions must be made in the present time period, but the income forthcoming is in the future. There needs to be a way to bring these two time periods together. The methods used to do this have their roots in the time-value-of-money and cash-flow accounting. It is the after-tax cash flow that is of interest to the investor. Thus, before considering several of the methods used to make future income investment decisions it is useful to illustrate the tabulation of after-tax income flows from a new investment opportunity. Be aware that these income flows are anticipated and not realized; they must be budgeted. Capital investment methods and procedures are illustrated using Example 10.21. These figures now can be used to illustrate the various decision-making criteria that could be used to justify the purchase.

EXAMPLE 10.21

Assume an investment with expected income flows as shown in column two (Cash Income). The new investment cost and associated information are shown.

Purchase price	$80,000
Depreciation life	8 years
Depreciation method	Straight line with no salvage value
Value of asset after 8 years	$20,000
Marginal tax rate	25%

Year	Cash Income	Annual Depreciation[a]	Taxable Income[b]	Income Tax[c]	After-Tax Cash Flow[d]
0	$ —	$ —	$ —	$ —	$-80,000
1	10,000	10,000	0	0	10,000
2	15,000	10,000	5,000	1,250	13,750
3	20,000	10,000	10,000	2,500	17,500
4	25,000	10,000	15,000	3,750	21,250
5	25,000	10,000	15,000	3,750	21,250
6	25,000	10,000	15,000	3,750	21,250
7	20,000	10,000	10,000	2,500	17,500
8	15,000	10,000	5,000	1,250	13,750
8	20,000[e]	—	20,000[f]	5,000	15,000
	Total after-tax income(0-8 years)				$ 71,250

Average income(1–8) from investment including salvage $18,906

Average income(1–8) from investment excluding salvage $17,031.

[a]Annual depreciation = $80,000/8 = $10,000.

[b]Taxable income = cash income – depreciation.

[c]Income tax = taxable income × tax rate.

[d]After-tax cash flow = cash income – income tax.

[e]Assumed the asset would be sold at the end of 8 years.

[f]The sale of the asset will be treated as depreciation recovery and will be fully taxed.

There are three capital budgeting methods that ignore the time value of money:

1. *Necessary and urgent.* This method pays little attention to future income flows. It is the method used for many consumer goods. The buyer feels that he or she must have it and cannot wait any longer. Everyone else has one and that is justification enough. It is sad, but true, that many business goods also are purchased on this basis. This method is highly speculative and very risky. There are no means for comparing results.
2. *Payback (PB).* Under this method the buyer considers how long before the asset will generate enough cash to recover the purchase cost. It certainly is better than necessary and urgent but fails in not considering the total economic life of the asset and the time value of money. It is really a measure of liquidity. It is illustrated below:

Formula: $PB = V_0/E$

where V_0 = cost of the new asset
E = annual additional after-tax earnings before depreciation

Note: If E is not uniform then an average can be used.

EXAMPLE 10.22

(See Ex.10.21 for data source.)

PB = \$80,000/\$18,906 = 4.2 years

or

5 years if each of the annual after-tax income flows is subtracted from the purchase cost of the asset until zero is reached.

3. *Return-on-investment (ROI).* Under this method the amount of the investment is divided into the annual income flows to determine the rate of return. The rate can be converted into a percentage by dividing by 100. In this case the annual income is after depreciation rather than before depreciation as with payback. This is because the asset does change in value over time as represented in the depreciation amount and should not be included with income. The change in value is a noncash cost. This method has some advantages over PB in that it considers the whole life of the asset and the ROI can be compared with other investments, although imperfectly. ROI is not the same as used in the business world and it too fails to take into account the time value of money. One of the problems is in deciding which investment to use in the denominator, the original cost or a half-life value. Both are illustrated below:

   Formula: $ROI = (E - D)/V_0$

   where D = average annual depreciation

EXAMPLE 10.23

(See Ex.10.21 for data source.)

ROI_1 = (\$18,906 – \$10,000)/\$80,000 = 0.111, or 11.1%

EXAMPLE 10.24

(See Ex.10.21 for data source.)

Average amount invested:

$(V_0 + V_n)/2$ = (\$80,000 + \$20,000)/2 = \$50,000
ROI_2 = (\$17,031[a] – \$10,000)/\$50,000 = 0.141, or 14.1%

[a] The after-tax salvage value of the asset was excluded when tabulating the average income because it was included when tabulating the average investment.

There are capital budgeting methods that consider the time value of money. The following methods discount future income flows to the present time. Methods that do not discount future income flows imply a discount rate of zero. Discounting and compounding methods were presented earlier in this chapter and illustrated in Examples 10.1 to 10.9. Two methods of capital budgeting using discounted income are illustrated. Both are in wide use.

1. *Net Present Value (NPV).* Under this method the discounted future income flows are compared with the original investment (V_0) to determine if there is a profit. Assume for the present that the discount rate is known. Its derivation is discussed later. There are two types of future income flows, those that change from year to year and those that are constant. Procedures are illustrated for both situations. Review the time value of money discussion if these procedures are not familiar.

Formula for a variable income

$$NPV = -V_0 + \frac{A_1}{(1+r)} + \frac{A_2}{(1+r)^2} + \ldots + \frac{A_n}{(1+r)^n}$$

EXAMPLE 10.25

Using the data in Example 10.21 and a discount rate of 12 percent, the NPV of this investment is as follows:

$$NPV = -\$80{,}000 + (\$10{,}000/1.12) + (\$13{,}750/1.12^2) + (\$17{,}500/1.12^3) + \ldots + (\$13{,}750/1.12^8) + (\$15{,}000/1.12^8) = \$8{,}202$$

It is assumed that if the NPV is positive, the investment is a profitable one and should be made if financing is available and the amount of risk is acceptable.

Formula for a uniform (constant) flow of income:

$$NPV = -V_0 + A[USPV_{r,n}] + (V_n/(1 + r)^n)$$

See Appendix Tables 10.2 and 10.4.

EXAMPLE 10.26

Assume that the income is constant at the average of the data in Example 10.21 at \$17,031 and the discount rate is 12 percent.

$$NPV = -\$80{,}000 + \$17{,}031\ [USPV_{.12,8}] + (\$15{,}000/1.12^8) = \$10{,}662$$

---

2. *Internal Rate of Return (IRR).* Under this method the discount rate is in question. The IRR seeks a rate at which the NPV would be zero, a break-even strategy. If the IRR is above the rate that the farm manager thinks is reasonably good then he or she proceeds with the investment. The procedure is somewhat of a guessing game in which different discount rates are tried until NPV = 0. The rate is arrived at by dividing the guess in half each time. (Actually it probably will be arrived at on a financial calculator or spreadsheet that plays the game for you.)

EXAMPLE 10.27

Consider the data in Example 10.21.

$$0 = -\$80{,}000 + \$10{,}000/(1 + r) + \$13{,}750/(1 + r)^2 + \ldots + \$13{,}750/(1 + r)^8 + \$15{,}000/(1 + r)^8$$

r = 14.5%

EXAMPLE 10.28

Assuming the income is constant at the average of the data used in Example 10.21 at $17,031, as used in Example 10.26.

$$0 = -\$80{,}000 + \$17{,}031\ (USPV)_{r,8} + \$15{,}000/(1 + r)^8$$

r = 15.5%

If management thinks this rate of return is satisfactory, can finance the purchase and is confident in the income stream, it will make the investment.

Selecting a Discount Rate

The mathematical procedures in discounting and compounding flows of funds are relatively simple compared with projecting future income flows and selecting discount rates. Anticipating future income flows was discussed in Chapter 5. It is appropriate to budget the expected return.

Selecting a discount rate is a "returns" concept and not a "cost" concept. The discount rate selected is more closely associated with the internal opportunity cost of capital than with an external rate of interest. The cost of borrowing can be built into the costs of financing the investment. When interest rates are used it is the past and projected interest rates that are more relevant than the current rate. Expectations are strongly influenced by past experiences.

There is an opportunity cost associated with selecting a discount rate. The appropriate comparison is with similar investments of like risk. Thus, the same discount rate should not be used for land, cattle, and machinery. Long-term land investments do not have as high of a discount rate as shorter-term machinery investments. The bond market, particularly municipals, would be a better guideline for the land market than returns on money markets. Those doing capital investment analyses should be acquainted with the returns from different kinds of investment opportunities. The farmland appraiser, for example, would want to study the returns to land that current owners are experiencing, and the rate necessary to give an NPV equal to that which represents current market sales. Appraisers are able to compare sales prices with their appraisals made before the sale, and thus arrive at a discount rate that will equate income to the price.

The weighted-average-cost-of-capital (WACC) has been a widely used approach to determine a discount rate. It is based upon the concept that both debt and equity are used when making new capital investments. Furthermore, even though the current purchase is made with 100 percent debt the balance sheet supports the purchase. If the asset is purchased with cash (100 percent equity) the same logic applies. The particular rate of return is circumstantial and part of a total investment package as defined in the liability side of the balance sheet. The WACC is defined as follows:

$$r = k_e W_e + k_d (1 - t) W_d$$

where r = discount rate
k_e = cost of equity (net cash rate of return on equity capital after taxes)
W_e = proportion of equity (long term) in the firm
k_d = cost of debt (before-tax interest rate)
W_d = proportion of debt (long term) in the firm
t = marginal tax rate

EXAMPLE 10.29

Assume the desired capital structure of the farm investor is to have a 60 percent equity level. If debt costs 12 percent before taxes and the after-tax return to equity is 10 percent and the marginal tax rate is 35 percent, what is the WACC?

$$r = (0.10)(0.60) + (0.12)(1 - 0.35)(0.40) = 0.091, \text{ or } 9.1\%$$

Incorporating Debt into Capitalization

The WACC incorporates debt into the discount rate. It is sometimes revealing to include the loan itself into the NPV determination as a return to equity. No new procedures are necessary, but the new principal and interest payments need to be incorporated into the after-tax cash-flow determination. The appropriate discount rate is the after-tax return to equity. An illustration shows the procedures.

EXAMPLE 10.30

Assume the variables are those in Example 10.21 except that an $80,000 asset is purchased with 20 percent down at 10 percent before-tax interest rate with annual standard amortized payments over 8 years. Loan payments are $11,996 per year. (Example 10.30 continues with the table on page 296.)

Tabulation of Present Values after Tax of an Investment Financed with Debt

Year	Cash Income	Annual Depreciation	Principal	Interest	Taxable Income[a]	Income Tax	After-Tax Income[b]	Discount Factor	Present Value
0	$ 0	$ 0	$16,000	$ 0	$ 0	$ 0	$(16,000)	1.000	$(16,000)
1	10,000	10,000	5,596	6,400	0	0	(1,996)	0.909	(1,815)
2	15,000	10,000	6,156	5,840	0	0	3,004	0.826	2,482
3	20,000	10,000	6,771	5,225	4,775	1,193	6,811	0.751	5,116
4	25,000	10,000	7,448	4,548	10,452	2,613	10,391	0.683	7,097
5	25,000	10,000	8,193	3,803	11,197	2,799	10,205	0.621	6,336
6	25,000	10,000	9,012	2,984	12,016	3,004	10,000	0.564	5,644
7	20,000	10,000	9,914	2,082	7,918	1,979	6,025	0.513	3,091
8	15,000	10,000	10,905	1,091	3,909	977	2,027	0.466	945
8	20,000	—	—	—	20,000	5,000	15,000	0.466	6,998
	Net present value (NPV)								$19,899
	Internal rate of return (IRR)								28%
If the losses in years 1 and 2 are used to offset other income then the cash flows are:									
1	10,000	10,000	5,596	6,400	(6,400)	(1,600)	(396)	0.909	(360)
2	15,000	10,000	6,156	5,840	(840)	(210)	3,214	0.826	2,656
	Net present value (NPV)								$21,526
	Internal rate of return (IRR)								30%

[a]Taxable income is cash income minus depreciation and interest.

[b]After-tax cash flow is cash income minus loan and tax payments.

Note: Where the after-tax interest cost is below the return to capital, the return to equity is leveraged to give a much higher return than to the total investment.

Adjusting for Growth

It is sometimes projected that the anticipated cash-flow income will grow or increase. If this growth is constant and if the base income is constant then the basic capitalization formula can be adjusted to accommodate for this. This situation is illustrated below:

$$\text{Formula: } V_0 = \frac{A_1}{(1+r)} + \frac{A_2}{(1+r)^2} + \ldots + \frac{A_n}{(1+r)^n}$$

where A_1 to A_n increases by a constant percentage; thus

$$V_0 = \frac{A_0(1+g)}{(1+r)} + \frac{A_0(1+g)^2}{(1+r)^2} + \ldots + \frac{A_0(1+g)^n}{(1+r)^n}$$

This formula reduces to the more simplified form:

$$V_0 = A_1\left[1 - \left(\frac{1+g}{1+r}\right)^n\right]$$

Note: If $g < r$ and N is large then $\frac{(1+g)^N}{(1+r)}$ approaches "0" and

$$V_0 = A_1/(r - g)$$

EXAMPLE 10.31

Assume $A_1 = \$1{,}000$, $r = 10\%$, $g = 5\%$, $n = 5$. What is the NPV?

$$V_0 = \frac{1{,}000\left[1 - \left(\frac{1+0.05}{1+0.10}\right)^5\right]}{(0.10 - 0.05)}$$

$$= \$4{,}151$$

EXAMPLE 10.32

Assume the same variables as in Example 10.31 except that n is large, say 50 years:

$$V_0 = \$1{,}000/0.05 = \$20{,}000$$

Adjusting for Risk and Inflation

Both risk and inflation tend to increase the discount rate. For example, if the real discount rate is 8 percent and an inflation rate of 4 percent is anticipated then the combined rate is near 12 percent. Risk can be accounted for by increasing the discount rate by a premium that reflects the risk associated with the investment. A procedure to estimate the risk premium is called the Capital Asset Pricing Model, not discussed here. (See Franz, Trisha and James Libbin, "The Single Index Model: Its Background and Application."1994. *Journal of American Society of Professional Farm Managers and Rural Appraisers* 58:79–54.)

Investing in Farmland[1]

This is a detailed illustration of capital budgeting. It is presented as a case study for those interested in capital investments. Even though the application is to farmland, the transition is not large to other investment opportunities. This illustration completes the chapter.

The land purchase decision is unique among the numerous business problems that farmers must resolve. It is unique because it usually involves considerable capital and happens infrequently. Farmers who have little or no experience in making land purchases with heavy capital commitments at stake should take a thorough look at the factors involved with making a land investment decision.

Four factors stand out: (1) the market price of the land; (2) the value of the land to the present business in terms of the stream of annual returns, liquidation value, and buildup of equity to support additional borrowing; (3) various financing constraints; and (4) risk of financial loss.

The market price of land, the first factor, is, of course, not determined until the land is actually sold. Prospective investors, however, should obtain an estimate of the market price prior to the sale. Such an estimate is valuable in formulating a bargaining strategy and in assessing the financial attractiveness of purchasing the property. Farmers may obtain market-price estimates by checking on comparable sales themselves or by hiring a professional appraiser.

The second important aspect of analyzing a land purchase is determining the value of the property to the farmer's business. Such a determination permits the consideration of all the variables that are specific to the farmer's business and personal judgment (e.g., enterprise selection; crop yields; cultural practices; product and input prices; rate of inflation; income taxes; length of planning period; and existing land, machinery, labor, and management resources, etc.). In addition, the valuation procedure requires that the farmer specify the particular after-tax rate of return desired for the land investment. The computed value of land thus indicates the price that can be paid for the property while yielding the prescribed rate of return. If the market price exceeds the value, the farmer must accept a return below that specified in the valuation process, or justify the purchase for various noneconomic reasons.

The third concern in a land investment analysis is the financial feasibility of the proposed purchase. Financial feasibility refers to ability of the business to finance the acquisition with equity or debt capital; and if the latter is involved, the ability of the business to service that debt. Because large amounts of debt are commonly used, it is important to make cash-flow projections to determine if sufficient funds will be available to make principal and interest payments on the real estate loan. If the amount of added real estate debt supportable by the farm's projected cash flow, plus the down payment, sum to more than the market price, the land investment is financially feasible. However, if cash-flow projections indicate difficulty in servicing the added debt, the investment will not be an attractive one.

A fourth consideration is how the land investment will affect the farmer's risk-bearing position. Typically, large amounts of debt capital are used to finance a land purchase. Thus, additional strain is imposed on the farm's cash flow as a result of making principle and interest payments on the land loan. Such fixed financial commitments, coupled with the uncertainty of future prices, yields, and land values, may increase the risk of financial loss beyond a level the farmer or the lender is willing to assume. The two critical questions farmers should raise about the purchase of additional farmlands are, What is the land worth to my particular business? and Can I pay for the land?

Land has a value because it entitles the owner to a set of future rights to that land. Such rights include a claim to land earnings and the right to sell property. Thus, the economic value of additional land to a farmer is dependent upon the stream of after-tax returns earned by the land and by the after-tax value of the land at the end of the farmer's planning period.

The value of land is the sum of these future benefits, discounted by the farmer's required rate of return. Discounted future benefits, thus, represent the maximum price the farmer is willing to pay for the land.

Variables in the valuation procedure include the following:

1. *Annual return to land and projected annual rate of change in the return.* The return to land is obtained by subtracting all production expenses, except interest on land investment, from gross receipts. To accurately estimate land returns, it is necessary to prepare detailed budgets for the crops to be produced on the land. Where the investor intends to rent out rather than farm the land, returns to land may be estimated by subtracting real property ownership costs (i.e., property taxes, improvement repairs, insurance and management fees, except interest on the land investment) from the rent received.

   Land returns should be estimated for the current year, based on yields, prices, and costs existing under normalized circumstances. In recognition that returns are likely to change over time, however, the valuation procedure gives the farmer an opportunity to indicate average annual percentage change in land returns anticipated for the planning period.

2. *Annual rate of change in land value.* Because the land's market value at the end of the ownership period is a component of the stream of returns on which the land's value is based, it is important that anticipated changes in the land's market value be considered in the valuation analysis. Such changes can be expected to result from variation in land use, production technology, price relationships, and availability and terms of financing. If land use is restricted to agricultural enterprises, the change in the land's market value should be similar to the rate of change estimated for annual agricultural returns. Where land is subject to a growing demand for industrial, urban, or recreational uses, however, the rate of change in market value will exceed that estimated on the basis of agricultural use. Thus, it is important that farmers study closely the factors potentially affecting the land's market value.

   The land's market price at the time of the proposed purchase is used as the starting point for computing annual changes in the value and the estimated sales price at the end of the ownership period. Farmers will, therefore, want to obtain an estimate of recent sales prices for nearby and comparable land.

3. *Annual rate of change in the general price level.* It is necessary to account for changes in the general price level when valuing land because these changes affect the purchasing power of the income realized from the land investment. Purchasing power and wealth are determined by real, not inflated, dollars. If, for example, a land investment yields a 6 percent return during a period of 6 percent general price inflation, the wealth position of the farmer has not changed. In effect, the farmer has received a zero real rate of return on the investment.

To ensure that valuation reflects the farmer's desired real rate of return, it is necessary to include in the analysis an estimate of the average annual rate of general price inflation occurring during the period of land ownership. A good measure of inflation would be the anticipated rate of change in the consumer price index or the implicit price deflator for the gross domestic product.

4. *Number of years in the farmer's planning horizon.* Because the value of the land is based on total earnings realized during the ownership period, the length of that period is an important determinant of value. The ownership period used in the analysis must be determined by each farmer in light of his or her age and business objectives. An older farmer about to retire and liquidate property may have less than a 5-year planning horizon. However, a 25-year-old farmer about to purchase his or her first piece of property could have a 40-year planning horizon.
5. *Income taxes paid on annual land returns and on capital gains when land is sold.* Federal and state income taxes paid on ordinary income generated annually by land and on capital gains when appreciated land is sold reduce the stream of earnings attributable to land. Thus, like production expenses, income taxes reduce the value of land to the farmer.

   To identify the impact of income taxes on land values, an estimate of the appropriate marginal tax rate is necessary. The marginal tax rate is the rate at which additional income from the land investment is taxed. Both federal and state taxes are important.
6. *Income tax savings from deductions of interest paid on debt capital.* Interest paid on money borrowed to finance land is a tax-deductible expense. Therefore, assuming a profitable business, interest deductions reduce income tax liabilities and increase the after-tax stream of earnings accruing to land. The size of these tax benefits can be substantial as illustrated by a $1,000 per acre loan with a 9 percent interest rate and a 20-year amortization period. Average annual interest for this loan would be $59.54 per acre. Assuming a 32 percent marginal tax rate, the annual income tax savings per acre are $19.05 ($59.54 × 0.32). The actual impact of interest tax savings on land values will, however, depend on the interest rate relative to the discount rate (discussed later) and the marginal tax rate.
7. *Required after-tax real rate of return on the land investment.* Once the after-tax stream of earnings accruing to land has been estimated, the earnings must be discounted to obtain the land's present value—the value of the land to the farmer at the time of the proposed purchase. Discounting (reducing in value) is necessary because the earnings are realized at future rather than current points in time. Future dollars are worth less to the farmer than current dollars, even assuming a zero rate of inflation and overlooking the added uncertainty of ever receiving those future dollars. The reason future returns are worth less is the opportunity one has to reinvest dollars on hand and realize returns accruing with the passage of time. Thus, future dollars should be discounted at a rate equivalent to the earnings sacrificed by not having the money on hand immediately to reinvest.

The discount rate used to value land is, in effect, the rate of return required on the land investment. Therefore, an important part of the valuation process is selection of a required real rate of return. A profit-oriented farmer will want the rate of return to be high enough to ensure a profitable land investment. This means the investment must yield a rate of return exceeding the cost of capital. Consequently, the cost of capital is commonly used as the minimal required real rate of return (discount rate) in valuing land.

Because the cost of capital (discount rate) has a sizable impact on land value, farmers should exercise care in estimating that cost. The cost should be at least as high as the after-tax cost of debt capital used to finance the land purchase. The after-tax cost of debt is calculated in the following manner:

After-tax cost of debt = Effective before-tax interest rate × (1 – marginal tax rate)

Thus, a loan with a 9 percent effective rate of interest, used by a farmer in the 24 percent marginal tax bracket, has an after-tax cost of 6.84 percent (i.e., 9% × (1 – 0.24). A tax adjustment is necessary to reflect the income tax deductibility of interest expenses.[2]

A procedure for valuing land according to its projected flow of income (annual earnings and liquidation value) is illustrated in Worksheet 10.1.

The Case Farm

Referring to Worksheet 10.1, assume a farmer is considering the purchase of a nearby 160 acres of cropland. The land, if purchased, will be used to grow a 4-year rotation of (winter) wheat–(spring) barley–(winter) wheat–dry peas. Preliminary figuring indicates the farmer can handle the added cropland with existing machinery resources.

One of the major problems confronting the farmer is determining the value of the land to the business, given the desired after-tax real rate of return. Once that value has been determined, it will be easier to assess the likelihood of acquiring the land. Also, the farmer can more effectively negotiate a price with the seller.

Before using the Worksheet, the farmer should prepare enterprise budgets for the three crops to be grown on the land. The farmer has grown wheat, barley, and dry peas for several years on land similar to that which he or she is considering buying. Consequently, the farmer's records and experiences will provide a good basis for estimating costs and returns for the crops. The budgets should be prepared assuming normalized current yields, costs, and returns. Estimates of future changes in land returns and general price inflation are made on the Worksheet.

Enterprise budgets prepared by the farmer are summarized in Worksheet 10.2. The cost estimates should include only those outlays expected to increase as a result of the land purchase. Thus, estimates should be made for such operating expenses as fuel, lubrication, machinery repairs, hired labor, seed, fertilizer, herbicides, insecticides, custom services, crop insurance, and interest on the added operating capital.

Certain overhead costs also will increase when additional land is purchased. If additional machinery is purchased, ownership costs should be estimated. Even if additional machinery is not needed, some increase in machinery overhead expenses can generally be expected because of the shortening of the machines useful life. Also, adjustments may be in order for changes in machine rentals or custom work. This increase, however, will be small compared with the added overhead cost when additional machinery is purchased. Excess machine capacity and cost economies realized through greater machine use is one of the leading reasons why many farmers are interested in buying additional land.

Other overhead costs increasing when more land is purchased are property taxes on the land and various miscellaneous items, including telephone, travel, accounting, utilities, etc. Because the added acreage normally requires more labor and management by the owner-operator, the cost of this effort also should be identified. Interest on the investment in added land should not be included because this cost is accounted for by the discounting procedure used on the Worksheet.

WORKSHEET 10.1

Analysis of Land Values Illustration

1.	Enter average annual before-tax gross receipts	$ 161.13
2.	Enter average annual before-tax costs. Do not include interest on land loan or investment	$ 93.11
3.	Subtract line 2 from line 1	$ 68.02
4.	Enter 1.00 minus your marginal income tax rate. Express as a decimal	0.63
5.	Multiply line 4 times line 3	$ 42.85
6.	Enter your required after-tax real rate of return on the land investment	7 %
7.	Enter your estimate of the average annual rate of general price inflation during the land investment planning period.	5 %
8.	Enter your estimate of the average annual rate of change in the land returns	6 %
9.	Add lines 6 and 7 and subtract line 8 from total	6 %
10.	Enter interest factor from Appendix Table 10.4 for interest rate on line 9and number of years in land investment planning period	12.783
11.	Multiply line 10 times line 5	$ 547.75
12.	Enter the proportion of the purchase price to be financed with debt (IF NO DEBT USED, SKIP TO LINE 20)	0.80
13.	Enter interest factor from Appendix Table 10.4 for interest rate equaling before-tax contractual rate of interest on loan and number of years in loan repayment period	9.129
14.	Divide line 12 by line 13	0.088
15.	Divide line 12 by number of years in loan repayment period	0.040
16.	Subtract line 15 from line 14	0.048
17.	Multiply line 16 times your marginal income tax rate	0.018
18.	Enter interest factor from Appendix Table 10.4 for interest rate equaling line 6 plus line 7 and number of years in loan repayment period	7.469
19.	Multiply line 18 times line 17	0.134
20.	Enter interest factor from Appendix Table 10.2 for interest rate equaling line 6 plus line 7 and number of years in planning period	0.059
21.	Enter interest factor from Appendix Table 10.1 for interest rate equaling your estimate of annual rate of increase in land market price and number of years in planning period	5.427
22.	Enter estimated market price of land	$ 1,200.00
23.	Multiply line 21 times line 22	$ 6,512.40
24.	Enter your capital gains tax rate (express as a decimal)*	0.20
25.	Multiply line 23 times line 24	$ 963.84
26.	Subtract line 25 from line 23 and multiply the answer times line 20	$ 327.37
27.	Multiply line 20 times line 24	0.009
28.	Add lines 19 and 27	0.143
29.	Enter 1.000 minus line 28	0.857
30.	Add lines 11 and 26	$ 875.12
31.	LAND VALUE (line 30 divided by line 29)	$ 1,021.14

*The capital gains tax rate can be estimated by using current IRS rates.

As indicated in Worksheet 10.2, by subtracting the added operating and overhead costs from the added gross receipts, the example farmer has estimated a return to land of about $68 per acre. This return is the farmer's estimate of what will be earned during the first year the 160 acres is farmed, assuming normal yields, product prices, and costs. All costs and returns are on a before-tax basis because income tax adjustments are made on the Worksheet.

WORKSHEET 10.2

Estimated Annual Receipts, Costs, and Returns for a 160-Acre Addition to the Example Farm, Wheat-Barley-Wheat-Dry Peas Rotation

	Crops in Rotation			Total	
Item	Wheat	Barley	Dry Peas	160 Acres	Per Rotation Acre
1. Gross receipts					
Acres	80	40	40	160	—
Yield per acre	65 bu	1.5t	1,300 lbs	—	—
Price	$ 3.60/bu	$ 70/t	$ 7.50/cwt	—	—
Gross receipts	$17,680	$4,200	$3,900	$25,780	$161.13
2. Costs					
A. Operating					
Machinery (fuel, oil, and repairs)	$ 1,028	$ 526	$ 555	$ 2,109	$ 13.18
Hired labor	454	215	236	905	5.66
Seed	680	304	986	1,970	12.31
Fertilizer	2,643	475	—	3,118	19.49
Herbicides and insectides	570	285	420	1,275	7.97
Custom services	260	130	25	415	2.59
Crop insurance	237	74	156	467	2.92
Interest on operating capital	222	32	30	284	1.78
Total	$ 6,094	$2,041	$2,408	$10,543	$ 65.90
B. Overhead					
Machinery depreciation	$ 736	$ 355	$ 402	$ 1,493	$ 9.33
Real estate taxes	480	240	240	960	6.00
Operator labor and management	910	285	195	1,390	8.69
Miscellaneous*	305	104	102	511	3.19
Total	$ 2,431	$ 984	$ 939	$ 4,354	$ 27.21
3. Total cost (A + B)	$ 8,525	$3,025	$3,347	$14,897	93.11
4. Returns to land (1–3)	$ 9,155	$1,175	$ 553	$10,883	$ 68.02

*Includes telephone, utilities, travel, accounting, legal costs, etc.

Other assumptions the example farmer must make before proceeding with the valuation analysis are as follows:

- Marginal income tax rate for the joint return filed with spouse = 37 percent.
- Required after-tax real rate of return on the potential land investment = 7 percent.
- Estimate of the average annual rate of general price inflation = 5 percent.
- Estimate of the average annual rate of change in returns to land = 6 percent.
- Length of the planning period = 25 years.
- Estimate of average annual rate of change in market price of land = 7 percent.
- Proportion of purchase price financed with debt = 80 percent.
- Interest rate on loan with a 20-year repayment period = 9 percent.
- Estimated market price for recent and comparable land sales = $1,200 per acre.

Estimates of the rates of change in the general price level, returns to land, and land values are made by studying past trends and giving careful thought as to how these trends might change over the 25-year planning period. The estimates reflect the farmer's judgment that the annual rate of increase in the land returns from farming, 6 percent, will stay slightly ahead of general price inflation, 5 percent. Also, because of the land's proximity to a growing community, it is expected that its value will increase at a slightly faster rate, 7 percent, than its return from farming activities, 6 percent.

Using the budget information and the previous assumptions, the farmer is now ready to complete Worksheet 10.1 on either a per-acre or 160-acre basis. Lines 1 through 5 of the Worksheet use information on land receipts (line 1), costs (line 2), and the farm's marginal tax rate to compute the annual after-tax land return (line 5). As indicated, that return is $42.85. Lines 6–8 request the farmer to designate the required after-tax real rate of return on the land investment, rate of general price inflation, and rate of change in land returns, respectively. From that information, the rate used to discount future land returns is computed on line 9 (6 percent for the case farm illustration). Next, the interest factor corresponding to the discount rate is inserted on line 10. Multiplication of the interest factor times the annual after-tax land returns (line 5) gives the after-tax present value of future land returns (line 11). Thus, the value of the land to the hypothetical farmer, excluding income tax benefits from interest and liquidation value, is $547.75 (line 11).

Information used to compute income tax savings from the deduction of interest paid on the land loan appears on lines 12 through 19. These steps can be omitted if no debt is used. Lines 20 through 26 request data for computing the after-tax liquidation value of the land. After completing the mathematical adjustments directed by lines 27 through 30, the final value of the land is computed on line 31.

As indicated on line 31 of the completed Worksheet, the land is worth $1,021 per acre to the hypothetical farmer. It can be concluded that if the farmer pays $1,021 for the land, an after-tax real rate of return of 7 percent will be realized on the investment. If more than $1,021 is paid for the land, a lower rate of return can be expected. Alternatively, a higher return is implied if the land is acquired for a price below $1,021. Because the market price was estimated to be $1,200, the farmer may have difficulty purchasing the land at a price providing the desired profit.

The farmer would be well advised to complete several worksheets to determine the impact of adopting different assumptions for such key variables as (1) required rate of return, (2) returns to land (e.g., wheat price and yield), (3) rate of appreciation in land value, and (4) level of general price inflation. This would provide a good indication of the risk associated with buying the land, as well as identifying flexibility in negotiating a price with the seller.

Determining the Financial Feasibility of a Land Purchase

After determining the value of a potential land purchase to the business, a farmer will want to consider the financial feasibility of the acquisition. More specifically, the question must be examined as to whether available equity capital reserves and unused borrowing capacity are sufficient to finance the purchase at the expected market price. Furthermore, where large amounts of debt capital are involved, the ability of the business to service that debt should be identified.

Worksheet 10.3 provides a procedure for determining the maximum financially feasible price the farm business will be able to pay for additional land.[3] The worksheet should be completed for the expanded (current-plus-added-land) business on a total farm basis. The worksheet assumes the maximum financially feasible price is determined by the equity funds available for a down payment, plus the maximum amount of debt the farm's cash flow can service. It is further assumed the latter is dependent upon (1) the amount of cash farm earnings retained in the business, (2) pre-expansion financial commitments that must be serviced, (3) the rate of interest on the new real estate loan, and (4) the number of years over which the loan is amortized.

WORKSHEET 10.3

Analysis of Ability to Pay for Land (Case Farm Illustration)

1. Enter average before-tax gross cash receipts from all enterprises in the expanded business	$ 133,000
2. Enter average annual before-tax cash costs for all enterprises in the expanded business*	$ 83,600
3. Subtract line 2 from line 1	$ 49,400
4. Enter average annual depreciation for income taxes paid on expanded business	$9,120
5. Enter income tax deduction for personal exemptions and zero bracket amount or itemized deduction	$ 7,000
6. Add lines 4 and 5	$ 16,120
7. Enter your average income tax rate for expanded business (express as a decimal)†	0.22
8. Multiply line 7 times line 6	$ 3,546
9. Enter 1.00 minus your average income tax rate (line 7)	0.78
10. Multiply line 9 times line 3	$ 38,532
11. Add lines 10 and 8	$ 42,078
12. Enter average principal payments on long-term debt (> 1 year). Do not include debt on land purchase	$ 7,000
13. Enter average annual depreciation reserve‡	$ 10,000
14. Enter social security taxes paid on self-employment income§	$ 2,098
15. Enter annual family living expenses	$ 14,000
16. Add lines 12, 13, 14, and 15	$ 33,098
17. Subtract line 16 from line 11	$ 8,980
18. Enter interest factor from Appendix Table 10.4 for interest rate equaling after-tax contractual rate of interest on loan and number of years in loan repayment period‖	10.594
19. Multiply line 18 times line 17	$ 95,134
20. Enter equity capital available for down payment on land purchase	$ 35,200
21. MAXIMUM PRICE THAT CAN BE PAID FOR TOTAL ACREAGE (line 19 plus line 20)	$ 130,334
22. MAXIMUM PER-ACRE PRICE (line 21 divided by number of acres in land purchase)	$ 815

*Cash costs should include hired labor, fuel, lubricants, repairs, all materials (e.g. seed, fertilizer, herbicides, pesticides), custom services, rent on machinery and land, crop insurance, property taxes, general overhead (e.g. telephone, utilities, travel, accounting, legal), and interest paid on all loans. Interest paid on the loan to finance the land purchase should not be included. If livestock enterprises are present, cash costs related to these enterprises should be included.

†Can be estimated by subtracting line 6 from line 3 to obtain taxable income and referring to tax tables for the average tax rate on that taxable income.

‡Estimate of annual capital needed to replace depreciable assets.

§Can be estimated by subtracting line 6 from line 3 and multiplying difference by the appropriate social security tax rate.

‖The after-tax interest rate equals the before-tax rate multiplied by 1 minus the average income tax rate.

Where the maximum price the business can pay for the land (line 22, Worksheet 10.2) exceeds the probable market price, the purchase is financially feasible. Alternatively, where the market price is likely to exceed the maximum price, there is a strong likelihood the business will experience cash-flow difficulties if the land is purchased.

The Case Farm Revisited

Thus far, the example farmer has determined that the 160 acres of land is worth approximately $1,021 per acre to the business. The farmer should now analyze the ability to pay for the land. That analysis can be made by completing Worksheet 10.2.

The analysis in Worksheet 10.3 is based on annual projections made by the farmer for the expanded (current and added land) business. The cash expense projection (line 2) does not include such noncash costs as depreciation, interest on equity capital investment, and operator and family labor. Also, interest paid on debt used to finance the land purchase is excluded at this point.

An estimate of depreciation (line 4) and personal exemptions and zero bracket amount or itemized deduction (line 5) is necessary for the determination of income tax savings.

After-tax earnings generated by the farmer's business are estimated to be $42,078 (line 11). Before determining the amount of additional debt the business can service, however, these earnings need to be reduced by several additional financial commitments (i.e., annual principal payments on long-term (over 1 year) debt assumed prior to the land purchase (line 12), an annual depreciation reserve (line 13), self-employed social security taxes (line 14), and family living expenses (line 15). After making these adjustments, the amount of cash available to make principle and interest payments on additional real estate debt is $8,980 (line 17).

The interest factor indicating how much debt a $1 annual payment will support can be found in the appendix tables. The interest factor should be selected for the interest rate equaling the after-tax interest rate on the land loan and for the number of years in the loan repayment period. For the case at hand, that would be 7.0 percent and 20 years (i.e., 9 percent × [1 – 0.22 average tax rate]). The corresponding interest factor is 10.594 (line 18), which, when multiplied times $8,980, gives the $95,134 of debt the expanded business can support (line 19). Assuming $35,200 of cash (equity capital) is available for a down payment (line 20), a total of $130,334 ($95,134 + $35,200) can be paid for the land (line 21). The maximum per-acre price is $130,334 divided by 160 acres, or $815 (line 22).

It is suggested that several worksheets be completed using different assumptions for such key variables as crop prices and yields, cash production costs, and real estate loan terms. By expanding the analysis in this manner, the farmer can obtain a better feel for the range of debt (and therefore, maximum land prices) that the farm can support under various business circumstances.

Interpretation of Analysis

Once Worksheets 10.1, 10.2, and 10.3 have been completed, the farmer is in a good position to assess the financial advisability of buying the land. If the analysis indicates the land's value and maximum financially feasible price both exceed the expected market price, the land purchase will appear attractive. Alternatively, if the value or maximum financially feasible price falls below the likely market price, the land investment will not yield the desired profit or present cash-flow problems.

The case farm analysis provides a good example of a financially unattractive land investment. Both the maximum financially feasible price ($815) and value ($1,021) lie below the expected market price ($1,200). The sizable difference between $815 and $1,200 indicates the farmer is in an extremely weak position to obtain the land. Under the adopted assumptions, purchase of the land at a competitive price (for example, $1,000 to $1,200) would involve considerable risk of being unable to service the added real estate debt.

Limitations of Analysis

Although the analysis outline is considerably more detailed than traditionally used, certain factors were, nevertheless, omitted in an attempt to keep the discussion manageable. The more important limitations are three:

1. The valuation procedure (Worksheet 10.1) does not consider the income tax benefits stemming from depreciation. Thus, where depreciable improvements (e.g., orchards, irrigation equipment, buildings) are included in the investment, the value computed by Worksheet 10.1 will be understated.
2. The possible recapture of depreciation when the property is sold is not considered.
3. Land values, returns to land, and general price levels are assumed either to remain constant or to increase at a constant annual compound rate. The valuation analysis, for example, does not allow a given rate of increase in land values for a certain number of years to be followed by either a zero or different rate of increase in later years.

APPENDIX TABLE 10.1

Future Value of $1 After n Periods Compounded at Various Interest Rates Expressed in Decimal Equivalents

n	0.005	0.008	0.010	0.015	0.020	0.030	0.040	0.050	0.060	0.070	0.080	0.090	0.100	0.120	0.140	0.160	0.180	0.200	0.250
1	1.0050	1.0080	1.0100	1.0150	1.0200	1.0300	1.0400	1.0500	1.0600	1.0700	1.0800	1.0900	1.1000	1.1200	1.1400	1.1600	1.1800	1.2000	1.2500
2	1.0100	1.0161	1.0201	1.0302	1.0404	1.0609	1.0816	1.1025	1.1236	1.1449	1.1664	1.1881	1.2100	1.2544	1.2996	1.3456	1.3924	1.4400	1.5625
3	1.0151	1.0242	1.0303	1.0457	1.0612	1.0927	1.1249	1.1576	1.1910	1.2250	1.2597	1.2950	1.3310	1.4049	1.4815	1.5609	1.6430	1.7280	1.9531
4	1.0202	1.0324	1.0406	1.0614	1.0824	1.1255	1.1699	1.2155	1.2625	1.3108	1.3605	1.4116	1.4641	1.5735	1.6890	1.8106	1.9388	2.0736	2.4414
5	1.0253	1.0406	1.0510	1.0773	1.1041	1.1593	1.2167	1.2763	1.3382	1.4026	1.4693	1.5386	1.6105	1.7623	1.9254	2.1003	2.2878	2.4883	3.0518
6	1.0304	1.0490	1.0615	1.0934	1.1262	1.1941	1.2653	1.3401	1.4185	1.5007	1.5869	1.6771	1.7716	1.9738	2.1950	2.4364	2.6996	2.9860	3.8147
7	1.0355	1.0574	1.0721	1.1098	1.1487	1.2299	1.3159	1.4071	1.5036	1.6058	1.7138	1.8280	1.9487	2.2107	2.5023	2.8262	3.1855	3.5832	4.7684
8	1.0407	1.0658	1.0829	1.1265	1.1717	1.2668	1.3686	1.4775	1.5938	1.7182	1.8509	1.9926	2.1436	2.4760	2.8526	3.2784	3.7589	4.2998	5.9605
9	1.0459	1.0743	1.0937	1.1434	1.1951	1.3048	1.4233	1.5513	1.6895	1.8385	1.9990	2.1719	2.3579	2.7731	3.2519	3.8030	4.4355	5.1598	7.4506
10	1.0511	1.0829	1.1046	1.1605	1.2190	1.3439	1.4802	1.6289	1.7908	1.9672	2.1589	2.3674	2.5937	3.1058	3.7072	4.4114	5.2338	6.1917	9.3132
12	1.0617	1.1003	1.1268	1.1956	1.2682	1.4258	1.6010	1.7959	2.0122	2.2522	2.5182	2.8127	3.1384	3.8960	4.8179	5.9360	7.2876	8.9161	14.552
14	1.0723	1.1180	1.1495	1.2318	1.3195	1.5126	1.7317	1.9799	2.2609	2.5785	2.9372	3.3417	3.7975	4.8871	6.2613	7.9875	10.147	12.839	22.737
16	1.0831	1.1360	1.1726	1.2690	1.3728	1.6047	1.8730	2.1829	2.5404	2.9522	3.4259	3.9703	4.5950	6.1304	8.1372	10.748	14.129	18.488	35.527
18	1.0939	1.1542	1.1961	1.3073	1.4282	1.7024	2.0258	2.4066	2.8543	3.3799	3.9960	4.7171	5.5599	7.6900	10.575	14.463	19.673	26.623	55.511
20	1.1049	1.1728	1.2202	1.3469	1.4859	1.8061	2.1911	2.6533	3.2071	3.8697	4.6610	5.6044	6.7275	9.6463	13.743	19.461	27.393	38.338	86.736
25	1.1328	1.2204	1.2824	1.4509	1.6406	2.0938	2.6658	3.3864	4.2919	5.4274	6.8485	8.6231	10.835	17.000	26.462	40.874	62.669	95.396	264.70
30	1.1614	1.2700	1.3478	1.5631	1.8114	2.4273	3.2434	4.3219	5.7435	7.6123	10.063	13.268	17.449	29.960	50.950	85.850	143.37	237.38	807.79
35	1.1907	1.3217	1.4166	1.6839	1.9999	2.8139	3.9461	5.5160	7.6861	10.677	14.785	20.414	28.102	52.800	98.100	180.31	328.00	590.67	2465.2
40	1.2208	1.3754	1.4889	1.8140	2.2080	3.2620	4.8010	7.0400	10.286	14.974	21.725	31.409	45.259	93.051	188.88	378.72	750.38	1469.8	7523.2

$V_n = \$1 \times (1 + i)^n$

APPENDIX TABLE 10.2

Present Value of $1 Received in Period n Discounted at Various Interest Rates Expressed in Decimal Equivalents

n	0.005	0.008	0.010	0.015	0.020	0.030	0.040	0.050	0.060	0.070	0.080	0.090	0.100	0.120	0.140	0.160	0.180	0.200	0.250
1	0.9950	0.9921	0.9901	0.9852	0.9804	0.9709	0.9615	0.9524	0.9434	0.9346	0.9259	0.9174	0.9091	0.8929	0.8772	0.8621	0.8475	0.8333	0.8000
2	0.9901	0.9842	0.9803	0.9707	0.9612	0.9426	0.9246	0.9070	0.8900	0.8734	0.8573	0.8417	0.8264	0.7972	0.7695	0.7432	0.7182	0.6944	0.6400
3	0.9851	0.9764	0.9706	0.9563	0.9423	0.9151	0.8890	0.8638	0.8396	0.8163	0.7938	0.7722	0.7513	0.7118	0.6750	0.6407	0.6086	0.5787	0.5120
4	0.9802	0.9686	0.9610	0.9422	0.9238	0.8885	0.8548	0.8227	0.7921	0.7629	0.7350	0.7084	0.6830	0.6355	0.5921	0.5523	0.5158	0.4823	0.4096
5	0.9754	0.9609	0.9515	0.9283	0.9057	0.8626	0.8219	0.7835	0.7473	0.7130	0.6806	0.6499	0.6209	0.5674	0.5194	0.4761	0.4371	0.4019	0.3277
6	0.9705	0.9533	0.9420	0.9145	0.8880	0.8375	0.7903	0.7462	0.7050	0.6663	0.6302	0.5963	0.5645	0.5066	0.4556	0.4104	0.3704	0.3349	0.2621
7	0.9657	0.9457	0.9327	0.9010	0.8706	0.8131	0.7599	0.7107	0.6651	0.6227	0.5835	0.5470	0.5132	0.4523	0.3996	0.3538	0.3139	0.2791	0.2097
8	0.9609	0.9382	0.9235	0.8877	0.8535	0.7894	0.7307	0.6768	0.6274	0.5820	0.5403	0.5019	0.4665	0.4039	0.3506	0.3050	0.2660	0.2326	0.1678
9	0.9561	0.9308	0.9143	0.8746	0.8368	0.7664	0.7026	0.6446	0.5919	0.5439	0.5002	0.4604	0.4241	0.3606	0.3075	0.2630	0.2255	0.1938	0.1342
10	0.9513	0.9234	0.9053	0.8617	0.8203	0.7441	0.6756	0.6139	0.5584	0.5083	0.4632	0.4224	0.3855	0.3220	0.2697	0.2267	0.1911	0.1615	0.1074
12	0.9419	0.9088	0.8874	0.8364	0.7885	0.7014	0.6246	0.5568	0.4970	0.4440	0.3971	0.3555	0.3186	0.2567	0.2076	0.1685	0.1372	0.1122	0.0687
14	0.9326	0.8944	0.8700	0.8118	0.7579	0.6611	0.5775	0.5051	0.4423	0.3878	0.3405	0.2992	0.2633	0.2046	0.1597	0.1252	0.0985	0.0779	0.0440
16	0.9233	0.8803	0.8528	0.7880	0.7284	0.6232	0.5339	0.4581	0.3936	0.3387	0.2919	0.2519	0.2176	0.1631	0.1229	0.0930	0.0708	0.0541	0.0281
18	0.9141	0.8664	0.8360	0.7649	0.7002	0.5874	0.4936	0.4155	0.3503	0.2959	0.2502	0.2120	0.1799	0.1300	0.0946	0.0691	0.0508	0.0376	0.0180
20	0.9051	0.8527	0.8195	0.7425	0.6730	0.5537	0.4564	0.3769	0.3118	0.2584	0.2145	0.1784	0.1486	0.1037	0.0728	0.0514	0.0365	0.0261	0.0115
25	0.8828	0.8194	0.7798	0.6892	0.6095	0.4776	0.3751	0.2953	0.2330	0.1842	0.1460	0.1160	0.0923	0.0588	0.0378	0.0245	0.0160	0.0105	0.0038
30	0.8610	0.7874	0.7419	0.6398	0.5521	0.4120	0.3083	0.2314	0.1741	0.1314	0.0994	0.0754	0.0573	0.0334	0.0196	0.0116	0.0070	0.0042	0.0012
35	0.8398	0.7566	0.7059	0.5939	0.5000	0.3554	0.2534	0.1813	0.1301	0.0937	0.0676	0.0490	0.0356	0.0189	0.0102	0.0055	0.0030	0.0017	0.0004
40	0.8191	0.7271	0.6717	0.5513	0.4529	0.3066	0.2083	0.1420	0.0972	0.0668	0.0460	0.0318	0.0221	0.0107	0.0053	0.0026	0.0013	0.0007	0.0001

$V_0 = \$1 / (1 + i)^n$

APPENDIX TABLE 10.3

Future Value of a Uniform Series of $1 at Various Interest Rates Expressed in Decimal Equivalents

n	0.005	0.008	0.010	0.015	0.020	0.030	0.040	0.050	0.060	0.070	0.080	0.090	0.100	0.120	0.140	0.160	0.180	0.200	0.250
1	1.0000	1.0000	1.0000	1.0000	1.0000	1.0000	1.0000	1.0000	1.0000	1.0000	1.0000	1.0000	1.0000	1.0000	1.0000	1.0000	1.0000	1.0000	1.0000
2	2.0050	2.0080	2.0100	2.0150	2.0200	2.0300	2.0400	2.0500	2.0600	2.0700	2.0800	2.0900	2.1000	2.1200	2.1400	2.1600	2.1800	2.2000	2.2500
3	3.0150	3.0241	3.0301	3.0452	3.0604	3.0909	3.1216	3.1525	3.1836	3.2149	3.2464	3.2781	3.3100	3.3744	3.4396	3.5056	3.5724	3.6400	3.8125
4	4.0301	4.0483	4.0604	4.0909	4.1216	4.1836	4.2465	4.3101	4.3746	4.4399	4.5061	4.5731	4.6410	4.7793	4.9211	5.0665	5.2154	5.3680	5.7656
5	5.0503	5.0806	5.1010	5.1523	5.2040	5.3091	5.4163	5.5256	5.6371	5.7507	5.8666	5.9847	6.1051	6.3528	6.6101	6.8771	7.1542	7.4416	8.2070
6	6.0755	6.1213	6.1520	6.2296	6.3081	6.4684	6.6330	6.8019	6.9753	7.1533	7.3359	7.5233	7.7156	8.1152	8.5355	8.9775	9.4420	9.9299	11.259
7	7.1059	7.1703	7.2135	7.3230	7.4343	7.6625	7.8983	8.1420	8.3938	8.6540	8.9228	9.2004	9.4872	10.089	10.730	11.414	12.142	12.916	15.073
8	8.1414	8.2276	8.2857	8.4328	8.5830	8.8923	9.2142	9.5491	9.8975	10.260	10.637	11.028	11.436	12.300	13.233	14.240	15.327	16.499	19.842
9	9.1821	9.2934	9.3685	9.5593	9.7546	10.159	10.583	11.027	11.491	11.978	12.488	13.021	13.579	14.776	16.085	17.519	19.086	20.799	25.802
10	10.228	10.368	10.462	10.703	10.950	11.464	12.006	12.578	13.181	13.816	14.487	15.193	15.937	17.549	19.337	21.321	23.521	25.959	33.253
12	12.336	12.542	12.683	13.041	13.412	14.192	15.026	15.917	16.870	17.888	18.977	20.141	21.384	24.133	27.271	30.850	34.931	39.581	54.208
14	14.464	14.752	14.947	15.450	15.974	17.086	18.292	19.599	21.015	22.550	24.215	26.019	27.975	32.393	37.581	43.672	50.818	59.196	86.949
16	16.614	16.997	17.258	17.932	18.639	20.157	21.825	23.657	25.673	27.888	30.324	33.003	35.950	42.753	50.980	60.925	72.939	87.442	138.11
18	18.786	19.278	19.615	20.489	21.412	23.414	25.645	28.132	30.906	33.999	37.450	41.301	45.599	55.750	68.394	84.141	103.74	128.12	218.04
20	20.979	21.596	22.019	23.124	24.297	26.870	29.778	33.066	36.786	40.995	45.762	51.160	57.275	72.052	91.025	115.38	146.63	186.69	342.94
25	26.559	27.554	28.243	30.063	32.030	36.459	41.646	47.727	54.865	63.249	73.106	84.701	98.347	133.33	181.87	249.21	342.60	471.98	1054.8
30	32.280	33.754	34.785	37.539	40.568	47.575	56.085	66.439	79.058	94.461	113.28	136.31	164.49	241.33	356.79	530.31	790.95	1181.9	3227.2
35	38.145	40.207	41.660	45.592	49.994	60.462	73.652	90.320	111.43	138.24	172.32	215.71	271.02	431.66	693.57	1120.7	1816.7	2948.3	9856.8
40	44.159	46.922	48.886	54.268	60.402	75.401	95.026	120.80	154.76	199.64	259.06	337.88	442.59	767.09	1342.0	2360.8	4163.2	7343.9	30089

$$V_n = \frac{\$1[(1+i)^n - 1]}{i}$$

APPENDIX TABLE 10.4

Present Value of a Uniform Series of $1 at Various Interest Rates Expressed in Decimal Equivalents

n	0.005	0.008	0.010	0.015	0.020	0.030	0.040	0.050	0.060	0.070	0.080	0.090	0.100	0.120	0.140	0.160	0.180	0.200	0.250
1	0.9950	0.9921	0.9901	0.9852	0.9804	0.9709	0.9615	0.9524	0.9434	0.9346	0.9259	0.9174	0.9091	0.8929	0.8772	0.8621	0.8475	0.8333	0.8000
2	1.9851	1.9763	1.9704	1.9559	1.9416	1.9135	1.8861	1.8594	1.8334	1.8080	1.7833	1.7591	1.7355	1.6901	1.6467	1.6052	1.5656	1.5278	1.4400
3	2.9702	2.9526	2.9410	2.9122	2.8839	2.8286	2.7751	2.7232	2.6730	2.6243	2.5771	2.5313	2.4869	2.4018	2.3216	2.2459	2.1743	2.1065	1.9520
4	3.9505	3.9213	3.9020	3.8544	3.8077	3.7171	3.6299	3.5460	3.4651	3.3872	3.3121	3.2397	3.1699	3.0373	2.9137	2.7982	2.6901	2.5887	2.3616
5	4.9259	4.8822	4.8534	4.7826	4.7135	4.5797	4.4518	4.3295	4.2124	4.1002	3.9927	3.8897	3.7908	3.6048	3.4331	3.2743	3.1272	2.9906	2.6893
6	5.8964	5.8355	5.7955	5.6972	5.6014	5.4172	5.2421	5.0757	4.9173	4.7665	4.6229	4.4859	4.3553	4.1114	3.8887	3.6847	3.4976	3.3255	2.9514
7	6.8621	6.7813	6.7282	6.5982	6.4720	6.2303	6.0021	5.7864	5.5824	5.3893	5.2064	5.0330	4.8684	4.5638	4.2883	4.0386	3.8115	3.6046	3.1611
8	7.8230	7.7195	7.6517	7.4859	7.3255	7.0197	6.7327	6.4632	6.2098	5.9713	5.7466	5.5348	5.3349	4.9676	4.6389	4.3436	4.0776	3.8372	3.3289
9	8.7791	8.6503	8.5660	8.3605	8.1622	7.7861	7.4353	7.1078	6.8017	6.5152	6.2469	5.9952	5.7590	5.3282	4.9464	4.6065	4.3030	4.0310	3.4631
10	9.7304	9.5737	9.4713	9.2222	8.9826	8.5302	8.1109	7.7217	7.3601	7.0236	6.7101	6.4177	6.1446	5.6502	5.2161	4.8332	4.4941	4.1925	3.5705
12	11.619	11.399	11.255	10.908	10.575	9.9540	9.3851	8.8633	8.3838	7.9427	7.5361	7.1607	6.8137	6.1944	5.6603	5.1971	4.7932	4.4392	3.7251
14	13.489	13.195	13.004	12.543	12.106	11.296	10.563	9.8986	9.2950	8.7455	8.2442	7.7862	7.3667	6.6282	6.0021	5.4675	5.0081	4.6106	3.8241
16	15.340	14.962	14.718	14.131	13.578	12.561	11.652	10.838	10.106	9.4466	8.8514	8.3126	7.8237	6.9740	6.2651	5.6685	5.1624	4.7296	3.8874
18	17.173	16.702	16.398	15.673	14.992	13.754	12.659	11.690	10.828	10.059	9.3719	8.7556	8.2014	7.2497	6.4674	5.8178	5.2732	4.8122	3.9279
20	18.987	18.414	18.046	17.169	16.351	14.877	13.590	12.462	11.470	10.594	9.8181	9.1285	8.5136	7.4694	6.6231	5.9288	5.3527	4.8696	3.9539
25	23.446	22.577	22.023	20.720	19.523	17.413	15.622	14.094	12.783	11.654	10.675	9.8226	9.0770	7.8431	6.8729	6.0971	5.4669	4.9476	3.9849
30	27.794	26.578	25.808	24.016	22.396	19.600	17.292	15.372	13.765	12.409	11.258	10.274	9.4269	8.0552	7.0027	6.1772	5.5168	4.9789	3.9950
35	32.035	30.422	29.409	27.076	24.999	21.487	18.665	16.374	14.498	12.948	11.655	10.567	9.6442	8.1755	7.0700	6.2153	5.5386	4.9915	3.9984
40	36.172	34.116	32.835	29.916	27.355	23.115	19.793	17.159	15.046	13.332	11.925	10.757	9.7791	8.2438	7.1050	6.2335	5.5482	4.9966	3.9995

$$V_n = \frac{\$1[1-(1+i)^{-n}]}{i}$$

APPENDIX TABLE 10.5

Annuity That $1 Will Buy at Various Interest Rates Expressed in Decimal Equivalents

n	0.005	0.008	0.010	0.015	0.020	0.030	0.040	0.050	0.060	0.070	0.080	0.090	0.100	0.120	0.140	0.160	0.180	0.200	0.250
1	1.0050	1.0080	1.0100	1.0150	1.0200	1.0300	1.0400	1.0500	1.0600	1.0700	1.0800	1.0900	1.1000	1.1200	1.1400	1.1600	1.1800	1.2000	1.2500
2	0.5038	0.5060	0.5075	0.5113	0.5150	0.5226	0.5302	0.5378	0.5454	0.5531	0.5608	0.5685	0.5762	0.5917	0.6073	0.6230	0.6387	0.6545	0.6944
3	0.3367	0.3387	0.3400	0.3434	0.3468	0.3535	0.3603	0.3672	0.3741	0.3811	0.3880	0.3951	0.4021	0.4163	0.4307	0.4453	0.4599	0.4747	0.5123
4	0.2531	0.2550	0.2563	0.2594	0.2626	0.2690	0.2755	0.2820	0.2886	0.2952	0.3019	0.3087	0.3155	0.3292	0.3432	0.3574	0.3717	0.3863	0.4234
5	0.2030	0.2048	0.2060	0.2091	0.2122	0.2184	0.2246	0.2310	0.2374	0.2439	0.2505	0.2571	0.2638	0.2774	0.2913	0.3054	0.3198	0.3344	0.3718
6	0.1696	0.1714	0.1725	0.1755	0.1785	0.1846	0.1908	0.1970	0.2034	0.2098	0.2163	0.2229	0.2296	0.2432	0.2572	0.2714	0.2859	0.3007	0.3388
7	0.1457	0.1475	0.1486	0.1516	0.1545	0.1605	0.1666	0.1728	0.1791	0.1856	0.1921	0.1987	0.2054	0.2191	0.2332	0.2476	0.2624	0.2774	0.3163
8	0.1278	0.1295	0.1307	0.1336	0.1365	0.1425	0.1485	0.1547	0.1610	0.1675	0.1740	0.1807	0.1874	0.2013	0.2156	0.2302	0.2452	0.2606	0.3004
9	0.1139	0.1156	0.1167	0.1196	0.1225	0.1284	0.1345	0.1407	0.1470	0.1535	0.1601	0.1668	0.1736	0.1877	0.2022	0.2171	0.2324	0.2481	0.2888
10	0.1028	0.1045	0.1056	0.1084	0.1113	0.1172	0.1233	0.1295	0.1359	0.1424	0.1490	0.1558	0.1627	0.1770	0.1917	0.2069	0.2225	0.2385	0.2801
12	0.0861	0.0877	0.0888	0.0917	0.0946	0.1005	0.1066	0.1128	0.1193	0.1259	0.1327	0.1397	0.1468	0.1614	0.1767	0.1924	0.2086	0.2253	0.2684
14	0.0741	0.0758	0.0769	0.0797	0.0826	0.0885	0.0947	0.1010	0.1076	0.1143	0.1213	0.1284	0.1357	0.1509	0.1666	0.1829	0.1997	0.2169	0.2615
16	0.0652	0.0668	0.0679	0.0708	0.0737	0.0796	0.0858	0.0923	0.0990	0.1059	0.1130	0.1203	0.1278	0.1434	0.1596	0.1764	0.1937	0.2114	0.2572
18	0.0582	0.0599	0.0610	0.0638	0.0667	0.0727	0.0790	0.0855	0.0924	0.0994	0.1067	0.1142	0.1219	0.1379	0.1546	0.1719	0.1896	0.2078	0.2546
20	0.0527	0.0543	0.0554	0.0582	0.0612	0.0672	0.0736	0.0802	0.0872	0.0944	0.1019	0.1095	0.1175	0.1339	0.1510	0.1687	0.1868	0.2054	0.2529
25	0.0427	0.0443	0.0454	0.0483	0.0512	0.0574	0.0640	0.0710	0.0782	0.0858	0.0937	0.1018	0.1102	0.1275	0.1455	0.1640	0.1829	0.2021	0.2509
30	0.0360	0.0376	0.0387	0.0416	0.0446	0.0510	0.0578	0.0651	0.0726	0.0806	0.0888	0.0973	0.1061	0.1241	0.1428	0.1619	0.1813	0.2008	0.2503
35	0.0312	0.0329	0.0340	0.0369	0.0400	0.0465	0.0536	0.0611	0.0690	0.0772	0.0858	0.0946	0.1037	0.1223	0.1414	0.1609	0.1806	0.2003	0.2501
40	0.0276	0.0293	0.0305	0.0334	0.0366	0.0433	0.0505	0.0583	0.0665	0.0750	0.0839	0.0930	0.1023	0.1213	0.1407	0.1604	0.1802	0.2001	0.2500

Annuity = $i / [1 - (1 + i)^{-n}]$ Note that these values are the reciprocals of Appendix Table 10.4.

APPENDIX TABLE 10.6

Principal and Interest Paid per $1 Borrowed by Term of Loan and Interest Rates Expressed in Decimal Equivalents

n	0.005	0.008	0.010	0.015	0.020	0.030	0.040	0.050	0.060	0.070	0.080	0.090	0.100	0.120	0.140	0.160	0.180	0.200	0.250
1	1.0050	1.0080	1.0100	1.0150	1.0200	1.0300	1.0400	1.0500	1.0600	1.0700	1.0800	1.0900	1.1000	1.1200	1.1400	1.1600	1.1800	1.2000	1.2500
2	1.0075	1.0120	1.0150	1.0226	1.0301	1.0452	1.0604	1.0756	1.0909	1.1062	1.1215	1.1369	1.1524	1.1834	1.2146	1.2459	1.2774	1.3091	1.3889
3	1.0100	1.0160	1.0201	1.0301	1.0403	1.0606	1.0810	1.1016	1.1223	1.1432	1.1641	1.1852	1.2063	1.2490	1.2922	1.3358	1.3798	1.4242	1.5369
4	1.0125	1.0201	1.0251	1.0378	1.0505	1.0761	1.1020	1.1280	1.1544	1.1809	1.2077	1.2347	1.2619	1.3169	1.3728	1.4295	1.4870	1.5452	1.6938
5	1.0150	1.0241	1.0302	1.0454	1.0608	1.0918	1.1231	1.1549	1.1870	1.2195	1.2523	1.2855	1.3190	1.3870	1.4564	1.5270	1.5989	1.6719	1.8592
6	1.0176	1.0282	1.0353	1.0532	1.0712	1.1076	1.1446	1.1821	1.2202	1.2588	1.2979	1.3375	1.3776	1.4594	1.5429	1.6283	1.7155	1.8042	2.0329
7	1.0201	1.0323	1.0404	1.0609	1.0816	1.1235	1.1663	1.2097	1.2539	1.2989	1.3445	1.3908	1.4378	1.5338	1.6323	1.7333	1.8365	1.9420	2.2144
8	1.0226	1.0363	1.0455	1.0687	1.0921	1.1397	1.1882	1.2378	1.2883	1.3397	1.3921	1.4454	1.4996	1.6104	1.7246	1.8418	1.9620	2.0849	2.4032
9	1.0252	1.0404	1.0507	1.0765	1.1026	1.1559	1.2104	1.2662	1.3232	1.3814	1.4407	1.5012	1.5628	1.6891	1.8195	1.9537	2.0916	2.2327	2.5988
10	1.0277	1.0445	1.0558	1.0843	1.1133	1.1723	1.2329	1.2950	1.3587	1.4238	1.4903	1.5582	1.6275	1.7698	1.9171	2.0690	2.2251	2.3852	2.8007
12	1.0328	1.0528	1.0662	1.1002	1.1347	1.2055	1.2786	1.3539	1.4313	1.5108	1.5923	1.6758	1.7612	1.9372	2.1200	2.3090	2.5035	2.7032	3.2214
14	1.0379	1.0610	1.0766	1.1161	1.1564	1.2394	1.3254	1.4143	1.5062	1.6008	1.6982	1.7981	1.9004	2.1122	2.3325	2.5606	2.7955	3.0365	3.6610
16	1.0430	1.0694	1.0871	1.1322	1.1784	1.2738	1.3731	1.4763	1.5832	1.6937	1.8076	1.9248	2.0451	2.2942	2.5538	2.8226	3.0994	3.3830	4.1159
18	1.0482	1.0777	1.0977	1.1485	1.2006	1.3088	1.4219	1.5398	1.6624	1.7854	1.9206	2.0558	2.1947	2.4829	2.7832	3.0939	3.4135	3.7405	4.5826
20	1.0533	1.0861	1.1083	1.1649	1.2231	1.3443	1.4716	1.6049	1.7437	1.8879	2.0370	2.1909	2.3492	2.6776	3.0197	3.3733	3.7364	4.1071	5.0583
25	1.0663	1.1073	1.1352	1.2066	1.2805	1.4357	1.6003	1.7738	1.9557	2.1453	2.3420	2.5452	2.7542	3.1875	3.6375	4.1003	4.5730	5.0530	6.2737
30	1.0794	1.1288	1.1624	1.2492	1.3395	1.5306	1.7349	1.9515	2.1795	2.4176	2.6648	2.9201	3.1824	3.7243	4.2841	4.8566	5.4379	6.0254	7.5093
35	1.0925	1.1505	1.1901	1.2927	1.4001	1.6289	1.8752	2.1375	2.4141	2.7032	3.0031	3.3123	3.6291	4.2811	4.9505	5.6312	6.3193	7.0119	8.7536
40	1.1058	1.1725	1.2182	1.3371	1.4622	1.7305	2.0209	2.3311	2.6585	3.0004	3.3544	3.7184	4.0904	4.8521	5.6298	6.4169	7.2096	8.0054	10.001

$$\text{Total } P + I = n \times \frac{i}{[1 - (1 + i)^{-n}]}$$

Notes

1. The discussion which follows was taken from Willett, Gayle S. and Myron E. Wirth, *How to Analyze an Investment in Farmland.* WREP 34, Washington State University.
2. Technically, the interest rate should also be reduced by the amount the lender has added to the real cost of capital to protect against loss of income due to general price inflation. This inflation premium, however, will be difficult to identify. Thus, since higher cost equity capital is commonly used with debt to finance land, basing the required rate of return on the after-tax cost of debt, unadjusted for inflation, should not result in a serious error.
3. The technical procedures used in the Worksheet were discussed earlier in this chapter.

CHAPTER 11

Acquisition of Farm Machinery Services

Introduction

Next to the investment in land, the acquisition of farm machinery is the most costly of all farm purchases, except for livestock on some operations. The investment in farm machinery in the United States totaled $78.6 billion in 1987, down from a high of $107.9 billion in 1982. This amount is about 15 percent of the real estate investment. Machinery investment quotations generally underestimate replacement costs because machinery values are usually listed using an adjusted tax basis. Depreciation schedules for tax purposes are more rapid than market-based depreciation, leaving this component undervalued.

The composition of farm machinery has changed significantly in recent years. For example, the number of tractors on farms has remained nearly constant at 4.6 million, whereas tractor horsepower (HP) has increased by over 50 percent since 1975. The shipment of moldboard plows decreased from near 60,000 in 1975 to less than 6,000 in 1985. Because all shipments of farm machinery decreased during this period, some of this decrease reflects the financial stress of the 1980s, but farming techniques are continually changing, as are farm size and organization, which also affect machinery choices. Not only has HP increased but also tractor conveniences (e.g., power steering and power brakes, automatic transmissions, adjustable seats, cabs with radios, TVs and computer controls).

The importance of farm machinery can be understood by considering Iowa crops.[1] In 1997 the average investment per acre was $179 ($178 for high-profit farms and $213 for low-profit farms). Machinery operating costs were $42 per acre of corn and $40 per acre of soybeans, with depreciation costing $23 for corn and $22 for soybeans. The costs for growing corn and soybeans in Iowa are shown in Table 11.1. To obtain $1 of additional profit per acre of corn, machinery costs would need to be reduced by only 1.3 percent for corn and 2.2 percent for soybeans. In contrast, it would be necessary to reduce fertilizer by 2.2 percent for corn and 4.1 percent for soybeans, or labor by 4.9 percent for corn and 6.4 percent for soybeans, to accomplish the same results.

TABLE 11.1

Costs of Producing Corn and Soybeans in Iowa[a]

Item of Cost	Corn (135 bu/acre) $/Acre	Corn (135 bu/acre) Percent	Soybeans (46 bu/acre) $/Acre	Soybeans (46 bu/acre) Percent
Machinery and power:				
Operating	31.36	9.5	14.81	5.9
Ownership	44.90	13.6	31.09	12.3
Subtotal	76.28	23.2	45.90	18.2
Seed	23.14	7.0	14.00	5.5
Fertilizer and lime	45.10	13.7	24.30	9.6
Pesticides	29.10	8.8	18.60	7.4
Labor @ $6.00/hr	20.40	6.2	15.60	6.2
Other variables	25.44	7.7	23.99	9.5
Land	110.00	33.4	110.00	43.6
Total	329.46	100.0	252.39	100.0

[a]*Estimated Costs of Crop Production In Iowa—1992,* Fm 1712, Cooperative Extension Service, Iowa State University.

Acquiring Farm Machinery Services

Acquiring farm machinery services means gaining control of a machine long enough to accomplish some farm business task. Some tasks need to be performed routinely, whereas other tasks need to be done only periodically. Feeding livestock is an example of the first and harvesting grain is an example of the latter. Whereas some machines can perform multiple tasks, other machines are specialized. Thus, it is important to identify the tasks that need to be performed and the amount of each before machine services are acquired.

Gaining machinery services may not mean ownership. Machine services can be obtained by sharing with neighbors, renting, leasing, and custom hiring. Each source needs to be evaluated and compared.

Ownership

Most long-term machinery services are acquired by purchase. Ownership provides complete control, but adds responsibility for financing; providing operator labor; making repairs; paying for fuel and taxes; acquiring insurance; and disposing of the machine when it is obsolete, worn out, or no longer needed. The owner benefits from being able to schedule services, but it is difficult to blame others for the quality of performance.

Machinery ownership might be profitable, yet it still may not be in the best interest of the business to purchase it. Most machines last 10 or more years, but financing their purchase through borrowing generally requires payback periods of 3 to 5 years. This forced saving (i.e., the loan payment period is shorter than the life of the machine) may cause cash-flow problems and create financial stress. Even if the machine is not debt financed, there is an opportunity cost of ownership that should be recognized when justifying a purchase. It was recommended in Chapter 10 that the weighted average cost of capital (WACC) be used to evaluate new investments. This rate combines the interest cost and the return-on-equity capital.

Taxes are another element to include when considering machinery ownership. It is the after-tax cost of ownership that should be weighed against the benefits. Depreciation and interest are both tax deductible, as are the operating costs of repair and fuel. The economic justification for acquiring machine services and the costs of ownership are discussed later in this chapter.

Joint ownership shares the cost of owning machinery and should be considered when annual use does not justify single ownership. The difficulty is in finding someone to be partners with and working out shared responsibilities of investment, repairs, and labor. The costs of ownership are present whether machines are used a lot or a little, and the per-unit costs of ownership diminish rapidly during the first stages of use. Thus, joint ownership may furnish enough total use to reduce these per-unit ownership costs to a level where ownership can be justified. Also, it may be easier for two persons to make payments. But, each party must trust the other implicitly and approve the other's work habits. A written agreement is useful to detail the responsibilities and cooperation of each party, including termination and death.

Used machinery ownership should be considered when there is insufficient cash flow to support the purchase of a new machine or if the annual use is low. Expect repair costs to be higher than for a new machine, offsetting in part the lower ownership costs. Also, the dependability may be less causing delays in work performance. The secret to successful ownership is to balance the higher operating costs with lower ownership costs by reducing the purchase price. Thus, it is important to have the used machine thoroughly checked for mechanical failure before purchase so as to reduce the probability of misjudging the machine's usefulness and total costs of ownership.

Exchange Work

Neighbor exchange is one of the oldest forms for obtaining farm machinery services. Two or more farmers working together to share their labor and equipment can reduce their individual investments in machinery and still have access to a complete system. Exchange work may be particularly attractive to young farmers starting their operations with older neighbors, one needing machinery and the other needing labor.

Because the parties involved in the exchange often have similar needs, scheduling may be a problem and priorities need to be set. Thus, the persons must be compatible and considerate of other's needs. There are things that should be set in advance, such as who pays for breakdowns, the hours of labor equivalent to an hour of machine use, and which jobs are cooperatively done. Rather than a total exchange of machinery or labor, it may be better to set up a rental and hiring system with a neighbor so each is compensated for services to the other.

Custom Hire

Custom hire is a common method for gaining short-term use of machinery services, particularly for fertilizing, applying pesticides, and harvesting. Generally, the operator's labor is included in the arrangement. Custom use of machinery can help both the provider and the purchaser of services. The custom operator, by increasing annual use, reduces the per-unit ownership costs. The farmer purchasing these services also may reduce ownership costs. The quality of service has always been an area of contention, but it is possible for custom operators to be very professional. Custom hiring is particularly useful for specialized machines that are expensive to purchase and used only seasonally (e.g., during harvesting). Some beginning farmers do custom work as a secondary business. Other beginning farmers find that using custom operators expands their limited capital resources. The following list contains some important considerations about custom work.

- The operator comes with the machine. He or she assumes the responsibility of operating the machine and its daily care. You are free to do other tasks, including the checking of service quality.
- You have no long-term capital commitment in the machine(s). All costs of custom hiring are paid out of gross income.
- All custom hiring costs are tax deductible as ordinary farm expenses.
- You know exactly what your costs will be for farm planning purposes.
- You have no ownership responsibility, including purchase, finance, and disposal when no longer needed.
- The hired machine probably will be nearly new and should perform efficiently with few breakdowns.
- Availability may be a problem, particularly if services are not contracted for in advance.
- The objectives of the custom operator may not coincide with yours as to the quality of the job performed.
- The custom operator, through conflicts of interest, over scheduling, breakdowns, and inclement weather may not get to your farm exactly when you want the job performed. Thus, a schedule needs to be worked out in advance.

Where the job is sufficiently large to attract the interest of several custom operators, bidding the job may be a useful means for ensuring the best price and service conditions. Even if competitive bidding is not used, it is important to work through a written contract where price, time, and quality considerations can be agreed upon in advance, thus removing some of the risk and uncertainty of having jobs performed by custom operators.

Rent or Lease

Rentals and leases are similar to ownership in that the renter or lessee furnishes the labor and pays some or all of the operating costs. Repair may be a negotiated cost item. Leases are normally of longer duration than rentals and take several forms, including purchase options.

Rentals are typically by the hour, day, week, month, season, or year, and occasionally by a unit of work performed, such as an acre. Rentals are particularly useful to obtain an infrequently used machine needed for a particular task. The cost per rented item may seem high, but still it is less costly than ownership, and the cheapest means of getting the job done.

Financial leases are long-term contracts, usually for the life of the machine. The lessee has exclusive use of the equipment or machine. Sometimes the contract is constructed to resemble complete ownership so closely that the Internal Revenue Service (IRS) considers the machine to be owned. Leasing with purchase options poses some complications with the IRS if not handled properly. The IRS may look at the agreement as tax evasion if the lessee has all of the benefits of ownership, including end ownership, but taxes are shifted and collections reduced. This detail needs to be worked out with the lessor in advance. Important considerations are as follows: Who pays the taxes and insurance? Who is responsible for repair and maintenance? Is the purchase optional or mandated and at what price? The tax difference is in whether the lease payments are ordinary expenses or must be capitalized.

The advantages of leasing farm machinery may include the following:

- Leasing is a hedge against inflation. Because payments are locked in at the time the lease is signed, future payments usually are not increased to adjust for inflation.
- Leasing transfers some of the risk of obsolescence and liquidation to the lessor.
- Leases may conserve investment capital for other parts of the business where financing is more difficult to obtain.
- You pay "as you go" out of earnings rather than buying a store of services to be rationed out in future years.
- Lease payments are fully tax deductible each year.
- Leases may be an effective way for a retiring farmer to transfer unneeded machinery to a family member or to the lessee of the farm.

Figure 11.1 summarizes the major characteristics of alternative methods of acquiring farm machinery services. The balance of this chapter is devoted to cost tabulation procedures and methods and means for selecting among the methods listed previously. Because all considerations cannot be discussed it is hoped that sufficient detail is given to take much of the guesswork out of the machinery decision process.

Economic Considerations in Farm Machinery Acquisitions

The decisions to purchase farm machines are not only among the most important ones made but also are the most difficult to analyze. The dollar size of the investment, financing, and other ownership considerations have already been alluded to. Investments of $50,000 to $100,000 required to purchase a tractor or harvest machine are among the largest made. Of course, all machines do not require this amount of cash outlay, but the combined investment for a system of machinery services is significant and could affect the success or failure of the business.

The costs of machinery acquisition are difficult to estimate but can be more directly approached than the benefits. The returns from machinery services are most often derived from the products they help to produce rather than directly measured from payments received for services rendered. Custom operations are the exceptions. For some machines, profitability is broadly determined over the total of all crops produced, as in the case of a farm tractor. Thus, there is no single procedure that can be applied in all situations. Refer to previous chapter discussions on budgeting, economic principles, and capital budgeting for useful theory, methods, and procedures. The discussion that follows uses this background for making farm machinery purchase decisions.

There are techniques for making machinery acquisition decisions in common practice that have served farm businesses well for a long time. Two of the most useful procedures are emphasized in the following discussion. The first relates to the time value of money, and the second considers cost or break-even analysis. Budgeting, including cash-flow accounting and partial budgets, is a necessary procedure in any investment decision analysis. The division of costs into ownership (fixed) and operating (variable) components is a must. The procedures for estimating these costs follows in the next section.

Machinery Investment Analysis

Investment analysis is an indirect measure of a machine's worth, unless the purchase is primarily for custom operations. The procedure involves estimating the flow of cash returns from a particular crop, or combination of crops, and similarly for livestock, to

FIGURE 11.1

Summary of Major Characteristics of Alternative Methods of Acquiring Farm Machinery Services

Alternative Acquisition Methods	Tax Considerations	Capital Outlay Required for Investment	Cash-Flow Requirement for Investment	Cash Flow for Operation and Acquiring Service	Repairs and Maintenance Cost	Operating Labor	Control over Use and Timeliness of Operation	Risk of Obsolescence
Ownership (a) Cash purchase	(a) Full tax benefits	(a) Full cash cost	(a) High in purchase year	Limited to operating costs	Full cost	Supplied by farm operator	Full control	Full risk
(b) Credit purchase	(b) Full tax benefits plus tax deductible interest expense	(b) Down payment	(b) High especially short-term loans					
Custom hire	Tax deductible expense	No investment capital required	None for investment	Full custom hire cost	No cost	Comes with machinery service	Limited control over timeliness and availability can be a problem	No risk
Renting short-term or operating lease (a) Rental	(a) Deductible expense	No investment capital required	None for investment	Operating cost plus full rental fee	Limited cost dependent on agreement	Supplied by farm operator	Limited control over timeliness and availability can be a problem	(a) No risk
(b) Conditional sales contract	(b) Rental fee considered as payment; investment must be capitalized							(b) Full risk
Long-term lease or financial lease (a) Lease with nonpurchase	(a) Lease payment deductible expense	No investment capital required	None for investment	Operating cost plus lease payment	Full cost	Supplied by farm operator	Full control	(a) Similar to ownership risk depending on length of the lease.
(b) Conditional sales contract	(b) Lease payment considered as payment; investment must be capitalized.							(b) Full risk

Source: George E. Ayres, extension agricultural engineer, and James M. McGrann, extension economist, Iowa State University, PM 787, 1977.

determine if the discounted values are greater than the machinery investment requirement. Example 11.1 illustrates this procedure.

EXAMPLE 11.1

Assume a 300-acre crop farm planted one-half in corn and one-half in soybeans. Corn yields 125 bushels per acre and sells for $2.60 per bushel. Soybeans yield 40 bushel per acre and sell for $5.75 per bushel. Cash production costs for seed, fertilizer, pesticides, machinery, fuel, and repair, hired labor, insurance, interest, and all other cash expenses are $250 per acre of corn and $175 per acre of soybeans. This provides a net cash farm income amount of $19,500 before taxes. The machinery investment requirement is $100,000. The planning horizon is 10 years, the discount rate is 10 percent, the marginal tax rate is 25 percent, and the growth in income rate is equal to the inflation rate. Is this a profitable machinery investment? The net present value (NPV) and internal rate of return (IRR) are detailed in Table 11.2.

Given the conditions and variables specified, this investment is profitable. But to be sure, determine if all of the variables were properly and accurately represented. For example, did the labor amount include the operator's labor, or is the $12,860 a return to unpaid labor? Was land compensated for its use? Was risk accounted for in the discount rate selected? (A 2.85% differential between the discount rate used and the IRR is not a large amount.) The answer to these questions must be given by the decision maker, the one making the $100,000 investment. Given the magnitude of machinery investments, great care should be used to ensure that the data used accurately reflects the situation. With these precautions, the tool is a good one for the purpose at hand.

Break-Even Analysis

Break even analysis is most useful when deciding how best to accomplish a specific task (i.e. by labor or machine, by custom or owned machine, by small or large machine). It does not substitute for investment analysis. It requires that the costs of owning and operating a machine be divided into its ownership and operating components. Cost of production procedures in Chapter 6 should be reviewed if these procedures are not fully understood. Some investigators have referred to this method as "The Depreciation Method."[2] The procedure is one of comparing average costs. Example 11.2 is used to illustrate the costs of owning and operating a tractor.

EXAMPLE 11.2

Assume the machine is a 100-HP tractor that costs $50,000 with a life of 10 years and a salvage value of $10,000. Depreciation is tabulated using the straight-line method. Taxes, housing, insurance, and interest combined (THII) cost $4,200 per year. Repairs and fuel and oil average $5.40 and $4.50 per hour, respectively. The marginal tax rate is 25 percent and will reduce each of these items by that amount. The per-year and per-hour costs for these items are shown in Table 11.3 and graphed in Figure 11.2 for annual use ranging from 100 to 1,000 hours.

Now suppose that a tractor could be rented or leased at a fixed rate of $10 per hour with the user paying all of the variable costs. Draw a horizontal line on Figure 11.2 at $10. Note that it intersects the fixed-cost line at about 600 hours. (The 600-hours figure can be verified by observing the ownership cost at 600 hours in Table 11.3, which equals $10.25.) Six hundred hours is a break-even level of use where the user is indifferent to ownership or rental agreement to obtain the use of the tractor. Below this level it is cheaper to rent it and above this level it is cheaper to own it. Suppose the rental

TABLE 11.2

Present Value Analysis for Acquiring Farm Machinery ($) (Example 11.2)

Year	Cash Receipts	Cash Expenditures	Net Cash Income	Depreciation (ACRS)	Taxable Income	Income Tax (25%)	After-Tax Cash Flow	Discount Rate (10%)	Net Present Value
0	0	100,000	(100,000)	—	—	—	(100,000)	1.000	(100,000)
1	83,250	63,750	19,500	10,710	8,790	2,198	17,303	0.909	15,728
2	83,250	63,750	19,500	19,130	370	93	19,408	0.826	16,031
3	83,250	63,750	19,500	15,030	4,470	1,118	18,383	0.751	13,805
4	83,250	63,750	19,500	12,250	7,250	1,813	17,688	0.683	12,081
5	83,250	63,750	19,500	12,250	7,250	1,813	17,688	0.621	10,984
6	83,250	63,750	19,500	12,250	7,250	1,813	17,688	0.564	9,976
7	83,250	63,750	19,500	12,250	7,250	1,813	17,688	0.513	9,074
8	83,250	63,750	19,500	6,130	13,370	3,343	16,158	0.467	7,546
9	83,250	63,750	19,500	0	19,500	4,875	14,625	0.424	6,201
10	83,250	63,750	19,500	0	19,500	4,875	14,625	0.386	5,645
10	20,000	0	20,000	0	20,000	5,000	15,000	0.386	5,790
Total net present value (NPV)									12,860
Internal rate of return (IRR) (%)									12.85

TABLE 11.3

Annual and per-Hour Costs of Owning a 100-HP Tractor (Cost = $50,000, Salvage = $10,000, Life = 10 yr, Tax = 0.25%)

Use (hr/year)	After-Tax Costs per Year ($) Depreciation	THII	Repair	Fuel and Oil	Total	After-Tax Costs per Hour ($) Depreciation + THII	Fuel/Oil + Repair	Total
100	3,000	3,150	405	338	6,893	61.50	7.43	68.93
200	3,000	3,150	810	676	7,636	30.75	7.43	38.18
300	3,000	3,150	1,215	1,014	8,379	20.50	7.43	27.93
400	3,000	3,150	1,620	1,352	9,122	15.38	7.43	22.81
500	3,000	3,150	2,025	1,690	9,865	12.30	7.43	19.73
600	3,000	3,150	2,430	2,028	10,608	10.25	7.43	17.68
700	3,000	3,150	2,835	2,366	11,351	8.79	7.43	16.22
800	3,000	3,150	3,240	2,704	12,094	7.69	7.43	15.12
900	3,000	3,150	3,645	3,042	12,837	6.83	7.43	14.26
1000	3,000	3,150	4,050	3,380	13,580	6.15	7.43	13.58

THII = Taxes, housing, insurance and interest.
Source: "The Costs of Owning and Operating Farm Machinery in the Pacific Northwest," PNW346, Cooperative Extension Service, Washington State University, 1989.

FIGURE 11.2

Average Cost per Hour for a 100-HP Tractor (Annual Use = 100 to 1,000 Hours)

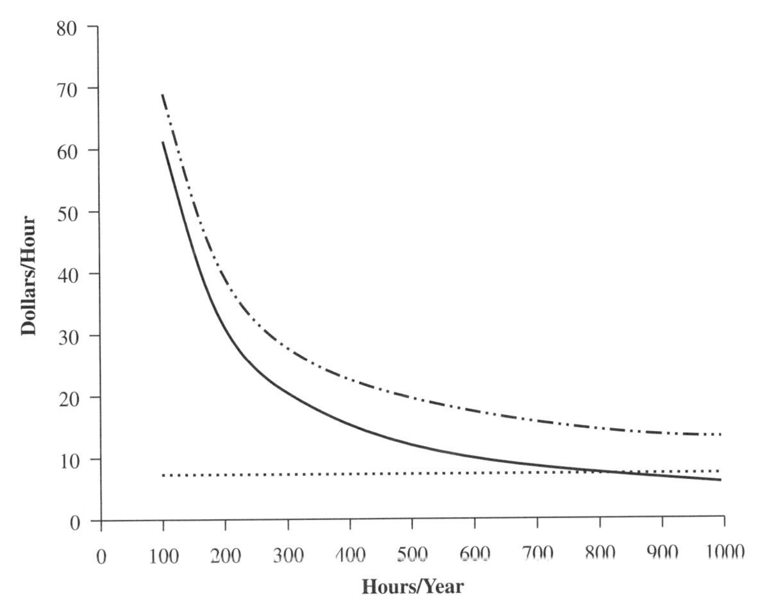

Cost type:
— Fixed Costs
···· Variable Costs
—·· Total Costs

cost is $20 per hour and the rental agency pays all of the costs. Now the user reads from the total average cost (TAC), and if another line is drawn it can be seen that it crosses the TAC line at about 500 hours of use. (The comparable figure from Table 11.3 is $19.73.) These observations can be summarized into a simple formula for determining the break-even units of use for a variety of situations as follows:

$$(VC_1 \times Z) + FC_1 = (VC_2 \times Z) + FC_2$$

where VC = operating (variable) cost per unit of use (the subscript signifies the alternative means of accomplishment)

Z = the break-even units of annual use measured in hours, acres, or other units

FC = ownership (fixed) cost per year (subscripts signify the alternative means of accomplishment)

This formula is now applied to the data presented above to obtain the break-even hours of use. The lessor of the machine pays all of the machine costs and the lessee pays a flat $20 per hour for its use. The tabulations for the lessee is as follows:

$$\$7.43Z + (\$3{,}000 + \$3{,}150) = \$20Z + 0$$
$$\$7.43Z + \$6{,}150 = \$20Z$$
$$\$12.57Z = \$6{,}150$$
$$Z = 489 \text{ hours}$$

Consider now the situation where the operator is deciding between two sizes of machines. The larger machine will have a higher ownership cost but due to the speed of operation the operating cost per unit is lower. What is the break-even annual units of use? Assume the machine is a field machine where the units are measured in acres:

$VC_1 = \$\ 8$; $FC_1 = \$8{,}000$ for the larger machine
$VC_2 = 10$; $FC_2 = \$6{,}000$ for the smaller machine

$$\$8Z + \$8{,}000 = \$10Z + \$6{,}000$$
$$\$2Z = \$2{,}000$$
$$Z = 1000 \text{ acres}$$

Thus, it can be seen that this method can be used to access the substitution of capital for labor (replacing a person with a machine), determining whether to rent or own, and what size of machine to buy. Example 11.3 is provided to give another illustration helpful in visualizing the tabulation procedure.

EXAMPLE 11.3

Assume a farmer is considering the purchase of a PTO hay baler. A new one costs $20,000 and carries with it ownership costs of $3,000 per year and operating costs of $4.00 per hour. He or she has the tractor to pull it with but will need to pay variable costs of $6.00 for the tractor and $5.00 for labor. Custom baling is by the bale but averages $25 per hour. What is the break-even hours of annual use? The problem and its solution can be seen in Figure 11.3.

Partial Budget Analysis

The partial budget should not be overlooked when selecting machinery or other investment items. The procedures for partial budgeting were presented in Chapter 5 and are illustrated again in Example 11.4.

FIGURE 11.3

Graphic Method for Finding Break-Even Units of Use

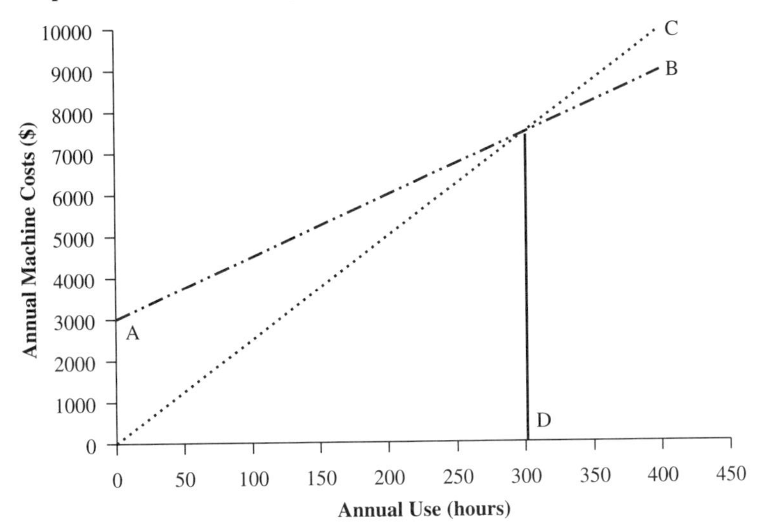

A = Annual ownership cost of owned machine at zero use.
B = Total (ownership and operating) annual cost of owned machine at a level of use above the anticipated amount.
C = Cost of custom services at the use level in B. (If two alternative owned machines are being use tested then A and B could be repeated for the second machine.)
D = Break-even units of annual use.

EXAMPLE 11.4

A small hog producer is considering automating his or her feed and watering systems. This automation will free labor from these chores so that he or she can take an off-farm job. The feeding system will cost $16,000 to install and the water system will cost $6,000, with annual repair and maintenance costs of $1,500 and $700, respectively. Hand feeding and watering costs will be reduced by $200. With the free labor an off-farm job can be obtained that pays $14.00 per hour for 2 hours per day, 5 days per week, 9 months of the year. The tabulations are shown in Table 11.4.

In this illustration, the advantage is not very large, but the investor will have more leisure time on weekends. (This will provide time to show off the new system to neighbors and friends.) The security of the alternative employments should be considered.

Machinery Reliability and Timeliness of Operations

There are limits on how much work a given set of machinery can accomplish per unit of time, and there are only so many hours in a day. Thus, only so much land can be farmed or livestock cared for with given sets of machinery. In addition, machines occasionally break down causing delays in the production process. (The pessimist would argue that a machine will select the busiest time to fall apart.) It has been shown, for

TABLE 11.4

Partial Budget Analysis of a Farm Investment

I. Added Costs		II. Added Income	
A. New costs		D. New income	
Depreciation		Employment	
Feed system	$1,600	$2 \times 5 \times 4.3 \times 9 \times 14$	$5,418
Water system	$ 600		
Total	$2,200		
Repair and maintenance			
Feed system	$1,500		
Water system	$ 700		
Total	$2,200		
Total	$4,400		
B. Reduced income		E. Reduced costs	
Added taxes on labor		Reduced taxes on investment	
$5,418 × 25%	$1,354	$4,400 × 25%	$1,100
		Reduced purchases	$ 200
C. Total	$5,754	F. Total	$6,718
		G. Net increase	$ 964

TABLE 11.5

Corn and Soybean Yield Losses Caused by Harvesting Dates[a]

	Planting Date					
	Apr. 25–May 4		May 5–14		May 15–24	
Harvest Date	% Yield Reduced	% Moisture	% Yield Reduced	% Moisture	% Yield Reduced	% Moisture
A. Corn						
Sept. 25–Oct. 5	0.3	28.5	0.4	30.8	0.9	34.1
Oct. 6–15	1.3	22.9	0.5	24.9	0.3	27.6
Oct. 16–25	3.1	19.5	2.1	20.3	1.4	22.4
Oct. 26–Nov. 4	4.3	18.3	3.6	18.7	2.8	20.1
Nov. 5–14	5.2	18.2	4.7	18.4	4.1	18.6
Nov. 15–24	6.2	18.0	5.5	18.2	5.0	18.4
B. Soybeans						
Oct. 16–25	3.0	—	3.0	—	3.0	—
Oct. 26–Nov. 4	10.0	—	10.0	—	8.0	—
Nov. 5–14	25.0	—	25.0	—	20.0	—

[a]*Acquiring, Selecting and Financing Farm Machinery,* Fm 1680, Cooperative Extension Service, Iowa State University, 1973.

example, that corn in Iowa planted between May 10 and 19 will yield 0.4 bushels per day less than if planted before this date. During the next 10 days yield losses per day will increase to 1.0 bushel and after June 1 will go to 1.6 bushels.[3] Also, there are losses associated with early and late harvest times. These losses are illustrated in Table 11.5 for Iowa for corn and soybeans.

TABLE 11.6

Estimated Number of Days Suitable for Fieldwork in Central Iowa at Probability Levels of 50%, 75%, and 90%

Dates	50%	75%	90%
April 12–25	7.0	5.9	5.7
April 26–May 5	9.0	7.4	6.6
May 10–23	10.8	6.9	6.3
May 24–June 6	9.8	8.0	7.3
June 7–July 4	20.5	17.7	16.2
July 5–Aug. 1	22.7	20.7	17.4
Aug. 2–Sept. 5	30.0	28.3	26.9
Sept. 6–19	11.8	9.2	6.7
Sept. 20–Oct. 3	10.8	8.6	7.0
Oct. 4–17	10.2	8.4	6.9
Oct. 18–31	11.9	9.2	9.0
Nov. 1–14	10.3	7.4	6.9

Source: *Fieldwork Days in Iowa,* Pm 695, Iowa State University Extension Service, 1976.

Coupled with the time it takes to do a job because of machinery capacity is the problem of weather and the number of days available to get into the field to prepare the land and plant and harvest the crop. Weather is a constant factor so a probability estimate must be made, based upon historical weather patterns and predictions. Estimates for central Iowa are shown in Table 11.6. Note that as the probability of getting into the field increases the number of days available becomes smaller.

The size of machines to purchase must take into account the days available for fieldwork, the capacity of the machines, and field losses from delayed planting and harvesting. A risk level of accomplishment must be selected suitable to the farmer's willingness and ability to withstand loss. Risk analysis was covered in Chapter 8. More is said about the field capacity later in this chapter. Most farmers select between the 75 and 90 percent probabilities of getting their crops produced without weather-induced machinery losses. These losses could be introduced into the partial budget and as a cost when developing machinery costs as illustrated in Table 11.3 and Figure 11.2. A crop loss factor can be introduced as the annual hours of use increases, effectively increasing operating costs. This produces an increasing average operating cost function rather than the horizontal one illustrated in Figure 11.2. The total average cost function then increases at some point in the annual use-per-year scale. This then affects the break-even points for selecting among alternative means of doing a farm machinery task. See Figure 11.5.

Few studies have been made of the down-time associated with a machines cumulative use and age. Agricultural engineers have represented failure rates with a "bathtub curve" in which there are decreasing failure rates early in the life of a machine due to manufacturing defects and increasing failure rates later in life. Failure rate in the period between is said to be caused by random factors. A California study could not substantiate the decreasing down-time rate but did show an increase in downtime hours as a machine becomes older. The results of this study are shown in Table 11.7.

TABLE 11.7

Estimated Wheel-Tractor and Grain-Combine Cumulative Costs and Downtime

Year	Cumulative Machine Hours	Annual Repair Costs	Cumulative Repair Costs per Hour Use	Annual Downtime Hours
Wheel tractors—1200 hours annually:				
1	1,200	5,026	4.19	22
2	2,400	7,585	5.25	28
3	3,600	9,650	6.18	35
4	4,800	11,448	7.02	43
5	6,000	13,070	7.80	54
6	7,200	14,565	8.52	68
7	8,400	15,961	9.20	85
8	9,600	17,279	9.85	106
9	10,800	18,531	10.47	133
10	12,000	19,727	11.07	167
Grain combines—500 hours annually:				
1	500	2,019	4.04	7
2	1,000	4,261	6.28	9
3	1,500	6,596	8.58	12
4	2,000	8,993	10.93	15
5	2,500	11,438	13.32	20
6	3,000	13,921	15.74	26
7	3,500	16,437	18.19	34
8	4,000	18,980	20.66	45
9	4,500	21,549	23.15	59
10	5,000	24,140	25.36	77

Source: Hardesty, Sermin D., and Hoy E. Carman, "A Case Study of California Farm Machinery: Repair Costs and Downtime," Giannini Series No. 88-2, University of California.

Estimating Machinery Costs and Capacity

The concept of operating and ownership costs was previously introduced. This section provides some guidelines for estimating these items. The estimating procedures presented here should not substitute for the user's actual situation and experience. But, it is often the case that the user's experience is limited. The longevity of life and the repair and fuel costs of one or two machines is insufficient to predict what these costs will be for a new machine. These guidelines may serve as an aid in new and replacement machinery purchase decisions. They also should be helpful when evaluating custom-rate charges and for renting and leasing decisions. Field–capacity-estimating procedures are included in this section because of the need to determine how large of a machine is needed to accomplish a task within the given allotted time as previously discussed.

Estimating Ownership Costs

Ownership costs (often referred to as fixed costs) include depreciation or replacement costs, interest on the investment, property taxes if they are levied, housing or other protection from the weather, and property and liability insurance. Each of these costs is discussed separately.

Depreciation. Depreciation was presented in Chapter 2 as a means of estimating the remaining or market value of depreciable assets in the balance sheet, for pricing a machine's services in the income statement, and for tabulating an income tax deduction. Depreciation is the reduction in value of a machine from wear and deterioration due to age and obsolescence. All machines are depreciable, as are most farm structures. Those procedures outlined in Chapter 2 are not repeated here. Recognize that the depreciation tabulated for income tax purposes is not the same as the depreciation tabulated for estimating the cost of owning and operating a farm machine. The Accelerated Cost Recovery System (ACRS or MACRS) rates used for tabulating income taxes are not appropriate to use in this case. Actually, it is impossible to determine the change in value of a machine over its useful life until that life is over and the machine has been sold or junked. But, the experience of others regarding a particular machine may provide reasonable guides to use.

The depreciation formula generally used is as follows:

$$\text{Depreciation per year} = \frac{\text{Cost} - \text{Salvage value}}{\text{Economic life}}$$

Normally, the straight-line method is used when estimating the depreciation charge for two reasons. First, it is assumed that the machine is purchased to be used over its whole economic life, and thus it is the average use cost that is wanted when budgeting its services. If a particular year is wanted, then perhaps more attention should be paid to the actual change in value of the machine during that year. Second, typically it is not just one machine that is being evaluated, but rather a package of machines to produce one or all of the crops and livestock on the farm. Because all of the machines will not be of the same age, it does not matter which depreciation method is used. The total depreciation claimed will be the same—an amount similar to that tabulated using the straight-line method.

The cost is generally the purchase price. If there was a trade, then it is the cash difference plus the value given in trade. (When tabulating depreciation for tax purposes the basis for the new machine is the cash difference plus the adjusted basis of the item traded.)

The depreciation amount used for estimating a machine's service cost probably will not be the same as the depreciation deduction for income taxes. Depreciation for income taxes reduces the remaining value to zero and is at an accelerated rate. Thus, the purpose of the tabulation needs to be specified before the depreciation cost can be computed.

The salvage value and the economic life are linked. Agricultural engineers have studied the remaining on-farm value (RFV) of various farm machines and came up with the guidelines shown in Table 11.8. These estimates are salvage value estimates based upon a machines purchase cost. Note that there is a schedule of values according to an asset's age. Thus, it isn't highly important that the exact economic life be specified, because the associated salvage value will adjust for this and give a similar depreciation amount. For planning purposes, the economic life selected is generally 8 to 10 years for most machines, tractors a little longer, and most other machines a little shorter.

TABLE 11.8

The Remaining On-Farm Value (RFV) of Machinery, Expressed as a Percentage of New Cost for Various Machines by Age

	RFV Group Number and Selected Machines in Each Group				
	1	2	3	4	5
Machinery Life (years)	Truck, Pickup	Tractors	Combines, Windrowers, Bale Wagons	Balers, Forage Harvesters	Potato Harvester, Hay Rake, Seeding Equipment, Tillage Equipment
	(% of new cost)				
1	86.64	62.56	56.64	49.56	53.10
2	79.02	57.56	50.13	43.86	46.99
3	72.06	52.95	44.36	38.82	41.59
4	65.72	48.71	39.26	34.35	36.81
5	59.94	44.82	34.75	30.40	32.57
6	54.66	41.23	30.75	26.91	28.83
7	49.85	37.93	27.21	23.81	25.51
8	45.47	34.90	24.08	21.07	22.58
9	41.46	32.11	21.31	18.65	19.98
10	37.82	29.54	18.86	16.51	17.68
11	34.49	27.18	16.69	14.61	15.65
12	31.45	25.00	14.77	12.93	13.85

Source: Derived from American Society of Agricultural Engineers, "Agricultural Engineer's Yearbook."

Consider, for example, the 100-HP tractor used in Example 11.3. Its cost was \$50,000. If a 12-year life is selected the corresponding RFV is 25 percent. The annual depreciation is as follows:

$$\frac{\$50,000-(\$50,000\times.25)}{12}=\frac{\$50,000-\$12,500}{12}=\$3,125$$

Interest on the investment. Interest is the investment cost and is either the cost of borrowed funds to purchase the machine or the opportunity cost of using equity capital to do so, or a combination of the two. The user's cost of capital was discussed in Chapter 10 where the weighted average cost of capital (WACC) was recommended as the rate to use. The typical procedure for budgeting an interest cost is as follows:

$$\text{Interest} = \text{Rate}\times\left(\frac{\text{Purchase price} + \text{Salvage value}}{2}\right)$$

For the tractor in the above illustration the interest amount would be as follows assuming the rate is 10 percent:

$$0.10\times\left(\frac{\$50,000+\$12,500}{2}\right)=\$3,125$$

Understand that when pricing machinery services, it is an after-tax amount that should be applied and compared. The tax adjustment can be applied to each element or as a composite to the total cost tabulated before taxes. Thus, this item is not specified when presenting each cost element. Furthermore, inflation may be a factor of concern when making long-term investments. Again, this consideration is not included when tabulating each item of cost. If the weighted average cost of capital (WACC) were 14.3 percent and the marginal tax rate 30 percent, then 10 percent would be the appropriate interest rate if the adjustment is made here.

The amortization method (AM) is an alternative procedure to estimating depreciation and interest costs. The formula used is a reciprocal of the uniform series present value (USPV) formula developed in Chapter 10. The amortization factors are tabulated by dividing one (1) by the USPV table values. Thus, we are looking for an annuity whose present value is equal to the purchase price of the new machine. The formula is as follows:

$$A = (P - S) \times AF + (S \times i)$$

Where A = Annual annuity required to recover depreciation plus interest on the machinery investment.

P = Purchase price of the new or replacement machine

S = Salvage value of machine traded, if there is one

AF = Amortization factor. This factor is found in Appendix Table 10.5 for an amount of $1.00. The formula given is as follows:

$$AF = 1/USPV_{i,n} = i/[1 - (1 + i)^{-n}]$$

i = compound interest rate

n = life of the machine

If there is a salvage value at the end of the replacement or planning period then this amount likewise commands an interest payment. Thus, the amount to be charged when figuring machinery costs should have the interest charge on salvage added to the A. Using the 100-HP tractor in another illustration, the depreciation and interest equivalent by using the AM procedure is as follows:

$$A = (\$50{,}000 - \$12{,}500) \times 0.1468) + (\$12{,}500 \times 0.10)$$
$$= \$5{,}280 + \$1{,}250 = \$6{,}530.$$

The AF amount can be found in Appendix Table 10.5 for n = 12 and i = 10 percent. This amount compares to $6,250 for depreciation plus interest illustrated previously. If there had been no salvage value the A would equal $7,340 ($50,000 × 0.1468 = $7,340). This method of pricing depreciation and interest is more accurate in specifying the interest charge because the half-life approach underestimates the sum of the interest payments on the unpaid balance amounts.

Using the future value of a uniform series (USFV) formula or table, the annual amount required to be set aside in an investment account, at a 10 percent return, to grow to be worth $37,500 at the end of 12 years is $1,754. For $50,000 the amount is $2,338. This procedure of planned replacement is called the sinking-fund method. A future value of a uniform series formula is used for this tabulation and is shown in Appendix Table 10.3.

Taxes, insurance, and housing costs. Taxes, insurance, and housing (TIH) costs are often listed together and a rule-of-thumb figure is used to estimate the amount to be charged. All of these items are not always present, thus sometimes are ignored in machinery cost tabulations. However, they can be significant for some machines and are usually applied when estimating fixed machinery costs. There is good justification for this because these costs, particularly insurance and housing, may be present in hidden ways. Property taxes on agricultural machinery are not levied in all states. For states with these taxes, an amount equal to 1 percent of the purchase price, or 1.5 percent of the average value, is often budgeted.

Insurance is often taken on machinery to protect the farmer from damage caused by accident, fire, vandalism, injury, wind, and other acts of nature. This cost is born by the farmer directly if insurance is not purchased. The cost of insurance is not the same for all machines. For example, it costs about 5 percent of value to insure a truck, 2 percent for a combine, and less than 1 percent for tillage machines.

Housing is perhaps the least standard of these items. Many machines are not housed, whereas others are housed in tightly fitted buildings. The cost of housing is directly associated with the square feet of space required. A compact machine like a tractor would cost less to house per dollar of value compared with a grain combine or forage harvester. One estimate used was 0.5 percent of the purchase price of the machine.[4] Another estimate linked the percentage charge to the specific machine.

Costs for taxes, insurance, and housing are shown in Table 11.9 for the Pacific Northwest states. These estimates can be modified to fit your particular situation. Applying these table estimates to the 100-HP tractor, the total percentage is 2.6 to be multiplied by the average value. The dollar amount per year to be budgeted is as follows:

$$\text{THI} = (\$50{,}000 + \$12{,}500)/2 \times .026 = \$812$$

The total ownership cost (TOC or TFC) is the sum of all the costs discussed above. For the 100-HP tractor, they are as follows:

$$\begin{aligned}\text{TOC} &= \text{Depreciation} + \text{Interest} + \text{THI}\\ &= \$3{,}125 + \$3{,}125 + \$812 = \$7{,}062 \text{ per year}\end{aligned}$$

TABLE 11.9

Percentage of Average Machine Investment Charged for Property Taxes, Housing, Interest, and Insurance (THII) for Selected Machines

	Cost Item (%)				
Machinery	**Taxes**	**Housing**	**Interest**	**Insurance**	**Total**
Wheel tractor	1.4	0.3	12.0	0.9	14.6
Crawler tractor	1.4	0.2	12.0	0.9	14.5
Combine	1.4	0.5	12.0	2.1	16.0
Potato harvester	1.4	1.4	12.0	0.6	15.4
Bean cutter	1.4	1.1	12.0	0.6	15.1
Self-propelled forage harvester	1.4	1.3	12.0	2.1	16.8
Pull-type forage harvester	1.4	1.3	12.0	2.6	15.3
Self-propelled windrower	1.4	1.1	12.0	2.1	16.6
Bean windrower	1.4	1.1	12.0	0.6	15.1
Hay rake	1.4	—	12.0	0.6	14.0
Hay baler	1.4	1.9	12.0	0.6	15.9
Self-propelled automatic bale wagon	1.4	1.0	12.0	2.1	16.5
Pull-type automatic bale wagon	1.4	1.0	12.0	0.6	15.0
Self-unloading forage wagon	1.4	—	12.0	0.6	14.0
Drills, planters	1.4	2.4	12.0	0.6	16.4
Tillage equipment	1.4	—	12.0	0.6	14.0
Sprayer	1.4	—	12.0	0.6	14.0
Pickup	1.4	1.2	12.0	5.2	19.8
Truck	1.4	1.2	12.0	8.5	23.1

If the tractor is used 1,000 hours per year the cost per hour is $7.06 but if it is used 500 hours the cost is $14.12.

Estimating Operating Costs

Operating costs (also referred to as variable costs), as the name implies, include those expenses that vary as a result of increasing or decreasing the hours the machine is operated. Included are repair, fuel and lubrication, labor, and dependability. The last item is the most intangible of all of the costs.

Repair and maintenance costs are extremely variable and difficult to predict. Your experience may be your best guide, and certainly you should keep machine cost data and know what it is costing to own and operate each machine. But, even if this information is available, it may not be a very accurate guide to the costs of a new or replacement machine. Some individuals are sure that they purchased a "lemon" when they experience higher-than-expected repair and fuel costs. This may be true and it is hoped that it is not typical performance. Repair costs do vary because of maintenance and operating conditions. Rough terrain, weather, overloading, and stress do affect a machine's performance.

In the absence of on-farm records, estimates have been made by the American Society of Agricultural Engineers (ASAE) for typical farm machines. The mathematical formulas used to estimate repair costs based upon the cost of the machine and the accumulated hours of use are as follows:

Annual repairs = Total annual repair (TAR)/Useful life

where $TAR = RF_1(X)^{RF_2}$

RF_1 = Repair factor #1

RF_2 = Repair factor #2

$$X = \frac{\text{Annual use} \times \text{Ownership period in years}}{1{,}000}$$

Factors for using this formula are shown in Table 11.10.

Repairs based upon this formula and in consideration of accumulated life give rise to the graphs in Figure 11.4. These graphs show the relationship between the sum of all repair costs for a machine and the total hours of use during its lifetime. Notice the shape of each curve. The slopes increase as the number of hours increase. This relationship indicates that repair costs per hour are low early in the life of a machine but increase as the machine accumulates more hours of operation. The different shapes indicate that all machines do not have the same schedule of repairs. Consider again the 100-HP tractor: assuming this tractor is used 600 hours per year, the accumulated 10-year use is 6,000 hours. Reading the graph for 2-wheel-drive tractors, the accumulated repair cost is 43 percent of the new cost. For the $50,000 tractor, the 10-year accumulated repair is $21,500. Dividing by 6,000 gives an hourly cost of $3.58. If the machine is kept for 12 years, the average repair cost rises to $4.17 [(60% × $50,000)/7,200 hours = $4.17].

Fuel and lubrication costs can best be obtained from your own experience. The second-best source is the manuals that come with the machine and company representatives. In the absence of these manuals, there are estimating procedures that may be helpful. Recognize that fuel consumption varies with the manufacturer, the kind of work performed, the running efficiency of the engine, the operator of the machine, etc. The formulas are meant to represent averages, so adjustments will need to be made for particular situations.

TABLE 11.10

Repair and Maintenance Cost Parameters (ASAE Standards)

	Estimated Life Machine Hours	Total Life Repairs Percent of List Price	Repair Factors RF_1	RF_2
TRACTORS				
2-wheel drive and stationary	10,000	120	0.012	2.0
4-wheel drive and crawler	10,000	100	0.010	2.0
TILLAGE				
Moldboard plow	2,000	150	0.43	1.8
Heavy-duty disk	2,000	60	0.18	1.7
Tandem disk harrow	2,000	60	0.18	1.7
Chisel plow	2,000	100	0.38	1.4
Field cultivator	2,000	80	0.30	1.4
Spring tooth harrow	2,000	80	0.30	1.4
Roller-packer	2,000	40	0.16	1.3
Mulcher-packer	2,000	40	0.16	1.3
Rotary hoe	2,000	60	0.23	1.4
Row crop cultivator	2,000	100	0.22	2.2
Rotary tiller	1,500	80	0.36	2.0
PLANTING				
Row crop planter				
No-till tillage	1,200	80	0.54	2.1
Conventional tillage	1,200	80	0.54	2.1
Grain drill	1,200	80	0.54	2.1
HARVESTING				
Corn picker-sheller	2,000	70	0.14	2.3
Combine				
Pull-type	2,000	90	0.18	2.3
Self-propelled	2,000	50	0.12	2.1
Mower	2,000	150	0.12	1.7
Mower-conditioner	2,000	80	0.26	1.6
Side delivery rake	2,000	100	0.38	1.4
Baler	2,000	80	0.23	1.8
Big bale baler	2,000	80	0.23	1.8
Long hay stacker	2,000	80	0.23	1.8
Forage harvester				
Pull-type	2,000	80	0.23	1.8
Self-propelled	2,500	60	0.12	1.8
Sugarbeet harvester	2,500	70	0.19	1.4
Potato harvester	2,500	70	0.19	1.4
Cotton picker or stripper	2,000	60	0.17	1.8
MISCELLANEOUS				
Fertilizer spreader	1,200	120	0.95	1.3
Boom-type sprayer	1,500	70	0.41	1.3
Air-carrier sprayer	2,000	60	0.20	1.6
Bean-puller-windrower	2,000	60	0.20	1.6
Beet topper-stalk chopper	2,000	60	0.23	1.4
Forage blower	2,000	50	0.14	1.8
Wagon	3,000	80	0.19	1.3

Accumulated Repair Costs Expressed as a Percentage of a Machines List Price

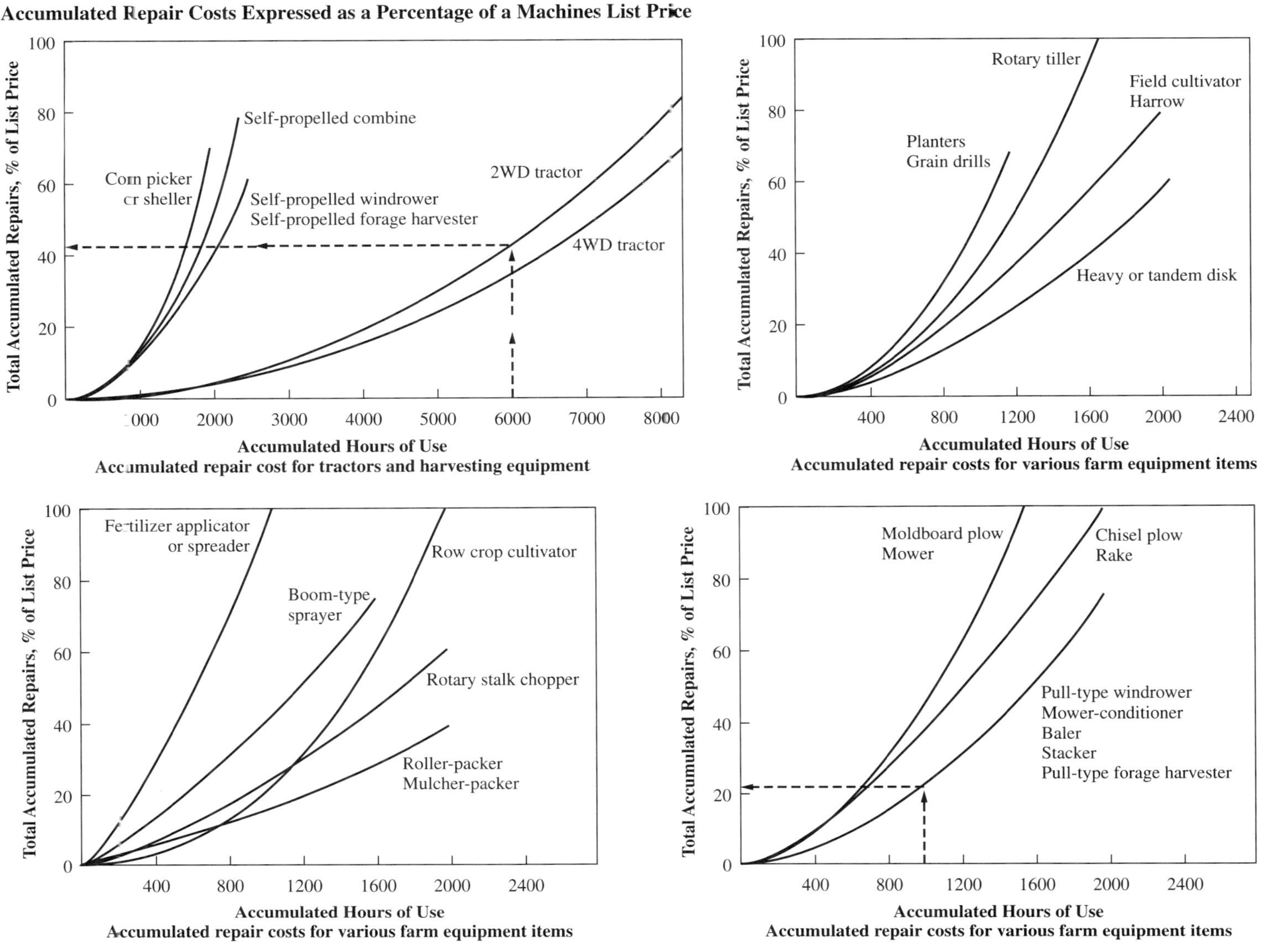

Source: Edwards, William, *Estimating Farm Machinery Costs*, Pm 710, Iowa State University Extension, 1989.

Fuel can be estimated using the following formula:

Annual fuel cost = f × 0.06 × PTO horsepower × fuel price × hours used annually

Fuel multiplier (f) = 1 for gasoline, 0.73 for diesel

For the 50-HP tractor the annual fuel cost for 600 hours of annual use is as follows, assuming the fuel is diesel and is priced at $0.95 per gallon:

0.73 × 0.06 × 50 × $0.95 × 600 = $1,248 or $2.08/hour

Lubrication is estimated at 15 percent of the fuel cost. For the 50-HP tractor this would be $.31 per hour and $187 per year.

Labor costs are associated costs and not a direct cost of the machine. Labor costs vary by machine and the job being performed. The labor hours for a single operator are usually greater than the hours the machine operates. In addition to the time for preparing the machine for operation, such as fueling and lubrication, is the time for getting to the field or workplace. Once in the field, there may be time required for loading of fertilizer or seed, unloading of product, and waiting for someone. These additional times have been estimated at 10 to 20 percent over the actual machine time. Your experience may give a better factor to use.

Total costs (TC) for owning and operating a machine are the sum of the ownership and operating costs. For the 100-HP tractor they are as follows:

TC = Depreciation + Interest + TIH + Repair + Fuel + Lubrication
= $3,125 + $3,125 + $812 + $2,502 + $1,248 + $187 = $10,999
Total cost per hour = $10,999/600 = $18.33

If labor costs are $7.00 per hour and it requires 1.2 times as many labor hours as machine hours, then the cost of operating the machine including labor is (1.2 × $7.00) + $18.33 = $26.73 per hour.

Adjusting for Inflation

Inflation was mentioned previously but it was suggested that this might be a consideration for the whole machine rather than treating the individual cost components. Inflation causes the replacement cost to increase, as well as prices of fuel and repair. If it is desirable to adjust for inflation, the factors in Table 11.11 are provided. An example illustrates how these factors can be used. If you wished to know the average cost of operating the 100-HP tractor in 5 years and the expected inflation rate is 4 percent, then you would multiply the $18.33 per hour cost by 1.22 (the corresponding factor from the table) to get a cost of $22.36. If labor is included, the cost is $32.61.

Adjusting machinery costs for inflation can help predict the custom rate needed to be charged each year to cover total costs, including the amount needed to replace the machine in future years. At the end of 12 years, this $50,000 tractor will have cost $80,000 ($50,000 × 1.6 = $80,000). Custom operators who do not take inflation into consideration will be underpricing their services.

Income Tax Considerations

The rates for charging machinery services should be an after-tax consideration. Taxes can affect measurably the amount of income remaining for personal use. The elements in the lists of ownership and operating costs outlined previously are not all affected the same. Depreciation is tax deductible but is tabulated differently from the average change in value tabulated here. Under the MACRS, depreciation is accelerated. The procedures for tabulating depreciation, including the MACRS, were discussed in

TABLE 11.11

Inflation Adjustment Factors for Machinery Costs

	Rate of Inflation (%)					
Years	3	4	5	6	7	8
1	1.03	1.04	1.05	1.06	1.07	1.08
2	1.06	1.08	1.10	1.12	1.15	1.17
3	1.09	1.12	1.16	1.19	1.23	1.26
4	1.12	1.17	1.22	1.26	1.31	1.36
5	1.16	1.22	1.28	1.34	1.40	1.47
6	1.19	1.27	1.34	1.42	1.50	1.59
7	1.23	1.32	1.41	1.50	1.61	1.71
8	1.27	1.37	1.48	1.59	1.72	1.85
9	1.30	1.42	1.55	1.69	1.84	2.00
10	1.34	1.48	1.63	1.79	1.97	2.16
11	1.38	1.54	1.71	1.90	2.11	2.33
12	1.43	1.60	1.80	2.01	2.25	2.52

Chapter 2 and are not repeated here. However, when pricing machinery services, tax adjustments should be included in the tabulations because they are used in practice to tabulate after-tax income. Decisions need to be made on an after-tax basis. Tax adjustments were illustrated in Table 11.2.

Interest is tax deductible only if it is borrowed and not if it is just an opportunity cost. Thus, it may be good practice to adjust interest for taxes directly when specifying the rate rather than to leave it as an item to be collectively adjusted. TIH is an even more variable item. If property taxes are paid, they will be tax deductible—the same for insurance. Housing is likely not to be closely associated with the machine, and thus there may be no tax adjustment.

Estimating Field Capacity[5]

The working capacity of a machine is the rate at which it performs its primary function; for example, the number of acres that can be disked per hour or the number of tons of hay that can be baled per hour. Measurements or estimates of machine capacities are used to schedule field operations, power units, and labor and to estimate machine operating costs.

The most common measure of capacity for agricultural machines is field capacity, expressed in acres covered per hour of operation. Theoretical field capacity (TFC) is dependent only on the full operating width of the machine and the average travel speed in the field. It represents the maximum possible field capacity that can be obtained at the given field speed when the full operating width of the machine is being used. It can be calculated from equation (1).

$$\text{TFC (A / h)} = \frac{\text{width (ft)} \times \text{speed (mph)}}{8.25} \qquad \textbf{(1)}$$

A machine cannot maintain its TFC for very long periods of time. Interruptions always occur to reduce its actual capacity below its TFC. The ratio of actual or effective field capacity (EFC) to TFC is called the machine's field efficiency (FE).

Field efficiency is expressed as the percentage of a machine's TFC actually achieved under real conditions. It accounts for failure to use the full operating width of the machine and many other time delays, such as turning, idle travel across headlands or to wagons, filling seed and pesticide hoppers, emptying grain tanks, cleaning a plugged machine, checking a machine's performance and making adjustments, waiting for wagons, and operator rest stops. Delay activities that occur outside the field, such as daily service, travel to and from the field, and major repairs, are not included in a field efficiency measurement.

A machine's EFC can be calculated easily. After completing a field of known size, divide the acres completed by the hours of actual field time to find its EFC. Record acres and hours for several fields to find the machine's average EFC over the whole season.

Average field speed also can also be easily measured. Mark off a distance of 88 feet in the field, place a stake at each end, and count the seconds it takes to drive between the stakes. Average field speed can be calculated from equation (2).

$$\text{Speed (mph)} = \frac{60}{\text{seconds to travel 88 feet}} \quad \textbf{(2)}$$

For example, if you traveled between stakes in 12 seconds, your average field speed was 5 mph.

After you have calculated the machine's average field speed, TOC (or TFC) can be calculated from equation (1) using the full width of the machine. The FE can be calculated from equation (3).

$$\text{FE (\%)} = \frac{\text{EFC}}{\text{TOC}} \times 100 \quad \textbf{(3)}$$

If you need to estimate a machine's EFC and have an estimate of FE, use equation (4).

$$\text{EFC (acre / hour)} = \text{TFC} \times \frac{\text{FE}}{100} = \frac{\text{width (ft)} \times \text{speed (mph)} \times \text{FE}}{8.25 \times 100} \quad \textbf{(4)}$$

The working capacity of harvesting machines is often measured by the quantity of material harvested per hour. This capacity is called the machine's material capacity (MC), expressed as bushels per hour or tons per hour. It is the product of the machine's EFC and the average yield of crop per acre, and can be calculated from equation (5).

$$\text{MC(bushels or tons/hour)} = \text{EFC(acre/hour)} \times \text{crop yield (bushels or tons/acre)} \quad \textbf{(5)}$$

For example, a baler with an EFC = 2.5 acre/hour working in a field yielding 2 tons of hay per acre would have an MC = 2.5 acre/hour × 2 tons/acre = 5 tons/hour.

The effective field capacities in Table 11.10 can be used as estimates of machine capacity under average field conditions. They were calculated from the speeds and field efficiencies listed, which were believed to be typical for many farmers. If your average travel speed or field efficiency differs much from those listed, calculate your machine's EFC from equation (4).

Choosing a Farm Machinery System

What Size of Machines to Buy?

Farm machines need to be large enough to perform the required task, and they also need to do the job in a timely manner. Yield losses from untimely planting and harvesting dates were presented earlier in this chapter, but their use was not illustrated. To draw upon the data presented, assume an Iowa farmer has 300 acres of row crops to be planted to corn and soybeans. The available machinery is sized for 6-row equipment. The 100-HP tractor used in previous illustrations would handle this size of equipment. A study of the days available to do fieldwork (Table 11.5) provides the hours available for seedbed preparation and planting. An 80 percent probability level was selected. Losses caused by late planting start May 5 for corn and May 15 for soybeans. (See Table 11.4 and related discussion.) There are 140 hours available before May 5, assuming 10 and 12 hour workdays. Following May 5, the average hours available per day are about 5. A study of the jobs to be performed and the effective field capacities (Table 11.12) reveals that it will require 0.5 hours per acre for seedbed preparation and 0.2 hours per acre for planting. Thus, by May 5, 280 acres (140/0.5) will be ready to plant. The farmer is now in the yield-reducing period, and it will take another 2 or 3 days to finish preparing the soil. Planting can proceed at the rate of 20 acres per day (0.2 acres per hour = 5 acres per hour; 5 acres per hour × 4 hours per day = 20 acres per day). Field capacity was reduced 20 percent to account for time required to fill seed and fertilizer boxes, etc. Assuming there are 150 acres of corn, it can be planted in 8 days. It will take another 8 days to plant the soybeans. The penalty for late planting of corn is 0.4 bushels per day per acre and for soybeans 3 percent of the yield. These outcomes are shown in the following table:

Crop	Acres	Penalty	Price	Total
Corn	150	(0.4 × 8)	$ 2.50	$1,200
Soybeans	150	(40 × 0.03)	$ 5.50	$ 990
Total				$2,190

This much loss may not justify a larger set of machinery. The NPV of $2,190 is only $15,382, assuming a 10 year discount period and a 7 percent discount rate. Now, assume that the acreage is 400. It would take 60 hours (about 2 weeks) to finish preparing the seedbed after May 5, and planting could not be completed until after June 5. It is easy to understand that yield losses would be much higher. Assume a loss of $8,760, four times as great as for the 300-acre farm. The discounted value is now $61,527, and most likely would justify trading up to 8-row equipment.

This analysis illustrates the increasing average cost curve mentioned previously. As the hours of annual use increases, as a result of using the same machinery set over an expanded farm acreage, the losses from untimely operations increase. These losses are illustrated in Table 11.13 and Figure 11.5. The losses due to untimely operations are fictitious but not unlike those tabulated in our example. The point where the farmer is indifferent to the size of machinery to purchase is at about 375 acres.

When Should Old Machinery Be Replaced with New Machinery?

There is an economic principle involved in deciding when to replace the old machine. It is derived for a study of the cost curves. The farmer should trade when the cost of keeping the present machine 1 more year (marginal cost consideration) is greater than the average cost of the replacement machine. As suggested, this is an after-tax consideration. This procedure is illustrated in Table 11.14. The cost adjustment for income taxes was figured using a 25 percent rate and was applied to depreciation, interest, property taxes, insurance, repairs, fuel, and lubrication. The total annual cost was the sum of these same factors less the tax. The cost per hour is obtained by dividing the total after-tax cost by 800, the annual hours of use. This cost is also the marginal cost per hour of use. The cumulative cost is obtained by summing the annual costs on an incremental basis over the 15 years. The average cost per hour is obtained by dividing the cumulative cost by the cumulative use. This procedure is the same procedure as illustrated earlier in this chapter. The break-even years are highlighted. If down-time is not considered the optimal trading year is 12. If there are losses due to delays in operations

TABLE 11.12

Average Field Speeds, Field Efficiencies, and Effective Field Capacities for Iowa Farm Machines

Machine	Speed (mph)	FE (%)	EFC (acre/hr)
Shredder			
6′ rotary	4.5	80	2.6
12′ flail	4.5	85	5.6
Fertilizer spreader, 4-ton	6.0	70	10.2
Anhydrous ammonia			
7 knife	4.5	60	5.7
Plow			
3–16″ bottoms	4.5	85	1.8
5–16″ bottoms	4.5	83	3.0
7–16″ bottoms	4.5	79	4.0
Chisel plow			
9′6″	5.0	85	4.9
13′6″	5.0	83	6.8
Offset disk			
10′	5.0	85	4.6
15′	5.0	81	6.6
Tandem disk			
14′	5.5	83	7.8
21′	5.5	80	11.2
Field cultivator			
15′	5.0	83	7.6
27′	5.0	80	13.1
Spring tooth, 14′	5.0	83	6.3
Peg tooth, 21′	6.0	80	12.2
Rolling cultivator			
4–38″	6.0	83	7.6
8–30″	6.0	78	11.4
Self-propelled windrower			
10′	6.0	85	6.2
14′	6.0	83	8.4
Combine, soybeans, and small grain			
10′	2.8	76	2.6
15′	2.8	73	3.7
20′	2.6	71	4.8
			Ton/Hr
Baler only and baler with accumulator			6.4
Baler with bale thrower, trailing wagon			4.8
Load-haul-stack bales, tractor with bale fork			4.0
Load-haul-stack bales, automatic balewagon			3.5
Stacker, 1 ton			5.0
Stackmover, field to roadside			6.0
Stackmover, haul and store or feed			4.0
Stacker, 6 ton			7.5
Stackmover, field to roadside			24.0
Stackmover, haul and store or feed			16.0
Forage harvester, corn silage			
2 row			24.0
3 row			32.0

Machine	Speed (mph)	FE (%)	EFC (acre/hr)
Rotary tiller			
80″	4.5	83	3.0
120″	4.5	81	4.4
160″	4.5	78	5.6
Sprayer, 12-row	5.0	65	11.8
Planter, seed only			
4–38″	5.0	76	5.8
4–30″	5.0	76	4.6
8–30″	5.0	72	8.7
12–30″	5.0	70	12.7
Rotary hoe			
4–38″	7.5	88	10.1
4–30″	7.5	88	8.0
8–30″	7.5	84	15.3
12–30″	7.5	82	22.4
Grain drill			
10′	5.0	72	4.4
14′	5.0	70	5.9
Broadcast seeder			
20′	5.0	75	9.1
Sweep cultivator			
4–38″	4.5	83	5.7
4–30″	4.5	83	4.5
8–30″	4.5	78	8.5
12–30″	4.5	76	12.4
Mower, 7′	5.0	80	3.4
Mower conditioner			
7′	5.0	85	3.6
12′	5.0	81	5.9
Rake			
7′	5.0	88	3.7
14′	5.0	83	7.2
Combine, corn			
2–38″	2.6	75	1.5
4-38″	2.6	71	2.9
4–30″	2.6	71	2.3
8–30″	2.5	65	3.9
			Ton/Hr
Large round baler			7.5
Bale mover, field to roadside			4.8
Bale mover, haul and store or feed			3.2
Stacker, 3 ton			6.5
Stackmover, field to roadside			12.0
Stackmover, haul and store or feed			8.0
Forage harvester, haulage or cornstalks,			
100 PTO HP tractor			14.0
150 PTO HP tractor			16.0

Source: "Estimating Field Capacity of Farm Machines," Pm 696, Iowa State University Extension Service.

TABLE 11.13

Costs per Hour and per Acre for 100-HP and 120-HP Tractors with Penalties for Untimely Operations

Acres	Hours	Fixed Cost	Variable Cost	Fixed + Variable Cost	Fixed + Variable Cost/Hour	Timeliness	All Costs	All Costs/ Hour	All Costs/ Acre
A. 100-HP tractor:									
100	200	7,426	1,062	8,488	42.44	0	8,488	42.44	84.88
150	300	7,426	1,755	9,181	30.60	0	9,181	30.60	61.21
200	400	7,426	2,556	9,982	24.96	0	9,982	24.96	49.91
250	500	7,426	3,466	10,892	21.78	0	10,892	21.78	43.57
300	600	7,426	4,483	11,909	19.85	0	11,909	19.85	39.70
350	700	7,426	5,608	13,034	18.62	0	13,034	18.62	37.24
400	800	7,426	6,841	14,267	17.83	1,000	15,267	19.08	38.17
450	900	7,426	8,182	15,608	17.34	3,000	18,608	20.68	41.35
500	1,000	7,426	9,631	17,057	17.06	6,000	23,057	23.06	46.11
550	1,100	7,528	10,594	18,122	16.47	10,000	28,122	25.57	51.13
600	1,200	7,628	11,557	19,185	15.99	16,000	35,185	29.32	58.64
B. 120-HP tractor:									
133	200	10,231	1,265	11,496	57.48	0	11,496	57.48	76.64
200	300	10,231	2,121	12,352	41.17	0	12,352	41.17	54.90
266	400	10,231	3,124	13,355	33.39	0	13,355	33.39	44.52
333	500	10,231	4,278	14,509	29.02	0	14,509	29.02	38.69
399	600	10,231	5,579	15,810	26.35	0	15,810	26.35	35.13
466	700	10,231	7,031	17,262	24.66	0	17,262	24.66	32.88
532	800	10,231	8,630	18,861	23.58	1,300	20,161	25.20	33.60
599	900	10,231	10,378	20,609	22.90	3,900	24,509	27.23	36.31
665	1,000	10,231	12,276	22,507	22.51	7,800	30,307	30.31	40.41
732	1,100	10,615	13,503	4,118	21.93	13,000	37,118	33.74	44.99
798	1,200	10,977	14,731	25,708	21.42	20,800	46,508	38.76	51.68

Costs were developed from "The Costs of Owning and Operating Farm Machinery in the Pacific Northwest," PNW346, Cooperative Extension Service, Washington State University, 1989.

because of repairs then this will shorten the trading year as illustrated. In this example the optimal trading year considering down-time costs is 11.

There are certainly other considerations than those illustrated for determining when to replace a machine. Because of these other factors the repair function is not a smooth one as illustrated. For example, repair costs normally do not happen in even increments and a decision may be made to replace an item rather than repair it. The comparison of the expected average costs of the replacement machine with the marginal cost of the old machine is still a valid one.

A major consideration in determining when to replace a machine is new technology. Within any 10- to 15-year period many changes take place that may cause a farmer to replace his or her machinery before its economic life is over. Also, within this time period additional land or livestock may be purchased. New machines may need to be obtained because of size efficiencies as previously illustrated and not because the old machines are worn out.

Is Leasing a Better Way to Obtain Machinery Services?

The financial lease is a long-term contractual arrangement in which the lessee acquires control of an asset in return for rental payments. The contract usually runs for several years and provisions may make cancellation difficult. Except for price variations, the

FIGURE 11.5

Average Cost per Acre as Acres Increase for Tractors with Penalties for Untimeliness

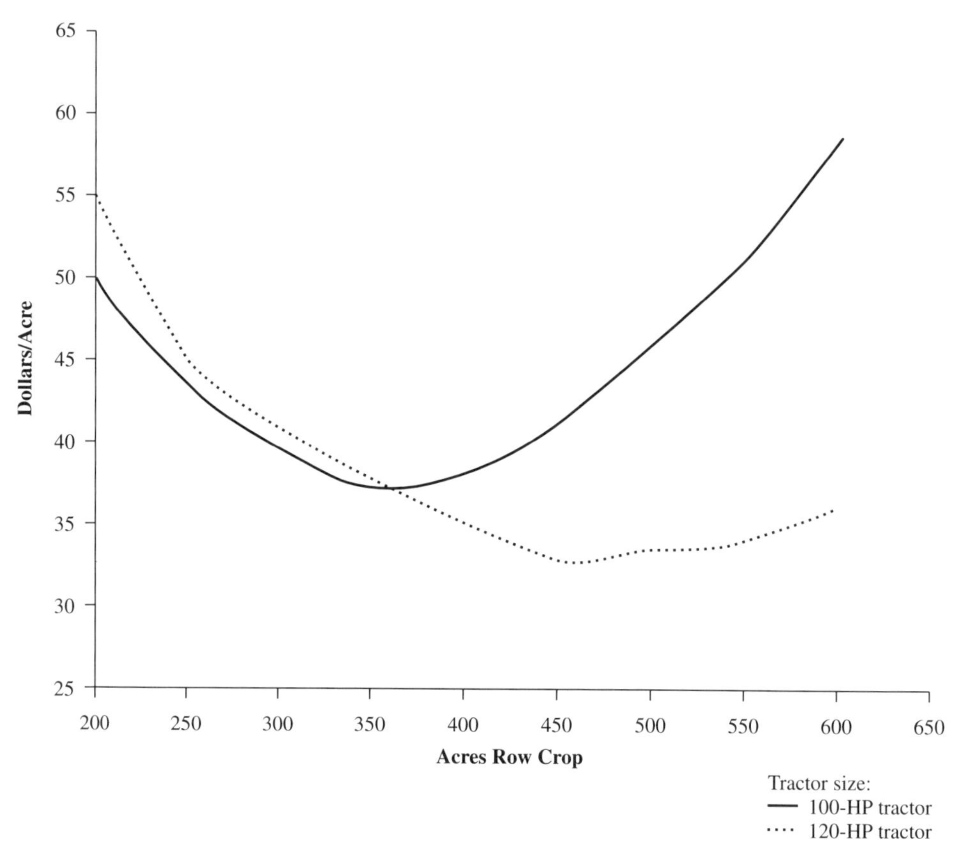

lessee acquires all of the benefits, risks, and operating costs of ownership without having to make the investment required for ownership. This type of lease is investigated here.

Leasing may be a lower-cost method of financing than obtaining the use of assets through buying and financing. There are many considerations in making this decision, but two of the major ones are profitability and cash flow. Financing may not be available for purchasing a new machine, and even if it is available, using limited funds to buy a new and expensive machine may not be in the best interest of the business. Adding a new machine through debt financing could reduce the borrowing capacity for other investments, which also could be profitable. This aspect was investigated in Chapter 10 and not further investigated here. Profitability depends on the comparison of the after-tax flows for the various financing methods. This procedure was illustrated in Chapter 10 but is further developed here.

An important consideration of the leasing arrangement is whether it qualifies as a deduction on the income tax statement. A major focus is on whether the lease is used solely for the transferring of tax benefits. A lease will be considered as a conditional sales contract if any of the following conditions exist:

- The agreement applies part of each payment toward an equity interest the lessee will receive.
- The lessee gets title after making a specified number of payments.

TABLE 11.14

Average and Marginal Cost Comparisons for a 100-HP Tractor

Year	Depreciation	Interest	Property Tax Insurance	Repairs	Fuel and Lubricants	Income Tax Reduction	Total Annual Cost	Marginal Cost/ Hour	Cumulative Hours of Use	Total Cumulative Cost	Average Cumulative Cost	Downtime Cost	Marginal Cost with Downtime	Average Cost with Downtime
	$/year	$/year	$/year	$/year	$/year	$/year	$/year	$/hour	Hours	$/life	$/hour	$/year	$/hour	$/hour
1	4,016	5 518	1,058	1,000	3,200	−3,698	11,094	13.87	800	11,094	13.87	500	14.49	14.49
2	7,174	4 657	893	1,700	3,200	−4,406	13,218	16.05	1,600	23,937	14.96	700	16.93	15.71
3	5,636	3,981	763	2,400	3,200	−3,995	11,985	14.14	2,400	35,247	14.69	900	15.26	15.56
4	4,594	3,430	657	3,100	3,200	−3,745	11,236	13.11	3,200	45,732	14.29	1,100	14.48	15.29
5	4,594	2,878	552	3,800	3,200	−3,756	11,268	13.05	4,000	56,175	14.04	1,300	14.68	15.17
6	4,594	2,327	446	4,500	3,200	−3,767	11,300	13.00	4,800	66,576	13.87	1,500	14.88	15.12
7	4,594	1,776	340	5,200	3,200	−3,778	11,333	12.95	5,600	76,933	13.74	1,700	15.07	15.11
8	2,298	1,500	288	5,900	3,200	−3,296	9,889	11.05	6,400	85,772	13.40	1,900	13.42	14.90
9	0	1,500	288	6,600	3,200	−2,897	8,691	9 46	7,200	93,338	12.96	2,100	12.08	14.59
10	0	1,500	288	7,300	3,200	−3,072	9,216	10 02	8,000	101,354	12.67	2,300	12.89	14.42
11	0	1,500	288	8,000	3,200	−3,247	9,741	10 58	8,800	109,819	12.48	2,500	**13.71**	**14.35**
12	0	1,500	288	8,700	3,200	−3,422	10,266	**11.14**	9,600	118,735	**12.37**	2,700	14.52	14.37
13	0	1,500	288	9,400	3,200	−3,597	10,791	13.49	10,400	129,525	12.45	2,900	17.11	14.58
14	0	1,500	288	10,100	3,200	−3,772	11,316	14.14	11,200	140,841	12.58	3,100	18.02	14.83
15	0	1,500	288	10,800	3,200	−3,947	11,841	14.80	12,000	152,682	12.72	3,300	18.93	15.10

Marginal and average cost figures are bold for the optimal replacement year for with and without downtime costs.

- The lessee must pay, over a relatively short period of time, an amount that is a large part of the purchase price of the machine.
- The lessee may have the option of purchasing the machine at the end of the contractual period for a preferential price.
- The lease designates some part of the payment as interest.

The following example illustrates the nature of the financial lease and aids in determining whether it is the most economic way for obtaining a machine's services.

EXAMPLE 11.5

Assume the 100-HP, $50,000 tractor used in previous illustrations and the budget procedures used in tabulating fixed costs.

If operating costs are paid by the lessee, and they usually are, then these need not be part of the tabulations. The marginal tax rate is 30 percent whether leased or owned. The lease payment is set at $10,000 and is fully tax deductible. If the tractor is purchased, depreciation and interest are fully tax deductible, but if the tractor is sold at the end of the decision period, then the amount received would be subject to depreciation recapture (MACRS).

In this illustration, assume that the purchase is financed with a 25 percent down payment at 12 percent interest with 6 annual payments of $9,121 each, and the machine is sold at the end of the leasing period for an amount equal to 34.9 percent of the purchase price. See Table 11.7. The balance of the procedures can be read from Table 11.14 where the tabulations are presented. The net present values are tabulated by discounting each of the annual cash flows by a rate of 10 percent and a time period equal to the number of years from the present to the end of the leasing period of 8 years. The tax benefit is lagged 1 year because of the filing procedures of most farmers.

In Example 11.5, as illustrated in Table 11.15, ownership has an advantage of $2,569 ($42,679 – $40,110 = $2,569). This is a present-value consideration based upon assumed tax rates, interest charges, and time periods. Thus, it is not a conclusive answer because there are other considerations. The lease could be fixed more in time than the purchase but may have financing advantages. A smaller down payment or different discount and interest rates would change costs and could reverse the choice. Even though there may be uncertainties over some of the variables, the procedure is a useful one and should be part of the comparisons used in making this kind of a decision. What-if questions should be investigated when making these kinds of decisions.

How Much Should Be Charged for Machinery Services?

This question serves not only the custom operator but also the farmer who wishes to allocate machinery costs to various farm enterprises (i.e. corn, soybeans, alfalfa, barley, beef cows, and feeder cattle). As suggested in Chapter 3, it is a useful procedure to establish a machinery service account. All machinery expenses are charged to this account, and in turn, this account charges the various production centers for machine services. The price by which these services are charged, as a professional custom operator, to extend farm operations or to allocate charges to farm profit centers from a machinery cost center, can be estimated using Form 11.1, *Worksheet for Estimating Farm Machinery Costs.*

This worksheet incorporates the procedures presented in the previous sections of this chapter for estimating fixed and variable costs. Where appropriate and when more accurate information is available, it should be used. Note that the interest cost is adjusted by an inflation factor to give a more accurate real fixed cost. Because this is a procedure for estimating average costs, the straight-line method of depreciation is used.

TABLE 11.15

Lease Versus Purchase of a Machine Using Net Cash-Flow Procedures

	Financial Lease				Purchase with Loan					
Year	Rent ($)	Tax Credit ($)	Net Cash Outflow ($)	Net Present Value ($)	Down and Principal Payments ($)	Interest ($)	Depreciation and Recovery ($)	Tax Credit ($)	Net Cash Outflow ($)	Net Present Value ($)
0	10,000		10,000	10,000	12,500	0	0	0	12,500	12,500
1	10,000	(3,000)	7,000	6,364	4,521	4,500	5,355	(2,957)	6,165	5,604
2	10,000	(3,000)	7,000	5,785	5,176	3,945	9,565	(4,053)	5,068	4,188
3	10,000	(3,000)	7,000	5,259	5,797	3,324	7,515	(3,252)	5,869	4,410
4	10,000	(3,000)	7,000	4,781	6,492	2,629	6,125	(2,626)	6,495	4,436
5	10,000	(3,000)	7,000	4,436	7,271	1,850	6,125	(2,392)	6,729	4,178
6	10,000	(3,000)	7,000	3,951	8,144	977	6,125	(2,131)	6,990	3,946
7	10,000	(3,000)	7,000	3,592	0	0	6,125	(1,838)	(1,838)	(943)
8	0	(3,000)	(3,000)	(1,400)	0	0	3,065	(920)	(920)	(429)
9	0	0	0	0	0	0	(17,450)	5,235	5,235	2,220
Totals	80,000	(24,000)	56,000	42,679	50,000	17,226	32,550	(14,933)	52,293	40,110

FORM 11.1

Worksheet for Estimating Farm Machinery Costs

Information		Tractor or Power Unit		Implement or Attachment
Machine		Tractor		Disk
a. Current list price of a comparable replacement machine		$ 50,000		$ 8,000
b. Purchase price or current used value of the machine		$ 45,000		$ 3,500
c. Accumulated hours to date (zero for a new machine)		0 hr		400 hr
d. Economic life, years of ownership remaining		10 yr		6 yr
e. Interest rate, %		13%		13%
f. Inflation rate, %		6%		6%
g. Annual use, acres				900 acres
h. Field capacity, acres/hr or tons/hr*				9 acres/hr
i. Annual use, hours ($\frac{(g)}{(h)}$ for implement)		600 hr		100 hr
j. Engine or PTO horsepower		125 HP		
k. Fuel type		Diesel		
Estimating fixed costs				
1. Salvage value (% from table 1) × list price (a)	29.5%	$ 14,750	17.7%	$ 1,416
2. Depreciation = {[(b) – (1)] × 1/(d)}		$ 3,025		$ 347
3. Average of beginning and ending investment = $\frac{(b)+(1)}{2}$		$ 29,875		$ 2,458
4. Interest = [(e) – (f)] × (3)		$ 2,091		$ 172
5. Taxes, insurance, and housing = 0.01 × (a)		$ 500		$ 80
6. Total fixed cost per year = (2) + (4) + (5)		$ 5,616		$ 599
Estimating variable costs				
7. Accumulated hours to date (c) and repair %, Figure 11.4	0 hr	0%	400 hr	5%
8. Total accumulated hours at end of life = [(d) × (i)] + (c), and %, Figure 11.4	6,000 hr	43%	1,000 hr	19%
9. Total accumulated repairs = (% from (8) – % from (7)) × (a)		$ 21,500		$ 1,120
10. Average repair cost/hour = $\frac{(9)}{(d)\times(i)}$		$ 3.58		$ 1.87
11. Fuel cost/hour = fuel price $1.25/gal × 0.44 (diesel) or 0.06 (gasoline) × (j)		$ 6.88		
12. Lubrication cost/hour = 0.15 × (11)		$ 1.03		
13. Labor cost/hour = wage rate $6.00/hr × 1.10		$ 6.60		
14. Total variable cost/hour = (10) + (11) + (12) + (13)		$ 18.09		$ 1.87
15. Fixed cost /hour = $\frac{(6)}{(i)}$		$ 9.36		$ 5.99
16. Total cost/hour = (14) + (15)		$ 27.45		$ 7.86
17. Total cost per hour for tractor and implement combined			$ 35.31	
18. Total cost/acre or ton = $\frac{(17)}{(h)}$			$ 3.92	

*Average hourly work rates for many farm machines are listed in Pm-696, *Estimating Field Capacity of Farm Machines.*
Source: Edwards, William, *Estimating Farm Machinery Costs,* Pm-710, Iowa State University Extension, 1989.

Ownership (fixed) costs are separated from operating (variable) costs. For custom operations the total cost should be charged. For on-farm allocations, it may be appropriate in some applications to only charge the operating costs. Increases in use affect directly the operating costs but may have little effect on the annual ownership costs. Thus, in considering an activity of short-term duration it may be more accurate to just charge the operating cost rate. However, in the longer run each machine will wear out, or be technologically replaced, and a new machine will need to be purchased.

The machinery costs estimated here are before-tax costs. This cost is appropriate for many decisions, including custom operations and internal cost allocations. If machinery costs are tax adjusted, then revenue also needs to be tax adjusted. It is appropriate to always raise the question of taxes to determine if the expense or revenue needed in the particular decision is a before- or after-tax consideration.

The following list provides generalities about machinery and equipment acquisitions:

- Small farms and operations requiring limited uses of machinery and equipment will favor owning used machines, renting equipment, and using custom operators.
- Higher discount rates for figuring the net present value of the acquisition costs for obtaining machinery services tend to favor credit purchases and financial leases over cash purchases.
- Higher discount rates tend to favor credit purchases over custom hire.
- Higher income tax brackets tend to favor financial leases and credit purchases over cash purchases.
- Higher wage rates tend to favor custom hire over credit purchases

Notes

1. *1996 Iowa Costs and Returns,* Iowa State Univeristy Extension, FM-1789.
2. Pasour, E. C., Jr., T. E. Nichols, Jr., and G. L. Bradford, *Applying Economic Principles in Replacing or Purchasing Agricultural Equipment.* EIR 10, North Carolina State University, 1969.
3. *Corn Management Research,* Pioneer Hi-Bred International, Inc., Des Moines, Iowa 50308, 1974.
4. Edwards, William, *Estimating Farm Machinery Costs,* Pm-710, Iowa State University Extension.
5. This topic was taken from *Estimating Field Capacity of Farm Machines,* PM-696, Cooperative Extension Service, Iowa State University.

CHAPTER 12

Acquiring and Managing Labor

Introduction

Humans are still the brain and brawn of agriculture. Even though machines and computers have taken over many of the functions previously performed by management and labor, people will always be a key element in agricultural production. But, the jobs performed by labor and the number of workers needed have changed dramatically. The farm population dropped from more than 12.5 million persons in 1965 to less than 3.8 million in 1995. During this same time, farm numbers declined from about 3.35 million to 2.1 million. Thus, farm population declined by 1.8 persons per farm. While the farm population was declining, the rest of the U.S. population increased from about 190 million people to 260 million. However, not all of the persons living on farms worked there, and some who lived in urban communities worked on farms. In July 1995, 3.67 million persons were working on farms. Of these, 1.63 million were self-employed, 0.63 million were unpaid, 1.07 million were hired with one-third of them working less than 150 days per year, and 0.35 million were service workers. The average wage rate for hired laborers was $6.44 per hour. It appears that the numbers of farmworkers has leveled off, with a slight increase in the 1990s.

The skill level required to operate today's farms has changed much from what it was in years past. Although there are jobs that must be done by hand, most tasks are performed with a machine or instrument. Many of these machines are complicated to operate and require operator training. In addition, the inputs of seed, fertilizer, water, pesticides, feeds, and health care must be applied by precise measurement and procedure. Training and supervision of operators and employees is a must and cannot be left to each employee's discretion and judgment.

In recent years there has been an increased awareness of the need to preserve the environment and protect the employee from job hazards. Thus, many federal and state laws and regulations have been legislated and are enforced. These regulations have been added to the job descriptions of workers and increased the requirement for training and supervision.

Agricultural workers, following the trend in nonfarm industries, demand benefits as part of the payment package. Some of these benefits are government mandated such as unemployment insurance (FUTA), and social and workman's compensation insurance. Many wage agreements include health and life insurance, retirement savings, overtime pay, and incentive plans in addition to income tax and social security (FICA) tax withholding.

The above considerations are the subject of this chapter. Everything about labor cannot be included. Many states have labor laws that strengthen those of the federal government and these obviously cannot be covered. Discussions include considerations in assessing labor needs, acquiring profitable employees, developing mutually beneficial payment packages, meeting government regulations, and providing supervision and management.

Sources of Supply for Agricultural Labor[1]

Operator and Family

The operator and his or her family have traditionally furnished the bulk of farm labor. In 1995 there were 2.1 million farms with 2.85 million workers. Of these workers, 1.89 million were family and 0.87 million hired. One-third of family workers were unpaid, meaning that they did not receive a wage. Although the majority of these workers are still male, an increasing number are female. Three factors account for this change: smaller families with fewer children to care for, more home conveniences relieving the drudgery of housework, and less physically demanding farm jobs. Improved technology has tended to reduce the physical burden of many farm tasks. However, the proportion of labor on commercial farms furnished by unpaid operators has been declining. This decline is not only true because of the increasing size of business but also by a changing farm structure. Farm corporations and some partnerships pay the entire workforce. It is probable that the future role of farm operators in supplying labor to the nations farms, particularly when measured on commercial farms requiring at least one full-time worker, will decline.

Hired Labor

Hired farmworkers, although less than 1 percent of all wage and salary workers, account for about one-third of the production workforce. Operators and their unpaid family members account for the remaining two-thirds. More importantly, hired farmworkers provide the labor at critical production times when operators and family members are unable to supply the necessary labor.

In 1996 hired farmworkers were more likely than all wage and salary workers to be male, younger, never married, and less educated. They were also more likely than all wage and salary workers to be Hispanic and foreign nationals who have citizenship in other countries.

About 704,000 workers (78 percent of all employed hired farmworkers) were primarily employed full-time (worked 35 hours or more per week), and 202,000 were primarily employed part-time. Part-time hired farmworkers were more likely than full-time ones to be female, white, younger (median age of 20 years compared with 37 years), never married, and born in the United States.

Hired farmworkers earn significantly less than most other workers. Full-time hired farmworkers received median weekly earnings of $280, or 58 percent of the $481 median of all wage and salary workers. Hired farmwork tends to be unsteady and seasonal. Some farmworkers find nonfarm jobs to supplement their incomes.

Employers should be aware of the many laws and regulations relating to living and working conditions of seasonal and migrant workers. Some of these laws are covered later in this chapter.

Custom Operators

Much of farm labor is furnished by custom operators. Often they are the cheapest source of temporary labor. They come with their machines and generally are well trained. They require no tax withholding or fringe benefit plans. The economics of using custom machine operators was covered in Chapter 11. This source should not be overlooked when considering the best and cheapest way of acquiring needed labor. Custom operators may release regular employees to perform other needed services.

Contract Service Workers

Many farm tasks are performed by contract persons on some farms. The worker may be self-employed or the employee of a contract provider. Typically, the jobs performed have been associated with crop harvesting and require large amounts of hand labor. This is an area of constant government surveillance to determine if there is tax avoidance or getting around some law or regulation. The fact that the employer and employee regard the agreement as contract work may have little weight in court if challenged by the federal government. Their guidebook reads, "If an employer–employee relationship exists, it does not matter what it is called. The employee may be called a partner, agent, or independent contractor. It does not matter how payments are measured or paid, what they are called, or whether the employee works full- or part-time."[2] The courts apply two tests to determine whether a worker is an employee or independent contractor. The "Control of Work Details Test" is generally used to determine independent contractor status for purposes of tax withholding and employer liability for negligence. The "Economic Realities of Dependence Test" is used to determine employee status under laws protecting workers (so-called social welfare legislation).

The degree to which a farm producer or employer has the right to control the details of work determines whether a worker is an employee or independent contractor. The more control the producer has, even if not exercised, the more likely it is that the worker is deemed an employee. In contrast, if a producer only has control over the work results, the worker is likely to be considered an independent contractor. The nature of the worker–producer relationship is ultimately decided case by case.

Generally, employees cannot waive the rights granted them under social welfare legislation. These rights include minimum wages, child labor, worker's compensation, and unemployment insurance. Because the independent contractor status waives employee rights, courts are reluctant to recognize workers as independent contractors. The "Employer's Tax Guide" provides conditions that help define if independent contractor status exists. Like many other matters one may not know for sure whether the worker is an employee or a contractor until a legal court of law rules.

Determining Labor Needs

The economics of labor utilization were covered in earlier sections of this book. Often labor is a component of some other consideration and not explicitly examined. This oversight could be a big mistake if the change in labor requirement upsets that which is already in place. The labor to meet an increased use may not be available, demand a much higher price than budgeted, or require special training. Replaced labor may not reduce the total labor cost if the labor payment cannot be reduced because the labor supply is unpaid (self-employed) or the workers are full-time employees. Thus, it is important to consider how the total labor of the business is affected when considering a change of products produced, adopting new production practices, or buying a new labor-reducing machine or building.

The discussions in earlier parts of this book are particularly relevant when assessing farm labor requirements. Attention is directed to the following:

- *Chapter 5.* Partial budgets can be used to evaluate changes that affected labor needs. A labor calendar and requirement chart are useful for assessing the total labor needs of the farm, including hired labor. Labor availability can be charted against projected needs. Both labor shortages and surpluses by month were charted. Linear programming can be used to evaluate the dollar contribution labor makes if in short supply during critical seasons of the year.
- *Chapter 6.* It was established that labor, a factor of production, should only be employed to the extent that it's marginal factor cost (payment rate) is just equal to the value of the added product it contributed. Furthermore, labor should be used to the extent that its marginal contribution to the income of the business is just the same as for all other factors of production.
- *Chapter 11.* Investments in machinery and equipment were presented as substitutes for labor, or vice versa. The substitution of capital for labor is still relevant. Break-even analysis was illustrated as a procedure for determining how far to go when making this substitution (Table 11.4 and Figure 11.3).

The labor requirements and availability assessment chart is again illustrated as a means of determining the needs for labor hiring. The procedure requires knowing the labor requirements of the profit and cost centers on the farm. Each of these requirements is listed with their monthly labor needs. If they are first listed on a per-unit basis, such as acre or head, they can be expanded by the number of units in each. The total need can then be compared with the labor supply. This procedure is illustrated in the following example.

EXAMPLE 12.1

A cattle rancher with 600 cows is having difficulty meeting labor requirements. Labor seems to be in short supply and hired laborers complain from being overworked and not being able to take a vacation. The rancher decides a labor requirements and availability chart will help identify shortages. First, a table is constructed listing the major farm activities and labor requirements on a monthly basis. This table requires a listing of the operations for each crop and livestock enterprise. Those things of a general nature, such as bookkeeping and marketing, are grouped together as overhead. A few things that do not need to be done on an exact timetable are not listed, but are estimated to take 400 hours. The crops are all custom harvested and this time requirement is excluded from the total. The labor requirements chart is shown in Table 12.1. The crop and livestock activities were grouped to create Figure 12.1.

TABLE 12.1

Monthly and Total Labor Requirements for a 600-Cow Ranch in Hours

	Overhead	Alfalfa (648 acres)		Small Grain (366 acres)		Irrigated Pasture (95 acres)		Dry Pasture (5000 acres)		Cows–Calves (600 head)		Feedlot Calves (560 head)		Farm
Month	Total	Hr/Acre	Total	Hr/Acre	Total	Hr/Acre	Total	Hr/Acre	Total	Hr/Head	Total	Hr/Head	Total	Total
JAN	40	0.0	0	0.0	0	0.0	0	0.00	0	0.5	300	0.4	224	584
FEB	40	0.0	0	0.0	0	0.0	0	0.00	0	0.7	420	0.4	224	684
MAR	40	0.0	0	0.0	0	0.0	0	0.00	0	2.0	600	0.4	224	864
APR	40	0.1	65	0.3	110	0.2	19	0.01	50	1.1	660	0.4	224	1,168
MAY	40	0.3	194	0.4	148	0.7	67	0.01	50	1.0	600	0.4	224	1,321
JUN	40	0.4	259	0.1	37	0.7	67	0.01	50	0.5	300	0.4	224	976
JUL	40	0.4	259	0.3	110	0.7	67	0.02	100	0.1	60	0.4	224	860
AUG	40	0.6	389	0.2	73	0.8	76	0.02	100	0.1	60	0.3	168	906
SEP	40	0.5	324	0.2	73	0.8	76	0.02	100	0.1	60	0.3	168	841
OCT	40	0.2	130	0.2	73	0.8	76	0.00	0	0.1	60	0.5	280	659
NOV	40	0.2	130	0.0	0	0.2	19	0.00	0	0.3	180	0.4	224	593
DEC	60	0.0	0	0.0	0	0.0	0	0.00	0	0.5	300	0.4	224	584
TOTAL	500	2.7	1,750	1.7	622	4.9	466	0.09	450	6.0	3,600	4.7	2,632	10,019

FIGURE 12.1
Labor Allocation Chart for a 600-Cow Ranch

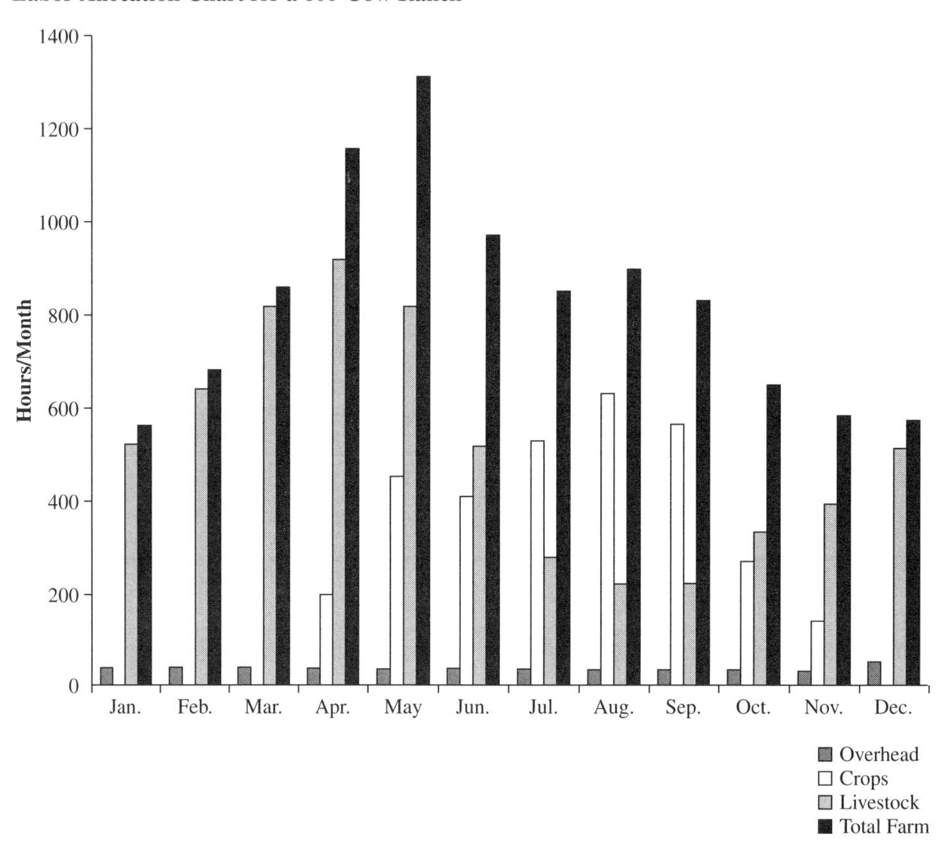

The current labor supply consists of the operator, his or her family, one full-time hired hand and undependable part-time employees as needed. The operator works from 200 to 240 hours per month, the hired hand 200 hours per month, and the family about 50 hours per month from September through May and 100 hours from June through August.

During 5 months of the year the labor supply is adequate but for the other 7 months it is in short supply. After studying this chart and plotting several labor scenarios, the manager decides to hire one more full-time employee and do most of the spring crop work with custom operators. There will be a surplus of labor during the fall and early winter months, but it will allow vacation time for all farmworkers and the manager will have more time for planning and marketing. The higher-quality work performance of the full-time employee over the part-time employees used in the past should more than pay the increased cost.

The effect of the weather was not mentioned in the above illustration, but it is an important consideration. In Chapter 11 weather was included when considering the size of machinery to purchase. Weather variables are no less important when planning labor needs and acquisitions.

In the above illustration, there are both crop and livestock activities, and it may be possible to do some trading to provide extra labor for one or the other when needed. Also, changing to custom operators may provide some help in this area. However, the availability, dependability, and efficiency of custom operators should be evaluated carefully before making this decision.

A labor estimate worksheet is provided in the appendix to this chapter (see Form A). Adaptations can be made to fit individual needs.

Acquiring Full- and Part-Time Employees[3]

Hired employees, like other resources, will make a farm business more profitable only if they are needed, qualified, and managed effectively. New employees should be complementary to existing personnel rather than duplicating their existing strengths. Therefore, planning before hiring is important. Planning provides a basis for seeking out employees who have the skills and personal characteristics that best meet the needs of the business. Remember that it is better to spend time in finding the right employee than in trying to improve less-effective employees and dealing with problems they have caused. So take planning and hiring tasks seriously, particularly if you are hiring key employees.

There are five basic steps in the personnel planning process (Figure 12.2).

1. Assess your supervisory skills, personnel needs, and working conditions.
2. Develop tentative job descriptions.
3. Match present workers with job descriptions.
4. Develop job descriptions for remaining tasks.
5. Hire employees who fit the job description.

Step 1. Assess your supervisory skills, personnel needs, and working conditions.

Assess your own characteristics and supervisory skills. Planning should start with a self-assessment by the farm manager and others who have major supervisory responsibilities. The experience, skills, and attitudes of each personnel supervisor or manager should be considered. Having an employer and employee who are compatible and complementary is essential but unlikely if left to luck in hiring.

Which questions are relevant in your self-assessment? The following suggest some possible lines of query:

- What are my strengths and weaknesses?
- Am I a good teacher?
- Do I have the patience to work with people with no farm background or little farm experience?
- Am I a good listener?
- Do I trust my employees?
- Which, if any, of my biases could get in the way of developing a good relationship with an employee (e.g., tobacco, alcohol, politics, sports, or breed of livestock)?
- Am I a perfectionist?
- Do the current employees respect and like me?
- Do I tend to be an optimist or a pessimist?
- Am I an effective delegator?

Your leadership style and attitude toward delegation will markedly affect your ability to hire and keep good employees. For example, to effectively manage others,

FIGURE 12.2

Flow Chart of the Personnel Planning Process

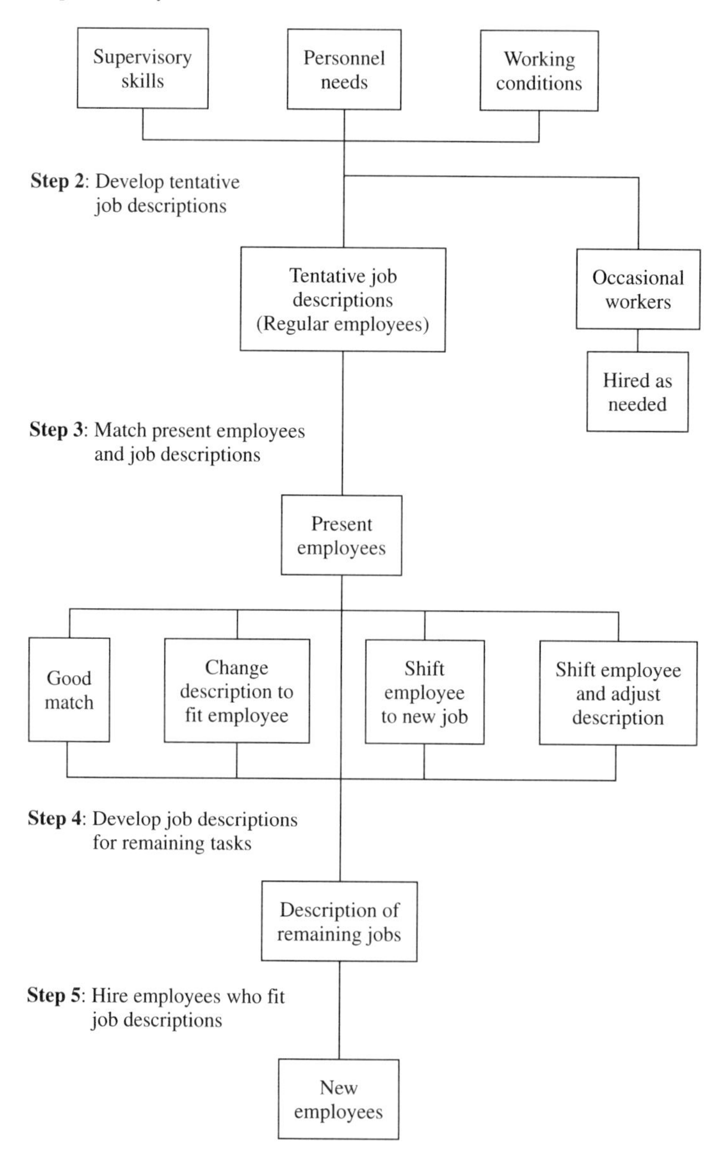

you must shift part of your time from being a "do-er" to being a manager. Helping others do the job for you is an essential part of leadership. Even though you may be able to do a job faster and better than employees, you must become their leader, teacher, and helper. If you are unwilling to accept leadership and training responsibilities and to delegate, then you should be cautious in building a business involving hired employees.

Your leadership style will directly affect the work environment on your farm and dictate some of the characteristics of employees you will be able to hire and keep. For example, if you are an autocratic sort of manager, you will need to hire followers and invest time in training to do things your way. At the other extreme is the manager with a loose or "free rein." If you are a "free-rein" manager, you need to hire self-starters who have the ability to work with others. Training will still be necessary. In the middle is the participative leader who hires the types of employees the job requires and adapts managerial style to their needs. With this approach, you need to be comfortable adapting your training program to the needs of individual employees.

Honest answers to these kinds of questions suggest desirable and undesirable characteristics in employees for your farm. It may become apparent that a person who would make an outstanding employee for a neighboring farm may present serious problems to you and your current employees.

Determine your personnel needs. How much labor do you need and when? How much do you need during various seasons? What can be done to reduce the amount of labor needed, for example, custom hire or bigger machines, shift enterprises, or adopt labor-saving practices? These questions were discussed in the previous section.

This analysis should result in a statement of the kind and amount of work to be done. Divide the total work into the kinds and numbers of jobs to be filled. Determine how many employees and what kind of employees are needed: regular or occasional, skilled or unskilled, technical or management. It is important to focus on the hiring of regular or key employees because occasional workers are usually available and hired on an as-needed basis.

Make an assessment of the amount and kind of labor needed, and when it is needed, by completing a worksheet as illustrated in the appendix (see Form A).

Define work environment and working conditions. To attract good employees, your business environment must have some advantages over your competitors. What are yours? Are you offering competitive wages? Are you offering a flexible work schedule, opportunity for training, or some type of incentive program? Is there a good work environment: good people to work with, good equipment in safe working condition, a good image as a place to work? What potential does the job hold for the employee: future role in the business, chance for growth, responsibility?

Who is the boss? Who reports to whom? What can each employee decide on his or her own? How have family members on the management team divided up responsibilities? Do employees think they are supervised or do they believe that everyone works together to get the job done? These questions all deal with organizational structure. The biblical precept that a man cannot serve two masters still holds true: it is a major reason why employees leave. If two masters are to be served effectively, these masters need to be exceptionally well coordinated and organized.

Organizational structure must be dealt with, particularly in multimanagement situations. Myths about organizational structure are rampant. Some examples follow:

- Only farms with many employees have organizational problems.
- Everyone knows who the boss is.
- Organizational structure is important only in dealing with big problems.
- We all work together so organization is unimportant.
- No one is better than anyone else on this farm.
- Having only family members involved in the farm eliminates organizational problems.

No two farms should be organized alike. Personnel managers, employees, families, and the work involved vary from farm to farm so the structures need to vary. Keys to effective organizing include proper planning of the organization, clear relationships among the people in the organization, delegation of authority, clear limits to the delegation, authority accompanied with responsibility, and neither over- nor underorganization.

Step 2. Develop tentative job descriptions. Next, you will need to develop a tentative set of job descriptions based on supervisory skills, personnel needs, and work environment as determined in step 1. However, these job descriptions should not be "cast in stone" because in step 3, you will need to match the proposed jobs with your present personnel. This matching may involve a change in the job description, a change in the employee's role in the business, or both.

Appendix Form B can be used as a guide in developing these tentative job descriptions. The descriptions should spell out the proposed duties and responsibilities associated with the job as well as the authority granted to the person, and the person's accountability. Details regarding accountability, job title, and compensation can be delayed until you are done matching proposed job descriptions with the present workforce.

As part of the personnel planning process, you also should develop a job description for yourself that includes statements as to how you plan to carry out the personnel management functions in light of your particular situation.

Step 3. Match present employees with tentative job descriptions. The next step involves a matching of tentative job descriptions with the present employees. To do this, first appraise the talents of your present employees, including their past experience and performance as well as their personalities, needs, and desires. Next compare the tentative job descriptions of step 2 with characteristics of your present employees. The four possible outcomes are as follows:

1. Employee continues in present job—satisfactory match between proposed job description and employee's skills.
2. Change tentative job description to fit present employee's skills.
3. Shift employee to a new job as described in the tentative job description.
4. Match by making adjustments in both the employee's responsibilities and the job description.

Of course, there could be a fifth adjustment—the dismissal of a present employee. This might be the result of a major business adjustment that no longer requires the employee's skills, or it could be the result of previous unsatisfactory employee performance.

This matching process should take into account both paid and unpaid family labor as well as present hired employees. Which of the four options should be used? This needs to be asked on an employee-by-employee basis. This should normally involve a discussion between you and the employee so that the employee's needs and desires are considered as well as those of the business. Following this discussion, the individual employee's job description should be firmed up with details such as job title, responsibility, authority, accountability, and compensation written into a job description and accompanying employment agreement.

Step 4. Develop job descriptions for remaining responsibilities. After the present employees have been assigned responsibilities, the remaining responsibilities not assigned to current employees and family members are for new employees. These remaining responsibilities need to be grouped to make sense from your business's standpoint as well as a potential employee's standpoint. This again involves development of job descriptions. The descriptions should be viewed as a fairly firm representation of the kinds of employees you need.

Again, you can use Appendix Form B as a guide in developing these descriptions. First, you need a job title or some way of differentiating jobs. Calling everyone a herdsman or milker defeats the purpose. Use job titles that are meaningful to your operation. You also need to spell out the duties and responsibilities associated with the job. Indicate the authority granted to that person, the person's accountability, and the results expected.

Qualifications, to include work experience and personal characteristics, should be specified for each position. Be clear about whether experience in a similar job is essential, preferred, or unimportant. If you have a thorough training program, previous experience and job knowledge may be unimportant. You also should note the starting wages and possible range in wages and benefits to be offered. Compensation needs to be in line with responsibilities and market competition.

Wages are often thought of as the prime motivational tool in a business, eliciting negative as well as positive reactions from employees. Care must be taken in establishing wage rates. Pay rates and benefits are the subject of the next section.

Step 5. Hire employees who fit job descriptions. The final step in the personnel planning process is noted here as a reminder that in hiring new employees you should normally be looking for people who fit the established job descriptions—a round peg for a round hole. The temptation may be strong to hire a square peg and attempt to round off the corners once the employee is on board. In a large firm the manager may be able to afford such a time-consuming, risky process. For most farm situations, however, finding a person who meets immediate as well as longer-term employment needs is the preferred strategy.

The Hiring Process

Hiring can be divided into three steps or phases: recruitment, interviewing, and selection. Each is separately discussed.

Recruitment and Application

Recruitment involves searching for prospective employees and enticing them to apply. Recruitment can be done in several ways: (1) asking neighbors and other farmers if they know of someone looking for employment; (2) asking local vocational ag teachers, extension agents, lenders, and others who may know of prospective employees; (3) referrals by current employees who know the job to be done and the type of person needed; (4) checking with a state job service agency or private agricultural employment agency; and (5) advertising.

You may want to advertise in a local newspaper or in a state or national publication, depending on the position being offered. Any such ad should include basic items from the job description (i.e., brief description of the operation; duties and responsibilities to be performed; special experience or skills required; salary range and benefits). The ad can instruct interested parties to reply to a post-office box rather than to a telephone number. Potential employees inquiring about the position should be asked to fill out an application for employment, such as Appendix Form D.

Interviewing Selected Applicants

The most promising applicants should be invited for an interview. Applicants' spouses also should be invited if it seems desirable. The purpose of the interview is threefold: (1) to give the applicant a clearer picture of the job and what is expected, (2) to allow the employer an opportunity to compare each applicants' qualifications and aptitudes, and (3) to provide opportunity for current employees to become acquainted with the applicants.

An employee interview form, such as Appendix Form E, should be developed for the interview process. The form needs to fit your situation. Key items in the employee interview form are likely to check the applicant's farm work experience and skills, reasons for leaving the previous job, the applicant's goals and aspirations relative to farming, and the reasons for applying for the job. Appendix Form E contains a proposed list of characteristics to look for in an employee.

Appraise the personal characteristics of candidates carefully. Employee values and attributes are more important and harder to change than present skills and knowledge. Look for a positive attitude, willingness to work, persistence, maturity, ability to get along with others, good judgment, and honesty.

Early in the interview, tell the potential employee about your business, yourself, other employees, and the job. Establish a good rapport and help the candidate relax. Present your job opportunity positively but do not oversell it. Describe in some detail your business and its long-term goals. Outline duties and responsibilities of the job and those of the manager. Show the prospective employee around and be willing to spend some time. Be mindful of the applicant's family: consider what the job offers them. Try to determine if the family will be satisfied with what is being offered and how supportive the family is of the applicant's interest in farm employment.

Ask the applicant to talk about him or herself and past job-related experiences. Be ready with some open-ended questions: What are your strengths and weaknesses? What do you like or dislike about farmwork? What are your most and least favorite things to do on a farm? What do you hope to be doing 5 years from now?

Having some other members of your family and one or two current employees interview the applicants is usually helpful. They are likely to have some observations and impressions different from yours. If there are substantial differences of opinion among the interviewers, it may be necessary to invite one or more applicants back for a second interview.

Remember that certain types of interview questions are illegal. In general, you are not allowed to ask about race, religion, age, or marital status. Employers may ask whether an applicant has any disability that would interfere with job performance.

You may or may not want to indicate what the wages and benefits would be. If you do mention them, be sure to do so near the end of the interview. Providing this information early in the interview may cause the applicant to gain or lose interest in the position on the basis of compensation alone.

When closing the interview, make no commitments. Assure the applicant that if he or she receives an offer, there will be ample opportunity for discussion of detailed questions about the job and conditions of employment. Emphasize that you will provide a written employment agreement. Tell the applicant what date you plan to make the decision. Be sure to call or write by that date and report the results: yes, no, on hold, or still interviewing.

Ending the interview on a positive note is very important. Even though you may not make an offer to the applicant or the applicant may not accept your offer, he or she might recommend someone else for the position. Applicants are likely to discuss the interview, you, your business, and the position with their friends and families. In those conversations, it is to your advantage to have each applicant be enthusiastic about you and your business.

Checking references may give you special insights regarding applicants. Watch out for vindictive references, particularly from previous employers. Answers to the question: "Would you hire this person back?" should give you some good insights. A more likely problem is fear of legal repercussions causing references to say little about the applicant and nothing that is negative. For this reason, telephone conversations and personal visits with references are likely to be more useful than letters of recommendation or letters addressed, To whom it may concern. You can expand the list of references when an applicant is a finalist for a position by asking those references the applicant lists to suggest other names as additional references. But this can only be done with the applicant's written consent or it is in violation of data privacy.

Selecting the Employee to Be Hired

As you go through the process of selecting a new employee, keep in mind that you may not be hiring the most qualified, the most experienced, or the one willing to take the position at the lowest wage. However, you must feel comfortable with the applicant.

Select the candidate who best fits the job description. Changing the job description to fit an employee you like is a common temptation. The planning process emphasizes recruitment based on the needs of the business. Changing a job description may be throwing away the benefits of planning. Also, do not hire an overqualified or underqualified employee. If you do, either you or the employee is apt to be dissatisfied. Also remember that you can train a potential employee, but you will find it next to impossible to change an employee's personality. New employees must fit well into the existing labor and management team.

For full-time, regular employees, a written employment agreement should be developed and signed by the parties involved. This agreement should include wages to be paid and benefits to be provided. Use Appendix Form F to document this part of the arrangement. The job description also should be reviewed, adjusted, and made part of the employment package.

Developing Pay and Benefits Packages

The importance of developing competitive pay and benefit packages was alluded to in the discussion about hiring. Even though employees are concerned about all aspects of a job, its pay rate is high on the list of considerations. Low pay is a common source of dissatisfaction. The employer often uses pay rates to attract top personnel. Thus, the total pay package is often an item to be bargained over. Use Appendix Form G.

The Wage Rate or Salary

There are few laws governing the minimum rate of pay on most farms. But, those laws that affect other parts of the economy have their effect on agriculture. Minimum-wage legislation does affect farmers with relatively large payrolls.

The most important comparative wage rate is that which prevails in the local farm community. It is not necessary to pay a rate above that which other farmers pay, but it is necessary for the employee to feel he or she is getting a fair wage. The total pay is not only in the stated wage but also in the benefits and working conditions present. Thus, list benefits provided with the dollar value of each shown, including the employer's share of the social security (FICA) tax. It is easy for an employee to undervalue the benefits from a house, garden, and produce provided, and forget entirely those benefits not used such as insurance and FICA.

Labor payments may be made monthly, biweekly, weekly, and daily. The labor agreement may call for the rate of pay to be established annually, monthly, daily, hourly, or on a piece rate basis. Only for the daily and hourly employees are the number of hours to be worked usually specified. For some situations overtime rates must be paid. The hours worked, payment schedule, and days off should be specified in the labor agreement. Working more hours may not be avoided during some seasons but they should be offset with compensatory time off, overtime pay, or some other recognized benefit.

Minimum-wage laws affects agricultural employers who used more than 500 person-days of labor in any calendar quarter of the preceding year. The minimum wage is currently $5.15 (1998) but is subject to future increases.[4]

Fringe Benefits

Fringe benefits include regular leave or paid vacation, sick leave, health and life insurance, retirement, training, housing, and farm produce. Paid vacations are common policy for regular employees. A 1991 Iowa study showed the average employee received 5 holidays plus 10 vacation days per year.[5] Even though vacation times should not disrupt important operations it cannot be expected that vacations must always be taken when farming activities are at their lowest. It is better that vacation time be planned in advance.

Sick Leave

Sick leave is another benefit often provided to employees. It probably will be provided only to full-time employees and then only after a period of time. The amount of time off with pay for illnesses should be specified along with the conditions for using it. For example sick leave may accrue at the rate of 1 day per month. In the event it is not used, the extent to which it can accrue or generate regular leave time needs to be specified. Some employers allow substitution of one-half day of regular vacation for each day of sick leave not used after 15 days of leave time has been amassed. What happens to unused sick leave in case of separation needs to be specified.

Health and Life Insurance

Insurance might be obtained at reduced costs through group coverage. The benefit package may include partial payment by the employer and the employee. The cost of the insurance is tax deductible to the employer and may not be taxed as income to the employee. The maximum amount of tax-free life insurance in 1997 was $50,000. Some states require employers with a minimum number of employees to pay worker's compensation insurance. This system is to provide protection to employees injured on the job and to employers from certain liability for such injuries. Where these laws are in effect the employer pays a fee based upon his or her total payroll. In case of injury the employee receives hospital and medical benefits plus unemployment pay while recuperating.

Unemployment Insurance (FUTA)

Unemployment insurance is required of some agricultural employers. This insurance is federally mandated of those who employed 10 or more workers in each of 20 or more weeks, or paid $20,000 or more in cash wages in any calendar quarter. This insurance temporarily replaces part of the employee's income lost due to unemployment. This benefit is financed by an employer payroll tax. The 1992 rate was 6.2 percent of the first $7,000 paid to each employee. Unemployment income is subject to be taxed.

Federal Withholding Taxes

Federal income taxes and social security (FICA), including Medicare, are required to be withheld from most employee's wages or salaries. These taxes must be withheld from employee wages and deposited with the employer's share in an approved bank or

Federal Reserve Bank if the farmer-employer is subject to FICA tax. With exceptions not detailed here farmer-employers must withhold FICA taxes if the employee was paid $150 or more in cash wages during the year, or if the wages paid to all employees totaled $2,500 or more during the year. The amount of income tax withheld is subject to several conditions that are not be detailed here.[6] FICA tax is paid by the employer but the employee share may be withheld from the total wage payment. For 1997 each share was 7.65 percent (6.20 percent FICA + 1.45 percent Medicare) of the taxable wage. The maximum wage subject to FICA was $65,400; the maximum for Medicare was unlimited. Expect changes!

Retirement

Many employee agreements include retirement plans. Most of these plans incorporate employer and employee contributions. Most retirement programs are not taxed to the employee until funds are withdrawn from the savings program. Three plans are available: Simplified Employee Pension (SEP-IRA), SIMPLE, and Keogh. SIMPLE is the last to be added and cannot be combined with other plans. It has a limit of $6,000 per year. Employers can participate with their employees. For SEP-IRAs the employer may contribute up to 15 percent of the employee's compensation or $30,000, whichever is less. If tax-deferred income, including any growth in value, is withdrawn before the recipient reaches a specified age (59 1/2 in 1997) there may be a penalty extracted, in addition to its being taxed.

Housing

Housing is often furnished to full-time employees. This might include the utilities. This cost is a tax deduction to the employer and may not be taxable to the employee. If the housing is required by the employer as a means of employment to care for livestock, secure the property, or other business reason then its value would not be taxable to the employee. Employer-provided housing does not always benefit the employee. Sometimes the employer will want to provide this benefit in lieu of cash and difficulties may arise over repairs, upkeep, and utilities. Thus, housing details should be included in the labor contract.

Education

Having an educated workforce almost always benefits the employer. Thus, this is another benefit good employers provide. Possible areas of instruction include bookkeeping, fertilizers and pesticides, irrigation, livestock breeding, livestock feeding, machinery repair and maintenance, and farm safety.

Bonus and Incentive Plans

Many farmer-employers like to tie part of an employee's compensation to volume, performance, longevity, or some other criterion. A study of Iowa farmers found nearly one-half of employers had such plans in effect at an average cost of about 10 percent of the total compensation. Motivation is one of the most frequent concerns voiced by farm employers. Because of the diversity of farming types, activities, and situations many types of plans are in use. Examples of these arrangements are shown in Appendix Form C.

Wage-incentive plans differ from bonuses in that they are tied to specific measures of performance. Wage-incentive plans are generally limited to skilled employees whose length of tenure and performance on the farm cannot be rewarded adequately with regular wages and benefits alone. If incentives are used, tie them to things the employee has control over, and limit the amount of benefits to what both you and the employee can live with. Incentive plans should not be substitutes for a fair wage plus fringe benefits package.

Perhaps the most common type of incentive is tied to volume (i.e., pigs weaned per litter, percentage calf crop, milk produced per cow, and grain harvested per acre). Other types might relate to overtime work such as tractor hours driven after 7 p.m. or cows detected in heat.

Performance may be paid at the end of the year after an overall evaluation of the employee's work. This arrangement is very subjective and probably relates more to the employer's income than to the performance of the employee. A more equitable reward would be a percentage of gross income.

A bonus should be just that. If each Thanksgiving the employer gives each employee a turkey, it is no longer a bonus but part of the employee's income, and he or she may like more control over what the income is used for. Bonuses should depend on how well the employee has performed and the profitability of the business rather than on tradition. These can be given at any time of the year. Incentive plans need to be part of the wage agreement and adhered to. The employee needs to know what is important and expected.

Share arrangements may be considered in certain circumstances. Examples include joint ownership of livestock or a mere sharing of profits. However, these arrangements need to be reserved for special employees and particular situations and require careful discussion. Share arrangements place the employee in a risk and management role that may cause his or her earnings to be less than without the plan. Some employees will be motivated, whereas others may become discouraged and frustrated by plans that could result in an income lower than that prior to the plan.

Government Regulations Affecting Farm Employees

In addition to minimum wages, unemployment insurance, and federal income tax withholding, government regulations include child labor laws, the Occupational Safety and Health Act (OSHA), and the Immigration Reform and Control Act. In addition, many states have regulations affecting employment.

Child labor regulations apply to employment on the farm. Age 16 is usually the minimum age for employment in agriculture during school hours. Those under 18 and who should be in school may be required to have a work permit from the school stating that the work will not interfere with the students' school performance. Outside school hours, the minimum age for employment is 14 with two exceptions: children 12 or 13 can be employed with written parental consent, and children under 12 can work on their parents' farms. There are still regulations against work considered unsafe or hazardous. Such jobs include working with agricultural chemicals; driving and operating farm machinery, including tractors, combines, and mowers; and working with breeding livestock, including bulls and boars. Children under age 14 cannot be employed in hazardous jobs. Employees 14 and 15 can be certified for certain hazardous jobs. Some state extension services and other agencies provide training for certification.[7]

OSHA regulations affect many farms and farm employees. The purpose of OSHA regulations is to provide safe working conditions and to eliminate accidents on the job. An example is the requirement that slow-moving farm equipment using public highways use safety signs. The handling of farm chemicals has become much more regulated in recent years. This is an area of increasing public concern.

Federal OSHA requires toilet and hand-washing facilities in fields where 11 or more workers perform hand labor for 3 or more hours, including travel time. Employers must provide drinking water and permit employees reasonable access to it. There are also similar regulations for migrant-worker housing.

The Immigration Reform and Control Act of 1986 requires employers to do five things regarding their employees:

- Have your employees fill their part of the Form I-9 when they start to work. This requires that the employee be a U.S. citizen or naturalized in the United States, a permanent alien, or an authorized alien. Persons hired before November 7, 1986, do not need to complete this form.
- Check documents establishing employees' identity and eligibility to work.
- Properly complete the employer's section of Form I-9.
- Retain the Form I-9 for at least 3 years. If you employ the person for more than 3 years, you must retain the form until 1 year after the person leaves your employ.
- If requested, present the Form I-9 for inspection by the Immigration Service of the Department of Labor.

The Equal Employment Opportunity Act prohibits job discrimination based on sex and prohibits sexual harassment at the worksite. Furthermore, a woman cannot be discriminated against because of pregnancy or pregnancy-related disability. As a minimum, a woman must be granted leave without pay for a period of up to 4 months to recover from the pregnancy.[8]

Principles of Labor Management

After employing the right people, they must be managed effectively to help make them productive. Remember that you, the employer, have entered into a contract that "cuts" both ways. You expect the employee to put in a full day's work; to grow and become more efficient; and to be honest, cooperative, and loyal. So, too, the employee expects you to provide guidance, support, training, encouragement, trust, and honesty, as well as compensation. If you do not fulfill your part of the bargain as an employer, you cannot expect employees to fulfill theirs.[9]

Good working relationships depend on the effectiveness with which the management performs major personnel functions. Effective communication skills are essential because communication pervades all personnel management functions.

Communication: A Key Element in Personnel Management

Communication influences the effectiveness of the hiring and training processes, the motivation of the employee, and an employee's willingness to provide you with useful feedback. You must be able to send understandable messages to your employees so that they are willing and able to follow the instructions and carry out the tasks. Many of the tasks in farming involve an "art" or degree of "husbandry" learned only through experience. This obviously complicates the communication process. You must be willing to listen to your employees and seek feedback to see if the messages are getting through and are being understood.

Communication involves five elements: (1) the sender or source of a message, (2) the message, (3) the channel used to send it, (4) the receiver, and (5) the effect the message has on the receiver. The message is whatever the sender wants to communicate to the receiver. It can be as simple as "time for a break." It can be as subtle as a smile in response to a job well done. It can be as complex as the technique for a difficult weld. The channel is the means—verbal or nonverbal—by which the message is sent. Verbal channels can be oral or written, live or recorded.

Communication problems can arise from any of these five elements. Obviously, perfect communication is when the sender transmits a clear message via the right channel, and the receiver acts on the message as the sender intended.

As an employer and manager of people, you need to work continually on improving your communication skills. The following guidelines should be useful:

- *Have clearly in mind what is to be communicated* before attempting to communicate it.
- *Use both formal and informal channels of communication.* The formal channel involves manager to employee (downward communication), employee to manager (upward communication), and communication among employees (horizontal communication). Informal communication also can be effective. The grapevine spreads information quickly and often reflects employee attitudes and levels of understanding.
- *Use various means of communication.* A face-to-face conversation between two or more people is probably the most common approach. Examples of other means of communication include written messages, posters, pictures, videos, and movies. Nonverbal communication such as facial expressions and gestures can often "say" more to an employee than spoken words. Your actions and inactions are also an important means of communication. Failure to discipline a negligent employee for tardiness will likely send the wrong message to the other employees about the importance of starting work on time.
- *Remove communication barriers.* Use language that is understood. Words that are common to management and older employees may be confusing to a new employee. Attitudes of senders and receivers can erect or tear down barriers. Employees tend to erect barriers against messages they feel are unimportant or threatening.
- *Use feedback techniques.* Be sensitive to the reactions of your employees. Are they puzzled by what is being said? Do they appear anxious to ask questions? One of the most effective feedback techniques is to ask the receiver to restate or summarize the information just received.
- *Be an effective listener.* A sincere interest in how much of the message has been understood is the beginning of effective listening. Making it clear to employees that you are willing to listen to their problems, concerns, or suggestions is also important.

Probably the two most important aspects of communication are to understand your employee better (i.e., have a greater empathy as to your employee's situation and needs) and to send more understandable and meaningful messages.

Functions of a Personnel Manager: Principles and Guidelines

There are five major personnel management functions in which you need to be proficient: (1) work scheduling, (2) training and coaching, (3) motivation and morale, (4) evaluation and compensation, and (5) discipline and termination. Competency in these functions, in combination with effective communication, greatly facilitates the building of effective employer–employee relationships. Each of these functions is briefly described.

Work scheduling. Planning the work to be done a day or week in advance leads to more efficient use of labor. As the number of employees and tasks increase, this planning and scheduling function increases in importance. Here is another reason that your role as a "do-er" may have to be cut back to ensure that your employees are able to do assigned tasks efficiently.

Effective work planning involves not only identifying the best timing and sequence of tasks but also providing the machinery, equipment, and supplies needed to accomplish the tasks on schedule. A task list identifies what needs to be done within the next period of time; a work schedule identifies the employees and machines to do them. Each task listed should include a deadline and priority. Instructions should be given so that employees know exactly what is expected of them as well as their authority and responsibility concerning the task. Continuous updating of schedules and instructions is necessary because of weather problems, machinery breakdowns, and unexpected days off for employees.

Training and coaching. Training and coaching your employees should be thought of as a continuous process. New personnel, new tasks, changes in the way specific tasks are done, and changes of responsibilities of experienced personnel suggest a continuous need for training and coaching.

A step-by-step procedure is likely to increase your effectiveness as a trainer. The following are suggested steps in the training process:

- *Break the job into key components.* What tasks is the employee expected to know how to do? Can learning some parts of the job be postponed until more important parts are mastered?
- *Prepare the employee for the training.* Preparation can include putting the employee at ease; finding out what the employee already knows; indicating why it is important to learn to do the task properly; and explaining that his or her productivity, satisfaction, and safety will be influenced by the training.
- *Teach the job by describing and performing the tasks involved and answering the employee's questions.* Active involvement of the learner in this step is important. Here, you might first explain the task, then perform it, and then have the employee describe how to do the task.
- *Follow-up.* You should periodically check the employee's progress to ensure that the task is being done correctly and efficiently.

Once the employee has learned the rudiments of the job and is settled in, then training can be replaced by coaching. The main thing to watch for is employee complacency or slippage in performance. Peak performance is reached only after substantial practice makes routine the various aspects of the task. Unneeded parts of the task are eliminated and remaining parts are integrated for maximum efficiency in time and effort. Good coaching increases the speed at which the employee moves through the rapid progress and false plateau phases and the overall level of peak performance achieved.

Motivation and morale. If you have done a good job of selecting employees and trained them well, they should be able to do an efficient job for you. The question is, Will they? Doing an effective job of motivating employees is usually necessary if this question is to be answered positively.

It is important that you first recognize that motivation comes primarily from within an employee. Employees generally do not change their behavior merely because someone tells them to do so, or how to do so. You might keep the apple tree in mind as you attempt to motivate your employees: the apple tree provides flowers and pollen so that the bees can make honey and at the same time help the tree produce apples.

Employees need to feel they are accepted not only by their employer but also by coworkers. The employee who does not feel accepted is likely to eventually seek employment elsewhere. After an employee feels accepted, the needs of esteem and status emerge. These are the external signs of being liked and appreciated by others. Examples include a promotion or handling new responsibilities effectively.

You must continually communicate with your employees about their needs and aspirations and how to increase their level of job satisfaction. Each employee will have a differing set of needs. Here again good communication is needed with each employee.

The way managers attempt to motivate their employees depends largely on their own beliefs or assumptions about people.[10] So-called Theory X Managers assume that the average person inherently dislikes work, prefers to be directed, avoids responsibility, has little ambition, and wants security above all. Such managers supervise employees closely to get adequate performance.

Theory Y Managers, however, believe employees will meet the high expectations of their employers. Work can be satisfying as a result of the pleasure of association and pride of achievement. These managers regard employees as responsible and hardworking. They try to make jobs more meaningful so that employees can develop their capabilities and satisfy the human need for self-fulfillment. Theory Y employers encourage involvement and innovation.

There are very few exclusively Theory X or Theory Y managers. Generally, those managers who are most successful in developing long-term relations with employees have adopted more Theory Y than Theory X assumptions. However, it must be recognized that employees vary widely in terms of their attitude toward work and responsibility as well as in terms of their length of employment and skill level. This means that you will need to be flexible in your management style as you deal with each of your employees and with each situation. However, you should strive to be consistent or predictable as to which management style you use in any given situation.

Evaluation and compensation. You also need to tell your employees how they are doing, where they have improved, and offer constructive suggestions for improvement where needed. Plans to improve job performance, to train, and to broaden responsibilities should be discussed. You should encourage your employees to indicate what might be done to make their jobs more productive and satisfying. In addition to oral communication, all evaluations and suggestions should be in writing and kept in the employee's file.

Even though evaluation is often only done at salary adjustment time evaluation and feedback are useful on an on-going basis. Plan to have informal conferences several times per year, preferably at the end of a particular activity.

Before conducting the evaluation interview, do your homework to gather performance information and personal characteristics. You should allow plenty of time for the evaluation sessions with no interruptions. First, point out the strengths of the employee, being as specific as possible. Second, point out areas of work where improvement is needed and suggest some possible avenues for doing so. Let the employee offer suggestions as to how he or she could improve performance. Ask for employee suggestions on how your supervision and overall organization could be improved.

The evaluation is often a good time to discuss the issue of compensation—wages, bonuses, incentive plans, and benefits. A wage rate or earnings level perceived to be unreasonable or unfair by the employee can be an important source of job satisfaction. However, any changes in compensation should be consistent with the evaluation you have just gone through. Employees who need considerable improvement in their work habits should not be rewarded with a large merit raise.

Make bonuses and incentive payments consistent with attainment of business goals. Payments beyond regular wages should be rewards for positive or desired behavior, not something that your employees expect to happen automatically at Christmas time.

Discipline and termination. Discipline is one of the least satisfying aspects of personnel management. You should therefore strive for self-discipline among employees. If there has been a careful recruitment of employees followed by a sound training program, and proper attention to human needs, discipline problems should be minimal. Nevertheless, disciplining an employee is needed on occasion.

Not disciplining when needed sends the wrong message to the employee, and more importantly to the other employees. Reasonableness in discipline is important. The penalty should be consistent with the rules violation. Permitting the employee to maintain self-respect is an important part of discipline. Therefore, you should discipline the employee's behavior or act—not berate the person.

Basic to effective discipline is an employee handbook or other form of written statement of policies, practices, and work rules. Employee orientation should include a review of the handbook with employees given ample opportunity to raise questions.

Progressive discipline is more effective than an "all-or-nothing" approach. Progressive discipline increases the severity of punishment as an offense is repeated. Typical levels in progressive discipline are as follows: (1) informal talk, (2) oral warning or reprimand, (3) written warning, (4) disciplinary layoff, and (5) discharge.

On occasion you may need to discharge an employee. This may be the result of discipline or just the failure of the employee to live up to expectations and job requirements. Discharge is a serious matter with potential for important legal and personal consequences. Therefore, any discharge should be preceded by notice, attempted rehabilitation, and consistency.

Notice includes assurance that each employee has knowledge of policies, practices, and work rules. Notice also includes periodic performance evaluations. The results of the evaluations should be discussed with the employee and made available in written form.

Rehabilitation includes efforts to correct conduct problems; for example, progressive discipline for being tardy. Efforts to improve performance problems are also important. Examples include additional training, change of jobs, and help of other employees. All rehabilitation efforts should be carefully documented.

Notice of termination is best given in writing and in a private setting with hopefully a calm discussion of the reasons of discharge. Many employers allow the employee time to look for a new job while still on the payroll; that is, after notice has been given and until the termination becomes effective. All matters related to a disciplinary problem and the steps leading to discharge should be carefully summarized in writing and put in the employee's file.

If an employee quits, you should conduct an exit interview to find out why. It is important that you evaluate reasons given and make any changes that would improve or correct the situation for current or future employees.

Measuring Employment Efficiency

It is difficult to measure the efficiency of labor if you do not know the time that is being spent at assigned tasks. Often the only gauge the employer has to measure the efficiency of an employee by is that which he or she perceives it takes him or her to do the same task. This may not be a very accurate or honest comparison method and could be unfair to the employee. Generally, the measurement is imprecise and inaccurate. It is possible that the employee could do the job more efficiently than the employer. To obtain good information about the time required to perform certain tasks such as plowing or feeding the cattle, or produce specific enterprises such as an acre of corn or head of livestock, it is necessary for the workforce, including management, to keep labor-use records.

Given that people's memories are inaccurate and subject to forgetfulness, daily labor records are almost a necessity. At one extreme are the clock-watchers who time every activity with precision. This is more detail than most farm businesses want or need. This is something that management pays and there are economic principles that govern the time spent in this type of data collection. Even here the benefits need to be compared with the costs at the margin. However, there are data-collection methods that do not require much time and that give reasonably accurate information. Computers with spreadsheet programs have taken much of the drudgery out of this type of work.

The daily labor record is a compromise between precision or accuracy and clock procedures. Figure 12.3 illustrates such a record. At the end of each day's activity the worker records how his or her time was spent at described tasks or for named profit or cost centers. It is assumed that if done on a daily basis the reporting is reasonably accurate. The time at work must equal the sum of the time spent on the individual enterprises or other defined activities being tracked.

If a more precise account of labor use is desired, a daily labor report could be kept. Such a report is illustrated in Figure 12.4. This record could be made of card stock to be carried in a shirt pocket, in a vehicle, or placed in a convenient location to the work being performed. The worker writes down the job being performed and the time taken to do it, and credits it to a specific enterprise of activity being tracked. These could be given numbers and listed on the back of the card for ready reference. This form could be used to provide the data for the monthly report shown in Figure 12.3. In addition it breaks labor performance into tasks that may be very useful to management, particularly if management is concerned with the efficiency of an enterprise or activity. This detail may be useful in evaluating the performance of an employee. After the particular problem is solved this labor detail may not be necessary and the less precise monthly report will be adequate.

FIGURE 12.3

Employee's Monthly Labor Activity Report

LABOR ACTIVITY REPORT

(Each employee to maintain his/her own report and give to supervisor.)

NAME: | MONTH: | YEAR:

Day:	Hours worked: Time in	Time out	Hr work	Cost centers: Over-head	Mach. oper.	Build. main.	Crops or fields: Corn Fld 1	SB Fld 2	Hay Fld 3	Livestock: Beef cow	Hogs	Feed lot	Projects: New build.	Drain Fld 7	Daily hours used
1															
2															
3															
4															
5															
6															
7															
8															
9															
10															
11															
30															
31															
Total															

FIGURE 12.4

Employee's Daily Labor Activity Report

DAILY LABOR REPORT Date ________ 20XX

Employee ______________________ Month ________ Day ________

Beginning Time ______________ Ending Time ______________ Total Hours Worked ______________

ACTIVITY	HOURS SPENT	PROFIT OR COST CENTER ALLOCATIONS					MACHINE(S)	HOURS USED

TABLE 12.2

Monthly Allocation of Labor to Profit and Cost Centers by Employee

	Worker											
	1	**2**	**3**	**4**	**5**	**6**	**7**	**8**	**9**	**10**		
$ Rate	**14.74**	**10.64**	**4.65**	**5.50**	**4.00**	**3.85**	**3.86**	**3.82**			**Total**	**Total**
Center					**(hours)**						**Hours**	**$Amount**
1	0.0	19.5	0.0	23.5	1.5	0.0	0.0	0.0	0.0	0.0	44.5	342.73
3	10.5	26.0	9.0	0.0	4.0	0.0	0.0	0.0	0.0	0.0	49.5	489.26
5	4.0	5.0	0.0	0.0	3.7	0.0	1.5	0.0	0.0	0.0	14.2	132.75
8	0.0	0.0	9.5	0.0	0.0	0.0	0.0	0.0	0.0	0.0	9.5	44.18
9	6.2	0.0	0.0	0.0	0.0	0.0	0.0	0.0	0.0	0.0	6.2	91.39
10	4.4	1.1	5.0	0.0	0.0	0.0	0.0	0.0	0.0	0.0	10.5	99.81
11	4.2	8.7	0.0	0.0	0.0	0.0	0.0	0.0	0.0	0.0	12.9	154.48
12	4.5	1.2	0.0	0.0	0.0	0.0	0.0	0.0	0.0	0.0	5.7	79.10
13	1.7	0.7	0.0	0.0	0.0	0.0	0.0	0.0	0.0	0.0	2.4	32.51
14	6.2	3.7	0.0	0.0	0.0	0.0	0.0	0.0	0.0	0.0	9.9	130.76
=											0.0	0.00
51	3.5	0.5	104.2	0.0	6.4	0.0	0.0	0.0	0.0	0.0	114.6	567.04
52	71.0	79.0	0.0	8.0	1.5	27.5	12.5	55.0	0.0	0.0	254.5	2,301.33
53	3.0	41.5	3.0	0.0	0.0	5.0	1.0	4.5	0.0	0.0	58.0	540.03
=											0.0	0.00
71	5.5	4.0	0.0	30.0	9.6	0.0	0.0	0.5	0.0	0.0	49.6	328.94
72	0.0	0.0	0.0	3.5	0.0	0.0	0.0	0.0	0.0	0.0	3.5	19.25
=											0.0	0.00
CP1	18.0	0.0	0.0	59.5	0.0	19.5	24.5	2.0	0.0	0.0	123.5	769.86
CP3	44.0	7.0	0.0	18.0	0.0	0.0	2.0	0.0	0.0	0.0	71.0	829.76
TOTAL	237.5	235.0	258.10	115.5	69.9	55.5	41.5	70.5	0.0	0.0	1,083.4	8,758.88

The monthly labor report for each employee can be accumulated into a report for all workers as illustrated in Table 12.2. In this report all workers and all farm activities and enterprises are represented. At the top of the report are the wage rates for each worker, shown as a number. On the left are listed the cost and profit centers shown as numbers. At the right side are the total hours spent for each enterprise or activity and the value or cost of that time prorated by each worker's wage rate. On the bottom of the form are the total monthly hours worked by each worker. This information can be used to allocate the total labor expense for the business to the various enterprises and activities. This can be done in the general ledger or in some other business accounting system. Once the labor data has been collected and summarized the information obtained should be compared with some guideline. Many agricultural cooperative extension services publish such information as illustrated in Table 12.3.

TABLE 12.3

Direct Labor Requirements for Crop and Livestock Enterprises

Enterprise	Unit	Annual Hours of Labor per Unit: Average	High Mechanization, Efficient Work Methods	Low Mechanization, Poor Work Methods
CORN, GRAIN	1 acre	3.5	2.0	7.0
SOYBEANS	1 acre	3.5	2.0	7.0
WHEAT	1 acre	1.5	0.7	4.5
OATS	1 acre	1.5	0.7	4.0
CORN SILAGE	1 acre	6.0	3.0	20.0
HAY HARVESTING				
0.5-1.2 tons per acre	1 ton	3.0	2.0	6.0
Over 1.2 tons per acre	1 ton	2.0	1.0	4.0
SILAGE HARVESTING				
1-7.4 tons per cutting	1 ton	0.6	0.3	2.0
Over 7.5 tons per cutting	1 ton	0.3	0.1	1.0
DAIRY HERD[a]				
10-24 cows	1 cow	115	90	140
25-29 cows	1 cow	90	65	115
50-99 cows	1 cow	75	55	100
BEEF COW HERD, CALF SOLD[a]				
1-15 cows	1 cow	52	20	40
15-39 cows	1 cow	15	12	25
40-100 cows	1 cow	10	8	16
BEEF COW HERD, CALF FED[a]				
1-15 cows	1 cow	40	30	60
15-39 cows	1 cow	25	20	40
40-100 cows	1 cow	20	15	30
FEEDER CATTLE, LONG FED[a]				
1-40 head	1 feeder	15	10	25
40-119 head	1 feeder	10	7	17
120-200 head	1 feeder	8	5	13
FEEDER CATTLE, SHORT FED[a]				
1-40 head	1 feeder	10	8	18
40-119 head	1 feeder	10	8	18
120-200 head	1 feeder	5	3	10
SHEEP, FARM FLOCK[a]				
1-25 ewes	1 ewe	7	5	10
25-49 ewes	1 ewe	5	3	7
50-100 ewes	1 ewe	4	2	6
HOGS				
15-39 litters	1 litter	23	15	35
40-99 litters	1 litter	18	10	30
100 litters or more	1 litter	15	8	25
FEEDER PIGS				
1-100 hogs	1 pig	2.2	1.0	4.5
100-249 hogs	1 pig	1.6	0.7	3.0
250-500 hogs	1 pig	1.4	0.5	2.7
POULTRY				
Commercial flocks				
> 2,000 hens	100 hens	40	20	80

[a]Includes time for harvesting hay and straw and hauling manure in addition to time for caring for livestock.

Source: *Farm Management Manual,* AE-4473, University of Illinois Cooperative Extension Service.

APPENDIX[11]

LABOR ESTIMATE WORKSHEET
(Example)

Form A

				Total hours for year	Distribution of hours			
					Dec. through March	April May June	July Aug.	Sept. Oct. Nov.
	(1)			(2)	(3)	(4)	(5)	(6)
1	Suggested hours for full-time worker			2400	600	675	450	675
2	My estimate for full-time worker							
3	LABOR HOURS AVAILABLE							
4	Operator (or Partner No. 1)							
5	Partner No. 2							
6	Family labor							
7	Hired labor							
8	Custom machine operators							
9	TOTAL LABOR HOURS AVAILABLE							
10	DIRECT LABOR HOURS NEEDED BY CROP AND ANIMAL ENTERPRISES							
11	Crop enterprises	Acres	Hr/Acre					
12								
13								
14								
15								
16								
17								
18								
19								
20	TOTAL LABOR HOURS NEEDED FOR CROPS							
21	Animal enterprises	No. Units	Hr/Unit					
22								
23								
24								
25								
26	TOTAL LABOR HOURS NEEDED FOR ANIMALS							
27	TOTAL HOURS NEEDED FOR CROPS AND ANIMALS							
28	Total Hours of Indirect Labor Needed							
29	TOTAL LABOR HOURS NEEDED (lines 27 and 28)							
30	TOTAL AVAILABLE (line 9)							
31	Additional Labor Hours Required (L. 29 minus L. 30)							
32	Excess Labor Hours Available (L. 30 minus L. 29)							

Source: Missouri Farm Planning Handbook, Manual 75; Feb. 1986, Department of Agricultural Economics, University of Missouri, Columbia.

Form B

JOB DESCRIPTION
(Example)

I. Job Title ______________________________

II. Work Duties, Authority, and Responsibilities:
____
____
____
____
____

III. Job Qualifications:
A. Formal Training: ______________________________
B. Special Training: ______________________________
C. Experience: ______________________________
____
____

D. Job Knowledge: ______________________________
____
____
____

E. Personal Characteristics: ______________________________
____
____

F. Physical Requirements: ______________________________
____
____

G. Flexibility (Time, Tasks, etc.): ______________________________
____
____

H. Other: ______________________________
____

IV. Supervision:
A. Amount: None ______ Minimal ______ Average ______ Close ______
B. Supervisor:

V. Job Advancement or Promotion Possibilities: ______________________________
____
____

VI. Wage Rate: Beginning $ _____ Per _____; Range __________

VII. Bonuses, Incentives, Benefits: ______________________________
____
____

VIII. Provisions for Time Off And Vacation: ______________________________
____
____

Examples of Bonus/Incentive/Share Arrangements[a]

Form C

The following examples of incentive programs should be used only as guides and be adapted to your situation. They should be tied to work responsibilities carried out by the employee and over which he or she has some control.

Suggested Incentives	TYPE OF EMPLOYEE STATUS		
	Semiskilled	Skilled	Supervisory/Management
Normal Incentive Should Equal To:	2–5% of cash wages	4–10% of cash wages	5–40% of cash wages
General Farm	End of year bonus = $100–$400 per year plus $50 for each year of service	End of year bonus = $200–$600 per year plus $75 for each year of service	Put emphasis on incentive plans
Small Farm	Pay 1½ times cash wage rate for each hour worked over 60 hours per week		2% of net cash income
Large Farm	Pay 1½ times cash wage rate for each hour worked over 48 hours per week		1–4% of net cash income
Crop Farm	$1–$2/hour tractor driven after 7:00 p.m. (paid weekly) $2–$3/hour tractor driven after 11:00 p.m. (paid weekly)	$1–3/hour tractor or combine driven after 7:00 p.m. (paid weekly) $2–$5/hour tractor or combine driven after 11:00 p.m. (paid weekly)	2–6 cents per bushel of corn produced over county average 5–15 cents per bushel of soybeans produced over county average
Dairy	$1–$3 for each cow detected in heat $3–$5 per calf weaned if death loss kept below 10% $5–$10 per calf weaned if death loss kept below 5%	Calving interval $100 = 13.5 months $200 = 13 months $400 = 12.5 months	Herd milk production average 14,000# = $100/year 16,000# = $400/year 18,000# = $800/year 20,000# = $1,600/year
Hogs	$0.50–$0.75 for each sow detected in heat	Pigs saved per litter 7.5 = $ 50 8.0 = $150 8.5 = $300 9.0 = $500 9.5 = $900	Feed conversion farrow to finish 400# = $100/year 375# = $200/year 350# = $400/year 325# = $700/year 300# = $1,100/year
Beef	$1-$2 for each feeder detected sick, treated and recovered	Calf crop sold 85% = $100 90% = $200 95% = $400 100% = $700	Same as other two categories

[a]These examples reflect Minnesota conditions. Contact your state extension services for arrangements for your area.

Form D

Date____________

APPLICATION FOR EMPLOYMENT
(Example)

____
NAME OF FARM OR EMPLOYER

1. PERSONAL INFORMATION

Name ________________________

Address________________________ Phone Number (__)____________

____

*Spouse's name (if any)____________ *Children/Ages (if any) ____________

Do you have any health problems or physical impairments that would interfere with your doing strenuous physical work?

If yes, please explain. ________________________

____

Education ________________________

____

Special training ________________________

____

2. WORK HISTORY (beginning with most recent employment)

Employer #1 ________________________

Address ________________________

________________________ Phone (__)____________

Job duties ________________________

____

Dates worked____________ Wages or salary received ____________

Reason for leaving ________________________

____

Employer #2 ________________________

Address ________________________

________________________ Phone (__)____________

Job duties ________________________

____

Dates worked____________ Wages or salary received ____________

Reason for leaving ________________________

____

3. LIST OF REFERENCES

Name____________ Address____________ Phone (__)____________

Name____________ Address____________ Phone (__)____________

Name____________ Address____________ Phone (__)____________

4. ON THE BACK OF THIS PAGE, PLEASE EXPLAIN YOUR EXPERIENCES, STRENGTHS, AND PERSONAL CHARACTERISTICS THAT YOU WOULD LIKE CONSIDERED AS PART OF YOUR APPLICATION.

*Optional--Applicant is not required to provide this information.

Form E

Date__________

Interviewer__________

EMPLOYEE INTERVIEW FORM
(Example)

1. Name of applicant __________

   Address __________ Phone No. (____) __________

2. Check of information on application form
   If different from application form:
   (Continue response on back of form if necessary)

   Exact nature of work experiences __________

   Present skills relative to this __________

   Reasons for leaving former __________

3. Personal goals and aspirations relative to farming?__________

4. Why are you applying for this job?__________

5. Evaluation of candidate relative to job description (select items from following list that relate to job description).

Characteristics	Low (Rating)				High
1. Leadership qualities	1	2	3	4	5
2. Ability to work with others	1	2	3	4	5
3. Receptiveness to receiving directions	1	2	3	4	5
4. Motivation to learn	1	2	3	4	5
5. Willingness to work (physical labor)	1	2	3	4	5
6. Training and background in:					
Animal nutrition	1	2	3	4	5
Animal health	1	2	3	4	5
Crop production	1	2	3	4	5
Livestock production	1	2	3	4	5
Mechanical skills	1	2	3	4	5
Management concepts	1	2	3	4	5
Finance	1	2	3	4	5
Marketing procedures	1	2	3	4	5
Machinery operation	1	2	3	4	5
Vehicle operation	1	2	3	4	5
7. Ability to manage others	1	2	3	4	5
8. Personal goals	1	2	3	4	5
9. Initiative and imagination	1	2	3	4	5
10. Motivation	1	2	3	4	5
11. Determination	1	2	3	4	5
12. Ability to compromise	1	2	3	4	5
13. Ability to identify problems	1	2	3	4	5
14. Ability to make a decision	1	2	3	4	5
15. Ability to understand directions	1	2	3	4	5
16. Willingness to ask questions	1	2	3	4	5

6. Comments, if any, about applicant's spouse and family (optional)__________

7. Key comments from references__________

8. Evaluations by current employees__________

9. Overall rating__________

10. Other comments__________

11. Agreed upon follow-up__________

Form F

EMPLOYMENT AGREEMENT
(Example)

Farmer Employer–Employee Agreement of Employment

I, ______________________________, agree to employ ____ to work on my farm located: ______________________________ beginning (date) ____________________ and continuing until such time as either wishes to terminate this agreement by a __________ day notice. ____________________________, the employers, and ____, the employee, agree to comply with the following conditions and actions:

1. To pay __________________________ $ __________ per ________ from which the employee's income tax (yes/no) and social security taxes will be withheld. Wages will be paid on __________ (day) of (week/bi-weekly/monthly).

2. To provide a house with utilities including heat and electricity. The maintenance is to be done by __________________ and paid for by ____________________. Any other agreements pertaining to the employee's house will be noted on the back of this page.

3. The normal working hours are from __________ A.M. to __________ P.M. with one hour off for breakfast and one hour off for lunch. Overtime will be paid for any work done after 7:00 P.M. at the rate of 1½ times the normal wage rate. Overtime will be paid after __________ hours are worked in any one week, Sunday through Saturday.

4. Time off shall be every other Sunday and holidays. The holidays for purposes of this agreement are New Year's Day, Easter, Memorial Day, Labor Day, Thanksgiving Day, and Christmas Day. On Sundays and holidays only the chore work will be done. The employer, ____________________, shall notify the employee, __________________________, at least 45 days before the holiday of what the time-off arrangements will be.

5. The employee is entitled to ________ weeks vacation with pay annually, which shall be taken during the nonheavy work season and agreed upon with the employer 30 days prior to beginning of vacation.

6. The employee is entitled to _______ days sick leave with pay annually for the time off due to actual illness.

7. The employee is entitled to _______ quarts of milk per day.

8. The employee is entitled to _______ pounds of beef and _______ pounds of pork per year.

9. The employee is entitled to a 15-minute break in midmorning and midafternoon.

10. The following insurance plans will be carried on the employee:
    ____
    ____
    ____

11. A bonus or incentive plan (is, is not) included. If included, the provisions are noted on Form G, attached.

12. Other provisions not included above are listed on the reverse side of this form.

______________________________________ Date ______________________________
Employer Signature

______________________________________ Social Security No. ______________________
Employee Signature

Form G

Date____________

Employee____________

Wage, Incentive, Benefits Agreement

Employee's Responsibilities:

Example

Son, John age 21, contributing only labor to the farm business. Provide labor where needed.

Your Plan

	Example: Cash Received Per Year	Example: Value Other Benefits Per Year	Your Plan: Cash Received Per Year	Your Plan: Value Other Benefits Per Year
Cash Wages	$7,200			
Bonus/Incentive Payments				
Crop____				
Livestock____				
Other/bonus____	200			
Fringe Benefits				
Housing, room and board		$2,400		
Utilities				
Meat, milk and other produce				
Other (car, gasoline, etc.)		240		
Insurance (health, accident, and life)		700		
Social security paid by employer		500		
Workman's compensation paid by employer				
Total Cash/Benefits Received	$7,400	$3,840		
Grand Total	$11,240			

Notes

1. Statistical data in this section and in the introduction were obtained from "1996 Current Population Survey," and other farm labor reports issued by the ERS, USDA.
2. Circular E, "Employer's Tax Guide," Publication 15, IRS.
3. The balance of this chapter is primarily from Thomas, Kenneth H. and Bernard L. Erven, "Farm Personnel Management." NCRP 329-1989, AG-BU 3613, Iowa State University.
4. Fair Labor Standards Act, USDL, WH Pub. 1288.
5. Edwards, William, *Farm Employee Management in Iowa.* FM 1841, Iowa State University Extension, 1991.
6. Farmer-employers should follow the guidelines in "Circular A," *Agricultural Employer's Tax Guide.*
7. Child Labor Requirements in Agriculture Under the Fair Labor Standards Act (Child Labor Bulletin No. 102), USDL, WH Pub. 1295.
8. Rosenberg, Howard R. and Daniel L. Egan, *Labor Management Laws in California Agriculture,* Cooperative Extension, University of California, 1990.
9. Note: AgriCareers of New Hampton, Iowa, has found that more than 80 percent of the reasons that employees gave for quitting a job could have been prevented. Reasons given included lack of achievement (12.9 percent), lack of training (7.9 percent), lack of responsibility (12.8 percent), lack of recognition (9.9 percent), low salary (13.8 percent), limited time off (6.9 percent), and problems with the boss and the boss's family (16.8 percent).
10. McGregor, D., *The Human Side of Enterprise,* McGraw Hill Co., New York, 1960.
11. These forms were taken from Thomas, Kenneth H. and Bernard Erven, "Farm Personnel Management." NCRP 329-1989, AG-BU 3613, available from Iowa State University.

CHAPTER 13

Income Tax Management

Tax Management—Part of Farm Management

Managing a modern farm business requires high capital investments and the handling of large sums of money annually. The tax consequences of farm business decisions have a greater impact on cash flow and net income as farm businesses become larger and farmers move into higher tax brackets. The farm manager is constantly making decisions during the year that affect the amount of income tax to be paid and the amount of cash available for operation of the business. State and federal income taxes and FICA taxes can amount to over one-third of net income. The tax consequences of various farm business transactions must be understood to make wise decisions in the framework of minimizing income tax while maximizing after-tax income. Thus, farmers must think about taxes throughout the year because tax management is a continuous process, not just a year-end endeavor.

Tax management assumes that the individual farmer can do a better job by planning investments and financing and by managing income and expenses than by unplanned business operations. The timing of transactions can play an important role in the short run in balancing income between years, thus minimizing yearly fluctuation in income and taxes. An attempt should be made to maintain an annual net income at least equal to the year's allowable nonbusiness deductions and personal exemptions and yet avoid extremely high taxable income in following years. It may be equally desirable to have a net income that approaches or equals the maximum earnings (from self-employment plus wages) eligible for social security credits. Tax management aims at the greatest after-tax income and net worth.

In the long run, however, tax management involves much more than merely timing transactions between years to even out income. The form of business organization under which the farm business is operated and the manner in which enterprises are organized within the farm business, as well as strategies used in financing the business, have a long-term effect on after-tax income.

Tax management is not concerned solely with minimizing taxes. If decisions are made and business transacted solely in an effort to reduce tax, net income after taxes may actually be lower.

For example, if a decision results in the saving of $100 in income tax but a larger amount is lost by a lower selling price of a farm product, the net income after taxes is reduced. Generally, there are no conflicts between wise tax decisions and good farm business decisions, but when a choice must be made, the one resulting in the larger net income after taxes should be followed.

Not all aspects of income tax management can be covered in a single chapter. Because of the complexity of income tax regulations, only major tax management considerations are discussed. Income tax regulations often change from year to year. Farmers should be aware of changes that may influence their management decisions. Tax management should be tailored to the organization and financial structure of the individual farm business. This discussion concentrates on those regulations that have been in place for several years and are expected to continue.

State income tax is not featured. But because states often follow federal guidelines most issues discussed are even more important than illustrated. Generally speaking, state income tax is less progressive than federal income tax and accounts for about 5 to 10 percent of taxable income.

Good Records—A Necessity[1]

Successful tax management depends on keeping complete records regularly and carefully throughout the year. Such records enable the taxpayer to determine approximate taxable income at any time. These records provide a basis for making business decisions that increase net after-tax income. If a preliminary check of expected total income and expenses indicates an unusually high taxable farm income, it may be wise to delay additional sales until after the end of the year or to increase deductible expenditures before the end of the year. Conversely, if the check reveals an unusually low net farm income, sales may be speeded up to include them in the current year, or expenditures or payment of accounts may be deferred until the next year.

Bank statements are often examined when tax returns are audited. Thus, the entries shown thereon should be supported with more detailed information in the account books. Nontaxable income by all means should be identified (e.g., borrowed money, bonds cashed, gifts, and inheritance). Account books suitable for keeping adequate records for tax purposes and farm business analysis are available at most county agricultural extension offices. There are good computer-based record systems suitable for farm management and tax reporting. See Chapter 3 for a discussion of important features of these systems.

Records and Tax Accounting Methods

Farmers may keep records and file their tax returns by using either the cash or accrual basis. A combination of methods may be acceptable if it clearly reflects income and expenses. The accounting method selected for business management should support income tax reporting. It was recommended in Chapter 3 that farmers keep an accrual set of business accounts, even if taxes are filed using the cash method.

Farmers make the election of filing their tax returns on the cash or accrual basis when filing the first tax return. A farmer who is part of a newly organized farm partnership or corporation or who files for a newly purchased farm operated as a separate business may file on the same basis or may change to the other. The chosen method must continue to be used unless written consent to change is obtained from the Commissioner of the Internal Revenue Service (IRS).

Depreciation of farm improvements; machinery and equipment; and purchased breeding, dairy, sporting, and work animals are allowable business deductions under both the accrual and cash methods. Farm records of such property should include a detailed record of the date of purchase, cost, trade-in property, years of life, method of depreciation, depreciation claimed to date, and any other tax-related items.

Farmers may report farm business income on the cash basis and other business or personal income on the accrual basis, or vice versa. Furthermore, a combination of methods may be used so long as income is clearly and accurately reported and consistently followed.

Cash Method

When the income tax return is filed on the cash basis, gross farm income includes income actually or constructively received from the sale of all crops and market livestock produced on the farm, gross profits (selling price less purchase price) on livestock and other items that were purchased for resale, and other farm income actually or constructively received. Allowable deductions include farm operating expenses paid during the year (regardless of when they were incurred) and depreciation expense allowable on farm improvements, machinery, equipment, and purchased dairy, breeding, sporting, and work stock.

There are certain advantages of filing on the cash basis in tax planning:

- *It is a simple method of reporting.* Fewer records are necessary because inventory accounts need not be kept. However, the inventory information may be needed for business analysis or for preparing a financial statement.
- *Taxes are postponed on increases in inventory.* Because inventories are not taxed until sold there is a clear tax advantage if tax rates remain constant or decline. The advantage is less if tax rates increase at sale time, which could be particularly troublesome if large inventories are sold.
- *There is flexibility* for adjusting income from year to year when wide variations may occur in prices and production rates, despite the tendency of the accrual method to make these types of adjustments automatically each year through beginning-of-year and end-of-year inventories.
- *Self-employment taxes may be decreased.* Income from the sale of draft, dairy, and breeding livestock is not reported on Schedule F and therefore does not affect self-employment income for cash-basis taxpayers. However, inventory changes for those same livestock are entered on Schedule F by accrual-basis taxpayers and do affect self-employment earnings. Some may wish for higher self-employment taxes to increase retirement benefits.

For example, with the cash method, part of the crop or livestock production in a good year may be held over for sale in years of lower production or sales may be speeded up in years of low production or low prices. Farmers also may delay expenditures, postpone payments, or conversely make certain cash purchases before actually needed, depending on the net income situation in a particular year.

Farmers may purchase deductible ordinary expense items such as feed, seed, fertilizer, and repairs in the latter part of the year, even though they are not used until the next taxable year. However, caution is necessary in making these purchases—the purchase must have a purpose other than simply to reduce taxes. It is often good management for a livestock feeder to purchase needed feed grain in the late fall when prices are generally lower than in the following spring, so there is often a reason other than tax-reduction for those purchases. The advantages depend on prices, cost of storage, and investment cost of borrowed money.

To deduct the cost of fertilizer, feed, chemicals, petroleum, or other annual operating supplies, they should be paid for and actually acquired in the year claimed. They must be bona fide irrevocable obligations, not advance payments on "orders to be placed" or on purchases that may be made in the succeeding year. The IRS has placed a limit on the amount of prepaid expenses to the extent that they cannot exceed 50 percent of the nonprepaid farming expenses. Prepayments must be made for a business purpose and not merely for tax avoidance. There should be a reasonable expectation of receiving some business benefit as a result of the prepayment. Examples of business benefits include fixing maximum prices, securing an assured feed supply, and securing preferential treatment in anticipation of a feed shortage.

Capital gains, including breeding livestock, are now treated as regular income for tax purposes and fully taxed, but this could change. There is a maximum tax rate on capital gains income that is lower than for regular income. Thus, there could still be some tax advantage even when tax rates are increasing.

Accrual Method

Gross farm income under the accrual method includes all income from sales made during the year regardless of when payment is received; all miscellaneous income regardless of source; and the inventory value of all livestock, crops, and supplies on hand and not sold at the end of the year.

To arrive at net income, the farmer subtracts the inventory value of livestock and products on hand at the beginning of the year, the cost of livestock and products purchased during the year (except livestock placed on the depreciation schedule), all operating costs or expenses incurred during the taxable year, and depreciation.

Costs of purchased feeder livestock are subtracted in the year purchased and then are included in the inventory at the end of the year if not yet sold. With the cash method these costs are not deductible until the year in which the animals are sold.

Some advantages of the accrual method include the following:

- *Income from the sale of stored crops may be leveled.* Crops stored for sale the following year are included in inventory and counted as income in the year produced as a credit against costs of production. When sold there is an inventory reduction debit against the sale. Thus, income is leveled between years. Examples include two calf crops and grain held over plus the current year's production. Farmers who report on the accrual method have their income tax paid more up to date than do farmers who use the cash method; thus, this method inherently results in a more even year-to-year taxable income.
- *Income taxes are paid on a more current basis.* Farmers starting with a small operation or with inadequate financing may desire to use the accrual method to keep their income and taxes on a more current basis, rather than postponing taxes until the year in which production is sold. Start-up costs may be so high that unsold production may be needed to bring taxable income up to the level of personal exemptions and standard deductions, or to avoid showing a net operating loss.
- *Social security benefits may be augmented.* The application of Section 1231[2] to sales of raised breeding, dairy, sporting, and work stock excludes proceeds from such sales from the self-employment income under the cash method. Furthermore, all costs of raising Section 1231 livestock are deducted from self-employment income. But with the accrual method the requirement of capitalization of "normal

costs" of raising these animals in the year of sale tends to increase self-employment income.

The requirement of capitalization of "normal costs" of raising Section 1231 animals can be a disadvantage of the accrual system. With more ordinary income from the sale of breeding stock, hog farmers, dairy farmers, and beef cattle breeders pay more taxes under the accrual method. Because both income taxes and social security taxes may increase, sound management calls for a careful comparison of these two methods of income tax reporting.

Comparisons of Methods

A significant difference between the cash and accrual methods of computing income is illustrated in the following example. A farmer purchases and takes delivery of feed for $500 on December 15 and charges the purchase, is billed for it January 2, and pays the bill in January. With the cash method, it is a farm expense in January when the bill is paid. With the accrual method, the $500 expense is a deduction in December when the obligation to pay was incurred or "accrued." Any feed on hand at the end of the year would be included in the ending inventory and the $500 debt in accounts payable. Therefore, it has no effect on net income because the added expense and added ending inventory offset one another. Income is similarly treated.

Increases in inventories are included in the income with the accrual method. For example, a farmer raises and feeds livestock during the year but does not sell any. With the cash method, there is no income until the livestock are actually sold. With the accrual method, there is income in the amount of any increase in the value of livestock and crops on hand at the end of the year compared with the value at the beginning of the year.

Table 13.1 shows the farm income subject to federal income tax under the cash and accrual methods of reporting for different farm situations (summarized below). In all situations except I, it is assumed that operating and ownership expenses are the same for the two methods. The purpose in A, B, and C is to illustrate what happens when a farm-produced product such as grain is added to, or reduced in, inventory. Situations A, D, and E illustrate the changes brought about by purchased feeder livestock. Situations F, G, and H are added to illustrate what happens when breeding livestock that are subject to capital gains are involved. In I the assumption of equal operating costs is relaxed to allow end-of-year purchases to reduced taxes. Situations H and I are not illustrated, but the reasons for the differences would be apparent after reviewing A-G.

A. *Crop and livestock-feeding farm.* All livestock sold were purchased. All crop and livestock increases were sold (i.e., beginning and ending inventories remain unchanged). The cost of livestock purchased remains unchanged from the year before.

B. Same as A except $2,000 more crops were sold, thus reducing ending crop and feed inventory by $2,000.

C. Same as A except $2,000 of the year's crop was added to the ending inventory; thus, $2,000 less crops were sold.

D. Same as A except that the cost of feeders was not the same as in the previous year. Cost of feeders purchased this year is less (i.e., $10,000 compared with $12,000 last year). The closing inventory difference from the purchase price was $2,000, the same as A.

E. Same as A except that the closing inventory value this year was $4,000 over the purchase price as opposed to $2,000 last year. The cost of the feeders was the same.

TABLE 13.1

Comparison of Net Taxable Income When Distributed by Cash and Accrual Methods Under Different Receipt and Expense Situations

Transactions	A Cash	A Accrual	B Cash	B Accrual	C Cash	C Accrual	D Cash	D Accrual	E Cash	E Accrual	F Cash	F Accrual	G Cash	G Accrual
Receipts:														
Miscellaneous receipts	$10,000	$10,000	$10,000	$10,000	$10,000	$10,000	$10,000	$10,000	$10,000	$10,000	$10,000	$10,000	$10,000	$10,000
Crops	20,000	20,000	22,000	22,000	18,000	18,000	20,000	20,000	20,000	20,000	20,000	20,000	20,000	20,000
Feeders purchased	20,000	20,000	20,000	20,000	20,000	20,000	20,000	20,000	20,000	20,000	0	0	0	0
Feeders raised	0	0	0	0	0	0	0	0	0	0	20,000	20,000	22,000	22,000
Breeding livestock (capital gains)	0	0	0	0	0	0	0	0	0	0	800	400	800	400
Ending inventory														
Crops and feed	—	10,000	—	8,000	—	12,000	—	10,000	—	10,000	—		—	10,000
Feeders purchased	—	12,000	—	12,000	—	12,000	—	12,000	—	14,000	—	0	—	0
Feeders raised	—	0	—	0	—	0	—	0	—	0	—	10,000	—	8,000
Breeding livestock	—	0	—	0	—	0	—	0	—	0	—	4,000	—	4,000
Total credits	$50,000	$72,000	$52,000	$72,000	$48,000	$72,000	$50,000	$72,000	$50,000	$74,000	$50,800	$74,400	$52,800	$74,400
Expenses:														
Operating	$20,000	$20,000	$20,000	$20,000	$20,000	$20,000	$20,000	$20,000	$20,000	$20,000	$20,000	$20,000	$20,000	$20,000
Fixed	2,000	2,000	2,000	2,000	2,000	2,000	2,000	2,000	2,000	2,000	2,000	2,000	2,000	2,000
Depreciation	2,000	2,000	2,000	2,000	2,000	2,000	2,000	2,000	2,000	2,000	2,000	2,000	2,000	2,000
Livestock purchased	10,000	10,000	10,000	10,000	10,000	10,000	12,000	10,000	10,000	10,000	0	0	0	0
Beginning inventory														
Crops and feed	—	10,000	—	10,000	—	10,000	—	10,000	—	10,000	—	10,000	—	10,000
Feeders purchased	—	12,000	—	12,000	—	12,000	—	14,000	—	12,000	—	0	—	0
Feeders raised	—	0	—	0	—	0	—	0	—	0	—	10,000	—	10,000
Breeding livestock	—	0	—	0	—	0	—	0	—	0	—	3,000	—	3,000
Total debits	$34,000	$56,000	$34,000	$56,000	$34,000	$56,000	$36,000	$58,000	$34,000	$56,000	$24,000	$47,000	$24,000	$47,000
Net taxable income	$16,000	$16,000	$18,000	$16,000	$14,000	$16,000	$14,000	$14,000	$16,000	$18,000	$26,800	$27,400	$28,800	$27,400
Difference		none		$2,000	$2,000			none	$2,000		$600			$1,400

H. Cash $25,000; accrual $27,500
$2,500

I. Cash $15,000; accrual $16,000
$1,000

Note: In situations F and G, beginning inventories were reduced by the inventory value of breeding stock sold. These inventories are subject to capital gains treatment. The taxable gain is included in receipts as sale of breeding livestock.

F. *Crop and livestock-raising farm.* All crop and livestock increases were sold (i.e., beginning and ending inventories remain unchanged). Normal replacements of breeding stock were sold. These animals had a beginning inventory value of $1,000 and were sold for $2,000.

G. Same as F except that inventories of feeders raised were reduced through sale by $2,000.

H. Same as F except that the breeding herd was increased by $2,000, thereby reducing the sale of feeders by $2,000.

I. Same as A except that $1,000 of feed was stockpiled (added to inventory) at the end of this year. It should be recognized that next year the advantage will be reversed.

It also should be observed that a difference in 1 year favoring one or the other of the methods may be reversed the following year. The important things to observe are the flexibilities and rigidities of each system. The level of taxable income over time will be the same under the two systems except for the effect of increased inventories. However, the amount of tax paid may be somewhat different through income averaging and the effects of capital gains.

Neither of these methods may provide the route to attaining all goals. However, if farmers are sure they will be faced with increasing inventories and a decreasing tax rate, the cash method would reduce their taxes. With decreasing inventories and an increasing tax rate, it is apparent that the accrual method would be desirable.

With other combinations of inventory change and rates of taxation the choice is not as clear-cut. For example, with increasing inventories and rates of taxation the individual using the cash method avoids paying taxes on the current year's increase in inventory but also loses out on paying on the previous year's accumulation of property at that year's lower tax rate. Similarly, when the accrual method with decreasing inventories and decreasing tax rate is used, the decreases in inventory operate to reduce taxable income, but taxes are paid on the basis of the current year's higher rate. However, compared with the cash method, taxable income and thus taxes would be reduced.

Using the method that delays or postpones taxes while taking advantage of all personal deductions and exemptions, even with rising rates of taxation, generally operates to the advantage of most taxpayers. Additional tax management alternatives are mentioned later in this section.

Income Tax Basis

An important term in income tax law is basis. In very simple terms, basis in an asset is the amount paid for that asset plus the cost of any improvements minus any depreciation claimed. For example, if $10,000 is paid for a corn planter, the basis in the corn planter is $10,000. If $1,500 of depreciation is claimed on the corn planter, the basis (or adjusted basis) is $8,500. If $1,000 is spent to improve the corn planter, the basis increases to $9,500.

It is important to know the basis in assets because the income tax that must be paid when an asset is sold is tabulated using the difference between the basis and the sales price. If an asset is sold for more than the basis in the asset, the difference between the selling price and the basis must be included in income. This difference is called gain and beginning in 1987 became fully taxable.

For certain assets that are sold for less than the basis, the difference between the basis and the selling price can be deducted as a loss. Again, the amount of the deduction depends on the type of asset and the length of time it was owned.

A cash-basis farmer has no basis in raised livestock, including those kept for breeding, because the costs of raising the livestock are claimed as expenses in the year they are incurred. Thus, when the animal is sold the sale is fully taxed. However, zero basis can appear to be detrimental in some cases. For example, if a raised dairy cow is killed by lightning, no casualty loss can be claimed because there is no basis in the cow. Furthermore, if the loss is covered by insurance, the proceeds from the insurance company must be included in income (although they may qualify for the long-term capital gains deduction). Although the denial of the loss deduction appears to be a detriment, it is simply the price paid for the advantage of deducting the cost of raising the livestock. Because the value of raised livestock has never been included in income, the loss of that livestock cannot be claimed as a deduction.

If money is borrowed to purchase an asset, the basis in the asset is likely to be more than the out-of-pocket purchase costs. For example, if $40,000 is borrowed to build a $60,000 livestock facility, the beginning basis is $60,000 even though the out-of-pocket cost is only $20,000.

Understanding basis is crucial to understanding income taxation and the management of taxes. Basis is referred to in publications on income taxation.

Maximizing After-Tax Income

The goal of farmers should be to maximize after-tax income. An individual's income is related to the resources at hand and how these resources are combined in doing business. The form of business organization chosen can have an effect on taxes paid. Some of these opportunities were discussed in Chapter 9. The farm business may be operated as a sole proprietorship, partnership, or corporation. The income of a sole proprietorship is reported on the sole proprietor's individual return. The income of a partnership is reported on a partnership return (Form 1065) but the partnership pays no taxes. Instead, each partner includes his or her share of the partnership income on his or her individual return. The income of a corporation is reported on a corporation return (Form 1120) and the corporation pays tax on the income, unless the tax-option corporation (Subchapter S) election is made. If the shareholders elect to be taxed as an S corporation, the income is not taxed to the corporation but is reported on the shareholder's individual returns and included in their taxable incomes. The regular corporation tax (Subchapter C) may be less than for individual shareholders. If earnings are held and not distributed as dividends taxes could be lower, particularly in the short run. Because the tax rates differ for corporations and individuals, taxes paid as a result of the operation of the farm business may differ considerably over a period of years, depending on which form of business organization is used. This is particularly true for higher taxable incomes. However, the form of business organization that best fits operational and estate-planning objectives should be considered. Income tax considerations may be an important factor in selecting a particular type of farm business organization. However, the discussion that follows concentrates on the sole proprietorship. Generally, those principles and regulations discussed have broad application.

For some, reduction of taxes may be accomplished by income averaging. For others, timing transactions to equalize income between years, particularly to level the allowable deductions in each tax year, reduces the total tax bill. Accelerated methods of depreciation used judiciously can assist in leveling income from year to year. The previously mentioned actions deal first with maximizing after-tax income through the type of income produced and the manner in which it is taxed and also with minimizing taxes by timing of income and deductions within the former framework. The dual action of both is to maximize after-tax income.

Some actions a manager may take do not necessarily reduce taxes but do tend to postpone them. Accelerated methods of depreciation, unless used to even income, tend to postpone taxes. This is true if soil- and water-conservation expenses are charged off as current expenses and disposed of within the time period that the new recapture provision allows. However, if the real estate is held for at least 10 years, it is possible to convert ordinary income to capital gains income and that could reduce tax liabilities.

Managing Income to Reduce Income Fluctuation

There are two reasons for reducing the fluctuation in income. First, progressive tax rates cause income in high-income years to be taxed at a higher rate. However, the flatter tax rates in the 1986 legislation reduced the impact of the more progressive tax rates of the past and the benefits from leveling income. The 1993 tax act again increased tax rates for high-income persons. Second, you cannot take advantage of the individual exemptions and deductions available on an annual basis if there is not income for them to offset in the low-income years.

The personal exemption was set at $2,000 in 1989 with increases for inflation in later years ($2,650 in 1997). The amount you can claim as exemptions is phased out once the taxable income goes above certain levels depending on the filing status ($181,800 for marrieds filing jointly in 1997). The standard deduction was set at $5,000 for joint taxpayers and $3,000 for single taxpayers with adjustments for age and blindness ($6,900 for marrieds filing jointly and $4,150 for singles in 1997). The level was not indexed but has been increased from these initial amounts. Itemized deductions may be used if the amounts exceed the standard deduction. If income is below the sum of these exemptions and deductions, the benefit of the excess deductions can never be realized. For example, incomes at or below the amount shown in Table 13.2 for 1997 are not subject to federal income tax. Persons may be required to file a tax return even though no taxes are due. Self-employment taxes are based upon income before adjusting for retirement, exemptions, and deductions.

When a preliminary check of income indicates a probable net taxable income less than the maximum income not subject to tax, consideration should be given to at least increasing income to the amount of the nontaxable income. Because personal exemptions, deductions, and credits are allowed annually, these deductions not absorbed by current income are automatically lost.

Unused exemption credits cannot be carried forward and applied against income of another year. The following example illustrates this principle. The John and Mary Jones

TABLE 13.2

Maximum Incomes, Not Subject to Income Tax, By Number of Exemptions, 1997

Number of Exemptions	1997
Single individual, under age 65	$ 6,800
Married couple, under age 65	12,200
Family of four, under age 65	17,500

and Jim and Jan Smith families each have two children. Their financial situations are shown in Table 13.3. The Joneses paid income tax but the Smiths did not, even though they had the same total net income for the 2 years. In the first year the Joneses paid no tax but could not use the $17,500 earnings that tax regulations permitted before payment of income tax. In the second year the Joneses' income exceeded exemptions so they had to pay taxes.

Reducing year-to-year variations in income also benefits from a progressive tax rate schedule. Consider for example two families with taxable incomes sufficient in all years to pay taxes and with the same average income but one family's income fluctuates more than the other. These incomes are shown in Table 13.4. The same average stable income resulted in a $3,586 lower total tax in 2 years than the fluctuating income.

The question is, How can farmers shift income from one year to another to take advantage of these two income-reducing opportunities? Receipts or expenses must be legally changed to fit the situation. This change is more easily done if taxes are paid using the cash method, but there are steps the accrual taxpayer can take to stabilize income. Assume for the sake of discussion that the marginal tax rate is higher in YRX1 than it is expected to be in YRX2; that is, it is desirable to shift income from this year to next year. The discussion centers on the cash method with comments where the accrual taxpayer can take advantage of the idea.

TABLE 13.3

Average Taxes Paid by Two Families with Different Income Streams

	Net Income			
Family	First Year	Second Year	Average Income	2-Year Tax
Jones	$ 0	$35,000	$17,500	$2,639[a]
Smith	17,500	17,500	17,500	0

[a]Using 1997 standard deductions and tax rate schedules.

TABLE 13.4

Federal Income Taxes Paid by Two Families with Equal Average Incomes but Differing Annual Incomes

Year	Taxable Income	Marginal Tax Rate	Federal Income Tax
Fluctuating incomes:			
YRX1	$ 87,000	28%	$14,097
YRX2	29,000	15%	1,721
Total	$116,000		$15,728
Stable incomes:			
YRX1	$ 58,000	15%	$ 6,071
YRX2	58,000	15%	6,071
Total	$116,000		$12,142
Difference:			$ 3,586

Methods for Shifting Income to Next Year to Level Taxable Income

- *Delay sales until next year.* Livestock feeders may have livestock about ready for market that can be sold either in December or January. Sales in January may be desirable from a tax standpoint, but should be weighed against losses from holding them too long. Other possibilities include postponement of selling grain or other farm products, culling dairy and breeding stock, payments for labor or custom work done, and selling of capital items not needed in the business and on which a gain can be realized.
- *Accelerate operating expenses.* Purchase and pay for operating inputs such as fuel, feed, seed, fertilizer, and chemicals. However, it is important to consider added costs of interest (paid or opportunity cost) and possibly lower costs in the spring. Advanced payments must meet IRS criteria to be valid so their rules need to be studied. In particular, prepayments must be done with care.
- *Accelerate depreciation.* There are three things that apply here: (1) the purchase of depreciable property, (2) Section 179 deduction, and (3) the depreciation method. The purchase of needed machinery, equipment, buildings, and breeding livestock surely will increase depreciation expenses. These were all discussed in Chapter 2 and are not detailed here. The Section 179 deduction allows the taxpayer to direct expense up to $18,000 in 1997 and will increase to $25,000 by year 2002. That is added expense in the year of purchase but will reduce the depreciation deduction in future years. Taxpayers also may select fast or slow methods of depreciation. The faster method uses the GDS or 150% declining balance system. The ADS system uses a straight-line depreciation with longer life periods than the GDS. The depreciation adjustment is equally applicable to the accrual method of accounting.
- *Expense rather then capitalize.* Section 179 deduction has already been mentioned. Other opportunities include classifying repairs as expenses rather than improvements, depreciating small tools rather than expensing them, expensing certain soil- and water-conservation expenses rather than capitalizing them, and a fruit-grower expensing preproductive costs. All of these must meet certain criteria and are not applicable in all situations. With repairs and small tools the method should not materially affect income. Both were discussed in Chapter 2. Not all conservation expenses can be expensed. Examples of those that can include the movement and treatment of earth, construction of drainage ditches and ponds, eradication of brush, and planting of windbreaks. Other rules apply. Generally, preproductive costs for crops with a preproductive period longer than 2 years must be capitalized. But if these crops are not citrus trees or almonds, it may be possible to elect out and expense these costs as they occur. Follow IRS guidelines in doing this. **Caution:** Once made, the decision to expense these items must be followed in future years. This adjustment also could benefit the accrual taxpayer.
- *Postpone taxation of crop insurance proceeds.* Crop insurance and federal disaster payments are taxable income, generally in the year received. However, if the failed crop normally would have been marketed in the following year then the payment may be declared in the following year.
- *Postpone taxation of Commodity Credit Corporation (CCC) loans.* Initially, CCC loans may be treated as income similar to a sale, or they may be treated as a

loan with the declaration shifted to the year when the grain is either forfeited or reclaimed and sold. IRS approval must be obtained to change this pattern once elected. This option could benefit accrual taxpayers.

- *Trade, don't sell, capital items with taxable gains.* There is a depreciation recapture if a depreciable asset is sold above it's adjusted basis. Included in this tabulation is all previous depreciation claimed and any Section 197 deductions. If the sale price is greater than the purchase price less these adjustments the difference is taxed as regular income. However, if this gain is transferred into the basis of a new replacement asset through trade the taxation is postponed and levied through the new item. Thus, current tax is moved into later years. Accrual taxpayers also can use this procedure.
- *Planning farm-operating expenditures.* Some expenditures may not be made every year but are nevertheless deductible in the year in which they occur. Such expenditures include painting buildings, minor repairs on improvements, many small shop tools, periodic seeding of legumes and grasses, and (within limits) costs of soil and water conservation. Farmers can manage to make many expenditures of this type in years of high gross income to reduce taxable income.

The above list is not exhaustive and other ideas to manage taxes are discussed later in this chapter. If the reverse income situation is anticipated (i.e., income is lower in the current year than it is anticipated to be next year), then income could be shifted into the current year to level income between years by reversing the strategies discussed above. This is a difficult thing to do because nobody wants to increase their current tax bill. The normal tendency is to try to shift taxes forward, and thus the opportunity for income leveling is forfeited. However, the benefits from income leveling should not be forgotten and not used because of an uncertain future.

Other Methods for Reducing Taxable Income

The list above primarily pertains to farm operations. The following list provides several other possible means for lowering taxable income and possibly leveling taxable income.

- *Pay cash wages to children.* Wages paid to children for farmwork actually done by them may be deducted as a farm business expense. To do this the wages should be reasonable and there should be a true employer–employee relationship. To establish this relationship, assign definite jobs, agree on wages ahead of time, and pay the children regularly as with any other employee. Wages paid to a child are included in his or her income and may result in the child having to file an income tax return. (See Table 13.2 for the maximum amount the child can earn before paying federal income taxes.) Parents, however, cannot claim a dependency exemption for the child if the child claims himself or herself as an exemption or if they do not pay half the child's support. Wages paid to children by parents are not subject to social security tax until the child reaches age 18.
- *Give income-producing property to children.* Gifts could include land, cattle, and machinery. Let the children report income from their work and capital. Family partnerships and farm corporations through stock transfers are sometimes used to do this. It is another way to spread family income over the lower tax brackets. Remember, gifts and partnerships must be legally sound to achieve tax savings. Gifts may be taxable if they exceed certain amounts.

- *Consider making a commodity (crop or livestock) gift to a spouse or child* The maximum amount of each gift must be less than $10,000. Use this option with great care and under the direction of a competent advisor.
- *Pay wages with a commodity.* Payment-in-kind wages (livestock, grain, etc.) are not subject to FICA taxes. The employer and the employee each save 7.65 percent on wages of $64,500 in 1997. In-kind wages are still subject to income tax once sold. The IRS will scrutinize these types of transactions very carefully so proceed with caution and under the direction of a competent tax practitioner. The employee needs to understand there may be a loss of social security benefits.
- *Contribute to a retirement plan.* Contributions made to a qualified retirement plan can reduce taxable income. Example plans for self-employed farmers include Keogh, employee pension plans (SEP and SIMPLE), and individual retirement accounts (IRA), including those for higher education. These plans all include payments of before-tax income, thereby reducing the amount of income taxes paid. At retirement when the funds are withdrawn the taxes are paid on the principal and earnings. There will be a penalty for early withdrawal except for death, disability, or certain other reasons such as medical.
- *Postpone itemized personalized deductions (Schedule A).* Benefits from itemizing deductions can only be realized when they total more than the standard deduction. See the discussion proceeding Table 13.2 regarding the standard deduction. Itemized deductions are medical and dental expenses exceeding 7.5 percent of adjusted gross income, state and local income tax, property tax, home mortgage interest, gifts to charity, and a few job-related expenditures. To illustrate assume two families who each have itemized deductions of $6,900 each year. Because the itemized deductions did not exceed the standard deduction, there is no benefit from filling out Schedule A. Now assume that the second family decides to only make charitable contributions every other year and is able to shift some medical costs, etc., such that their itemizations are $3,000 and $10,800 in alternating years. When itemizations are $3,000 the standard deduction is used, and when $10,800 Schedule A is submitted. Taxable income is reduced by $3,900 more than the standard deduction every other year for a tax savings of $585 if the marginal tax rate is 15 percent and $1,092 if the marginal tax rate is 28 percent. If itemizations can be used to level taxable income then even a greater savings is possible.

Investment Credit

Investment tax credit (ITC) has been one of the most effective tax-reduction tools. However, the Tax Reform Act of 1986 did away with most applications of it. Certain energy-saving devices, reforestation, and rehabilitating old buildings remained, but farm machinery and equipment and farm structures were dropped. It could be revived if the country were to go into recession as a means of stimulating investment. Investment tax credit allows the taxpayer a portion of the cost (historically 10 percent) as a credit against any income tax liability on a dollar-for-dollar basis. Energy credit could include solar heating. Only reforestation is discussed.

Reforestation Expenditures

Since 1980, certain reforestation expenditures have qualified for amortization and the regular investment credit. It is now possible to amortize qualified reforestation expenses over a 7-year period. Under the old law, such expenses had to be capitalized and could not be deducted until the timber was sold. The limit on the amount of expenses that qualify for this provision each year is $10,000 on a joint return. Any qualifying expenditure greater than $10,000 in a tax year must still be capitalized.

To qualify, an expenditure must be to forest or reforest property located in the United States and held for the commercial production of timber. The following types of reforestation expenditures qualify for the amortization provision: (1) cost of site preparation, (2) cost of seed or seedlings, (3) labor, (4) tools, and (5) depreciation on equipment (tractors, trucks, tree planters, etc.). Depreciation is only treated as a direct cost for the period the equipment is used in reforestation.

Soil and Water Conservation

One may elect to deduct certain expenditures for soil and water conservation. Land-clearing costs are no longer deductible. Otherwise, these expenditures must be capitalized and added to the cost or other basis of the land.

Deductible soil- and water-conservation expenditures include the treatment or movement of earth, such as leveling, conditioning, grading, terracing, contour farming, or restoration of fertility. In addition, the construction, control, and protection of diversion channels, drainage ditches, earthen dams, watercourses, and ponds; the eradication of brush; and the planting of windbreaks qualify for this deduction. There is a limitation on the amount of soil- and water-conservation expenditures that can be deducted in any tax year. All expenditure for conservation purposes must be consistent with a plan approved by the National Resources Conservation Service.

Expenditures for structures or facilities such as water wells, pipe, tile, and dams of wood, masonry, metal, or concrete construction cannot be included in soil- and water-conservation deductions. The cost of these items must be capitalized and the investment recovered through annual allowances for depreciation. Expenses for maintaining completed soil- and water-conservation structures, such as annual removal of sediment from a drainage ditch, are standard farm business expenses.

The maximum deduction in any year can not exceed 25 percent of gross income from farming, including the gains from the sale of breeding livestock. If land is disposed of within less than 10 years after acquiring it, the law provides for the recapture of a part or all of the soil- and water-conservation and land-clearing expenditures. These expenditures are recaptured by treating the portion of the gain resulting from deductions allowed for these expenditures and ordinary income rather than as a capital gain. Any remaining portion of the gain will be a long-term capital gain, provided the land is owned long enough for long-term treatment.

Sales or Trades of Property

When a farm is sold or traded for another farm or business property, it is necessary to establish the cost basis of the farm sold to compute the actual gain or loss on the transaction and to ensure that no unnecessary tax is paid. To establish this cost basis, it is necessary to have a record of original cost, costs of all improvements made on the property since it was acquired, and all depreciation claimed. Some states have special depreciation and investment record books available at county extension offices.

Three types of improvements are made on farmland:

1. Improvements subject to depreciation include farm buildings, silos, fences, tile drains, etc., that are made of wood, concrete, brick, masonry, or metal.
2. Improvements that are not depreciable include construction of ditches, soil- and water-conservation expenditures, and cost of clearing land not previously used for farming if it was decided to capitalize them instead of deducting them as farm

operating expenses. Costs of such improvements are added to the original cost. This total investment is then reduced by the amount of all depreciation previously deducted or allowable. Any item deducted as an expense, such as soil- and water-conservation expenditures, cannot be included in the cost basis. Thus, it is important to have a complete record of all depreciation and capital expenditures during the entire period of ownership.

3. Improvements to the farmer's personal dwelling are not depreciable for tax purposes, except for a part used for the farm business office or for hired labor. Their costs should be added to the original investment to determine the basis for gain or loss when sold. New legislation beginning after May 6, 1997, allows married couples to exclude $500,000 ($250,000 if single) of gain on the sale or exchange of their principal residence if occupied at least 2 of the last 5 years. There is no restriction by age and the exclusion may be used every 2 years. The time lived in prior residences may be counted if the unrecognized gain was rolled over into the house now being sold. There are a few rules related to divorce, surviving spouses, nursing homes, and use of the home as a business. A prorated share may be claimed if the time owned and occupied is less than 2 years and the reason for sale is due to unforseen circumstances.

Installment sales. In selling a farm, the tax liability on the gain can be spread over a period of years and in many cases can be reduced by the use of the installment sales method in the year of sale. The income above cost (gain) is taxable as received. The installment method may be used for any sale of real property and for casual sales of personal property. Note that if the sale includes different classes of property there may be separate rules for each class. For example the recapture is different for buildings than machinery and soil- and water-conservation expenses may need to be capitalized.

Trading a farm. There is frequently a tax advantage in trading a farm for another farm or other business property. In case of trade all or part of the tax liability is postponed. No gain is recognized for tax purposes unless a difference in cash or nonbusiness property is received in the transaction.

However, in some cases it is desirable in the long run to sell and pay taxes on the gains to get a higher cost basis, including a higher basis for depreciation, on the new farm or business property. This could apply in a situation where an unimproved farm, or a farm on which the depreciation basis has been exhausted or nearly exhausted, is exchanged for a well-improved (with a high depreciation basis) farm or business property. The following example is an illustration. The Browns own Farm A in which they have a cost basis of $50,000. They have exhausted their depreciation allowances. Farm A has a market value of $100,000. The Browns can purchase Farm B for $100,000 or trade Farm A for it. Farm B has a fine set of improvements that can be reasonably valued at one-half the total value of the farm of $50,000, with the other $50,000 allocated to the land. If they trade Farm A for Farm B, they have no tax to pay but retain their cost basis of $50,000 (with no depreciation basis) in the new property.

If they sell Farm A and buy Farm B, the Browns will have a $50,000 long-term capital gain ($100,000 sale price minus $50,000 cost basis), 100 percent taxable beginning in 1987. If sold on the installment basis, the gain may be spread over a period of years. If sold with a down payment of $20,000 on the principal and principal payments of $20,000 per year for 4 additional years, they would have $10,000 long-term capital gain per year. Moreover, by purchasing Farm B they have obtained a $50,000 depreciation basis. Because the buildings are not new and obsolescence is a factor, they could logically be depreciated on a 10-year straight-line basis, or $5,000 per year. Thus, in the first 5 years, depreciation would more than offset the taxable gain and the Browns would be better off than if they had traded. In the next 5 years, they would have $5,000 per year depreciation that, if they had traded, they would not have had to offset ordinary income.

It is also conceivable to assume that the Browns' interest income from the sale of Farm A would be offset or nearly offset by interest expense in purchasing Farm B. Thus, by sale and purchase, $25,000 depreciation would be gained to offset ordinary income in addition to the $25,000 depreciation that would affect gain from the sale of Farm A.

Capital Gains and Losses

From an income tax standpoint, there are two types of income: ordinary income and capital gains income. Beginning in 1987, both are taxed as ordinary income with limits on the maximum rate at which capital gains income is taxed. Some of the favorable tax treatment was restored by congress in 1997. The maximum rates were reduced in 1997 to 20 percent for qualifying assets held for 18 months if regular income is taxed at 28 percent and 10 percent if regular income is taxed at 15 percent. The 20 percent rates are reduced to 18 percent in the year 2000. The holding period is 24 months for dairy and breeding livestock.

Suppose, for example, an unproven bull is purchased for $2,000. After 3 years the bull is shown to produce superior offspring and sells for $10,000. Depreciation aside, there has been a capital gain of $8,000. If the taxpayer's taxable income is under $35,800 the marginal tax rate is 15 percent, and the tax would be $1,200. But, it the taxpayer's income is over this amount the marginal tax rate is 28 percent and the taxpayer would owe $2,240.

Rules for Droughts and Disasters

For farmers who use cash accounting, there is an exception to the general rule that payments must be reported in the year they are received. That exception applies to crop insurance and disaster payments received from the federal government when crops cannot be planted or are damaged or destroyed by a natural disaster such as a drought or a flood. The exception does not apply where the crop insurance indemnity is due to a low price, such as may occur with the Crop Revenue Coverage (CRC) or Income Protection Plan (IPP) programs.

The exception lets the farmer postpone reporting such a payment by 1 year. That is, if a payment was received this year, it can be reported on next year's taxes. To qualify for the exception, it must be shown that the income from the crop for which the payment is received would have been reported in a year following the receipt of the payment.

If the taxpayer qualifies for this exception, there is the option of reporting the payment as income in the year it is received or as income in the following year. The election to postpone reporting the payment as income covers all crops from a farm. A separate election for each separate farming business must be made. For purposes of this provision, separate businesses are defined as those for which separate books are kept and are allowed to use different methods of accounting. In general, that requires the business to be separate and distinct.

To make the election, a statement must be attached to the return showing the following:

- Name and address.
- Declaration that the election is under Section 451(d).

- Identification of the specific crop or crops destroyed or damaged.
- Declaration that under your normal business practice the income derived from the crops that were destroyed or damaged would have been included in gross income for a taxable year following the taxable year of such destruction or damage.
- Cause of destruction or damage of crops and the date or dates on which such destruction or damage occurred.
- Total amount of payments received from insurance carriers, itemized with respect to each specific crop and with respect to the date each payment was received.
- Name(s) of the insurance carrier or carriers from whom payments were received.

If livestock are sold because of a shortage of water, grazing, or other requirements due to a drought, the recognition of the proceeds from the sale may be postponed. If the drought forced the sale of livestock (other than poultry) that were held for any length of time for draft, breeding or dairy (no sporting) purposes, the gain realized on the sale does not have to be recognized (reported as income) if the proceeds are used to buy replacement livestock within 2 years of the end of the tax year of the sale. The new livestock must be used for the same purpose as the livestock sold. Dairy cows must be replaced with dairy cows, for example. It must be shown that the drought forced the sale of more livestock than would have sold if there had been no drought. For example, if one-fifth of the herd was normally sold each year, only the sales in excess of one-fifth qualify for this provision. There is no requirement that the drought conditions caused an area to be declared a disaster area by the federal government.

The basis in the replacement livestock is equal to the basis in the livestock sold plus any amount invested in the replacement livestock that exceeds the proceeds from the sale.

If drought forces the sale of livestock, including draft, breeding, or dairy livestock, the taxpayer may be eligible for another exception to the general rule that livestock-sale proceeds must be reported the year they're received. This exception allows for the postponement of reporting the income by 1 year. To qualify, it must be shown that the livestock normally would have been sold in a subsequent year, and the drought that forced the sale must have caused some area to be declared a disaster area. But it's not necessary that the livestock have been raised or sold in that disaster area. The sale can take place before or after an area is declared a disaster area as long as the same drought caused the sale.

The election must be made by the due date of the return, including extensions, for the tax year in which the drought sale occurred. To make this election, attach to the return a statement with the following information:

- A declaration that an election is being made under Section 451(e).
- Evidence of the existence of the drought conditions that forced the early sale or exchange of the livestock and the date, if known, on which an area was designated as eligible for assistance by the federal government as a result of the drought conditions.
- A statement explaining the relationship of the drought area to the early sale or exchange of the livestock.
- The total number of animals sold in each of the 3 preceding years.
- The number of animals that would have been sold in the taxable year under normal business practices in the absence of drought.

- The total number of animals sold and the number sold on account of drought during the taxable year.
- A computation, pursuant to Reg. Section 1.451-79(e) (the computation shown above), of the amount of income to be deferred for each such classification.

Tax Planning When Buying a Farm

To ensure maximum tax savings at the time of purchasing a farm, the buyer should allocate the total cost of the farm to any growing crops, depreciable improvements, dwelling, timber and mineral resources, and land. From a tax management viewpoint the amounts allocated to the various items are handled differently. The cost assigned to the growing crops is an offset (shown on Schedule F) against the selling price of the crop in the year of sale. The cost basis of the farm is reduced by the amount allocated to the growing crop.

The part of the "cost" allocated to land cannot be recovered until the farm is sold because land cannot be depreciated. So, too, the portion allocated to the dwelling is not depreciable if used solely as the buyer's personal residence. A tenant house is depreciable for tax purposes. Cost allocated to depreciable improvements is recovered through depreciation. Recovery of cost is faster on short-lived improvements than on long-lived ones. Investment credit can be claimed only on the part of the investment allocated to assets that qualify for the credit, such as reforestation, rehabilitation of certain buildings, and energy conservation property.

For management and tax purposes, the "cost" must be broken down and allocated to each particular structure or improvement. In allocating cost to depreciable improvements, the following procedure may be helpful.

- Figure the present cost of replacing the improvement.
- Establish the years of normal useful life.
- Determine the age of the present improvement.
- Determine the remaining years the improvement will be used.
- Compute the value by dividing the remaining years the improvement will be used by the total years of normal use and multiplying the result by the replacement cost.

The following is an example of this procedure:

- Replacement cost of barn = $20,000
- Useful life of new barn = 25 years
- Age of present barn = 15 years
- Remaining life of present barn = 10 years
- Value of present barn = 10/25 of $20,000 = $8,000. No salvage value need be claimed, as cost of removing old buildings are generally about equal to salvage value.

Another guide in allocating costs is the reasonable insurance values of insurable property. Care should be taken that in the final allocation the amount allocated to the bare land represents a reasonable value for similar land in the community.

The proper allocation of cost may help determine the price a buyer will pay for the farm, particularly when the buyer is looking to future farm income after taxes to pay off the purchase price.

Closely related is the manner of payment of the purchase price. In computing taxable income, the buyer deducts interest payments but not payments on principal. The seller treats interest as ordinary income, whereas principal payments in excess of the seller's cost basis are capital gains.

Net Operating Losses

Because of depressed farm product prices, increasing production costs, and high interest rates, many farmers suffer losses in some years. If there is a loss in the operation of the farm during the year, the farmer may have a net operating loss, which can be carried back or forward to offset income in other years. Many farmers pay more taxes over a period of years than required by law because they fail to take advantage of the net operating loss provisions. If there is a net operating loss in a given year, the net operating loss deductions may be carried to certain other tax years. This feature allows the farmer to obtain a refund on taxes previously paid or to reduce tax liability in future years. A net operating loss is the excess of allowable deductions over gross income after certain adjustments are applied to that excess. Business losses are usually the dominant factors in operating losses, but other types of income and deductions must be considered in determining the net operating loss deduction.

The computation of a net operating loss carryback or carryover is complex because several adjustments must usually be made. Anyone who has a loss in the operation of his or her farm business should consult a qualified tax consultant for help in computing the carryback or carryover.

A net operating loss may be carried back 2 years and carried forward up to 20 years or until it is used up, whichever comes first. One can elect to forego the carryback and just carry the loss forward. The carryback of a net operating loss deduction can result in a refund of tax paid in prior years. The claim for refund may be filed any time within 3 years after the due date, including extensions, of the tax return for the year of the net operating loss.

Alternative Minimum Tax

The Alternative Minimum Tax (AMT) is a second tax system that runs parallel to our regular federal income tax system. The idea is to prevent taxpayers, particularly those in high-income brackets, from using tax deductions so as to pay little or no tax. By eliminating many tax incentives, broadening the definition of taxable income, and applying a set percentage to the resulting taxable income, the AMT system often results in paying a larger tax. AMT applies to both individuals and corporations. (If the total of itemized or standard deductions, plus personal exemptions, plus adjustments and tax preference items is greater than $45,000 for the married filing jointly or $33,750 for singles in 1997 the taxpayer must fill out Form 6251 to determine if AMT applies.)

The AMT calculation starts with taxable income for regular tax. Any net operating loss taken to come up with that income is first added back in. Tabulations of the alternative minimum tax are shown in Table 13.5. Note that there are three major considerations for tabulating the AMT from taxable income: adjustments, preferences, and exemption. Under AMT the standard deduction, personal exemption, and itemized deductions are not allowed. Also there are a different set of rules for tabulating depreciation. These, plus a few lesser items, constitute the adjustments. Preference items include some tax-exempt bonds, accelerated depreciation on real property, intangible drilling costs, and some other lesser items. Only the accelerated depreciation is of major concern to farmers. The exemption is a standard amount. The tax rate applied to this remainder income is 26 or 28 percent, and the amount paid is the difference between the regular tax and AMT. The 26 percent applies to incomes less than $175,000 for married couples.

TABLE 13.5

Tabulating the Alternative Minimum Tax for the FABA Family (Assume a Family of 4 and Taxable Income Equal to the Cash Method Income in Table 3.2)

Taxable income as computed for regular tax	$53,581
Less itemized or standard deductions	(6,900)
Plus ATM adjustments and preferences	6,900
Add back regular tax net operating loss (NOL)	0
ATM taxable income (AMTI)	$53,581
AMTI	$53,581
Exemption	(45,000)
AMTI less exemption	$ 8,581
Net AMTI	$ 8,581
AMT tax rate (26%)	× 0.26
Tentative minimum tax (TMT) before credits	$ 2,231
AMT tax credit	0
TMT after credits	$ 2,231
Regular tax	(5,411)
Alternative minimum tax	$ 0

Managing Income for Maximum Social Security Benefits

In some cases, farmers may wish to increase their net farm income to the maximum amount subject to self-employment tax to secure larger social security benefits when they retire. The maximum amount of income subject to self-employment tax was $65,400 in 1997 and scheduled to increase in future years. Increasing taxable income to boost social security benefits increases the amount of income tax and self-employment tax. It may be desirable to do this to gain additional retirement benefits. Each individual must weigh the increased cost of obtaining the higher benefits against the value of the benefits.

A higher taxable income can be had by increasing sales and other receipts or by decreasing expenses. Both of these methods have been mentioned in this chapter and throughout the book and are not repeated here.

Farm cash rental payments are generally not considered to be self-employment income. However, if the landlord "materially participates" in the management of the farm, the income qualifies for self-employment. So if a landlord desires to increase social security coverage, the lease arrangement should be designed so that he or she "materially participates" in the management of the farm.

For the years when farm income is low, there is an optional method of determining net earnings from self-employment. Under this method, a farmer whose gross income is more than $2,400 and whose net farm income is less than $1,600 may treat $1,600 as net earnings from farm self-employment. This could help to increase social security coverage in a low-income year or a year with a net operating loss. Also, under the optional method, a farmer whose gross income from farming is less than $2,400 may treat two-thirds of gross income as net earnings from farm self-employment.

Tax Management Reminders

- Do not include in income any indemnity for diseased animals if payment has been or will be used to buy like or similar animals within 2 years.
- If using the cash method, deduct cost of purchased livestock lost, stolen, or died during the year.
- If using the accrual method, make an inventory check to make sure numbers reflect livestock purchased, born, sold, died, and slaughtered.
- Deduct as many auto, utilities, telephone, and family shared expenses as actually used in the farm business (half may not be accurate). Make certain this use is well documented.
- Keep records to insure deduction of easily overlooked items such as farm magazines, farm organization dues, bank service charges, business trips, the portion of the dwelling used for farm business, household supplies used for hired help, and cash outlay to board hired workers.
- Use installment sales of property to spread income over a period of years to avoid high income in one year.
- If age 63 or 64, postpone income to age 65 to take advantage of the double personal exemption. Persons approaching retirement, however, may want to maintain income as near as possible to the maximum social security ($65,400 in 1997) to increase retirement income in later years.
- Do not report capital gains on the sale of a dwelling as income if planning within 2 years to buy and occupy another dwelling. A once-in-a-lifetime exclusion of up to $500,000 of the gain on the sale of the personal dwelling may be allowed if more than 55 years old, even if it is not reinvested in another house.
- Itemize on bank deposit slips all gifts, borrowings, and nontax deposits so that they will not be considered taxable income.
- Keep records of all medical, dental, and hospital bills, including premiums for accident and health insurance to determine if these can be itemized above the standard deduction.
- Use charge accounts and pay by check to help ensure that all expenses are tabulated. This prevents omitting many small items that might be overlooked if paid by cash.
- Keep exact records of the date of purchase, cost, and date of sale of all items purchased for resale.
- Keep all paid receipts, invoices, canceled checks, etc., for a minimum of 3 years (5 years recommended), including checks for payment of income taxes. For machinery and improvements, vouchers should be kept for as long as the property is held plus at least 3 years.
- File an income tax estimate by January 15 to relieve the pressure of filing a final return by March 1. This may improve the completeness and accuracy of the return.

Tax Estimate Worksheet

The tax estimate worksheet shown on Form 13.1 can be used throughout the year in planning farm business and tax management strategies. Even if it is not used throughout the year, its use in November to plan tax savings in December is strongly advised.

FORM 13.1

Federal Income Tax Management for Farmers Worksheet, (NCR-2, MWPS, Iowa State University)

Use this worksheet throughout the year in planning farm business and tax management strategies. If you do not use it throughout the year, use it in November to plan tax savings in December.

		Amount Year to Date	Estimated Rest of Year	Estimated Year's Total
FARM RECEIPTS:				
Sales of product raised[a] and miscellaneous receipts:		$	$	$
Cattle, hogs, sheep, and wool, etc.		$	$	$
Poultry, eggs, and dairy products		$	$	$
All crop sales		$	$	$
Custom work, prorations and refunds, agriculture program payments		$	$	$
Total sales and other farm income	(1)	$	$	$
Sales of purchased market livestock[b]		$	$	$
Purchase cost (subtract)[c]		$	$	$
Gross profits on sale of purchased livestock	(2)	$	$	$
Gross farm profits (Item 1 + 2)	(3)	$	$	$

FARM EXPENSES:

Car and truck	$	Labor (incl. benefits)	$
Chemicals	$	Rent or lease	$
Conservation expenses	$	Repairs, maintenance	$
Custom hire	$	Seeds, plants purchased	$
Feed purchased	$	Storage, warehousing	$
Fertilizer and lime	$	Supplies purchased	$
Freight and trucking	$	Taxes	$
Gasoline, fuel, oil	$	Utilities	$
Insurance	$	Vet, breeding, medical	$
Interest	$	Other	$

Total cash farm expenses	(4)	$	$	$
Depreciation on machinery improvements, dairy and breeding stock	(5)	$	$	$
Total deductions (Item 4 + 5)	(6)	$	$	$
Self-employment farm income (Item 3 less item 6)	(7)	$	$	$
OTHER INCOME:				
Net taxable gain from Schedule D (Sales of dairy and breeding stock, machinery, and other capital exchanges)	(8)	$	$	$
Taxable nonfarm income	(9)	$	$	$
Adjusted gross income (Item 7 + 8 + 9)	(10)	$	$	$
Less: standard deduction or itemized deductions[d]		$	$	$
$2,650 × ____ personal exemptions[e]		$	$	$
Total nonbusiness deductions and exemptions	(11)	$	$	$
Taxable income (Item 10 less item 11)	(12)	$	$	$
Estimated income tax (calculated from applicable tax computation table or rates)	(13)	$	$	$
Estimated self-employment tax (Item 7 × .9235 × .153)[f]	(14)	$	$	$
TOTAL TAX (Item 13 + 14)	(15)	$	$	$
Less Credits: gas tax, income tax withheld, and estimated tax paid	(16)	$	$	$
Estimated tax due (Item 15 less item 16)	(17)	$	$	$

Last year's marginal tax bracket____________%

This year's estimated marginal tax bracket____________%

Next year's expected marginal tax bracket____________%

[a]For accrual method include sales of all livestock.

[b]Omit for accrual method.

[c]For accrual method adjust for change in inventory and new livestock purchases.

[d]Use itemized deductions if larger.

[e]Exemption for 1997, see current tax regulation for subsequent years.

[f]1997 rate is 15.3 % on up to $65,400 and 2.9% on amounts over $65,400.

Notes

1. If the reader is not familiar with accounting methods and procedures, he or she should study Chapters 2 and 3 of this book.
2. Section 1231 transactions include gains or losses from sales of breeding livestock, depreciable machinery, real estate, and other property used in your farm or business. Gains or losses are determined by the difference between the sale price and the basis or adjusted basis of the property. There are other rules that may affect this, such as depreciation recapture of Section 1245 property.

INDEX